Culture, Class, and Politics in Modern Appalachia

Essays in Honor of Ronald L. Lewis

WEST VIRGINIA AND APPALACHIA

A SERIES EDITED BY RONALD L. LEWIS
AND KEN FONES-WOLF
VOLUME 10

Other books in the series:

Transnational West Virginia
Edited by Ken Fones-Wolf and Ronald L. Lewis

The Blackwater Chronicle
By Philip Pendleton Kennedy; Edited by Timothy Sweet

Clash of Loyalties
By John Shaffer

Afflicting the Comfortable
By Thomas F. Stafford

Bringing Down the Mountains
By Shirley Stewart Burns

Monongah: The Tragic Story of the 1907 Monongah Mine Disaster
By Davitt McAteer

Sectionalism in Virginia from 1776 to 1881, Second Edition
By Charles Ambler;
Introduction to the second edition by Barbara Rasmussen

Matewan Before the Massacre
By Rebecca J. Bailey

Governor William E. Glasscock and Progressive Politics in West Virginia
By Gary Jackson Tucker

Culture, Class, and Politics in Modern Appalachia

Essays in Honor of Ronald L. Lewis

edited by Jennifer Egolf, Ken Fones-Wolf, and Louis C. Martin

MORGANTOWN 2009

West Virginia University Press, Morgantown 26506

© 2009 by West Virginia University Press

All rights reserved

First edition published 2009 by West Virginia University Press
Printed in the United States of America

16 15 14 13 12 11 10 09   1 2 3 4 5 6 7 8 9

ISBN-10: 1-933202-39-4
ISBN-13: 978-1-933202-39-6
(alk. paper)

Library of Congress Cataloguing-in-Publication Data

Culture, class, and politics in modern Appalachia : essays in honor of Ronald L. Lewis / edited by Jennifer Egolf, Ken Fones-Wolf, and Louis C. Martin. -- 1st ed.
    p. cm. -- (West Virginia and Appalachia ; 10)
  ISBN 978-1-933202-39-6 (alk. paper)
  1. Coal trade--Appalachian Region. 2. Coal mines and mining--Appalachian Region. 3. Appalachian Region--Social conditions. 4. Appalachian Region--Economic conditions. I. Lewis, Ronald L., 1940- II. Egolf, Jennifer. III. Fones-Wolf, Ken. IV. Martin, Louis C.
  HD9547.A127C85 2009
  338.2'72409754--dc22

Library of Congress Control Number: 2009921863

Cover Design by Than Saffel

Table of Contents

vii   LOUIS C. MARTIN AND KEN FONES-WOLF
         Preface

1   DWIGHT B. BILLINGS
         Introduction: Writing Appalachia: Old Ways, New Ways, and WVU Ways

Section I: Culture

31   DEBORAH WEINER
         'Scrip Was a Way of Life': Company Stores, Jewish Merchants, and the Coalfield Retail Economy

56   PAUL RAKES
         A Combat Scenario: Early Coal Mining and the Culture of Danger

88   JENNIFER EGOLF
         Radical Challenge and Conservative Triumph: The Struggle to Define American Identity in the Somerset County Coal Strike, 1922–1923

118   CONNIE PARK RICE
         Separate But Never Equal: Dewey W. Fox and the Struggle for Black Equality in the Age of Jim Crow

Section II: Class

141 MICHAEL E. WORKMAN
'Sadly in need of organization':
Labor Relations in the Fairmont Field, 1890 to 1918

166 REBECCA BAILEY
The Matewan Massacre: Before and After

204 RICHARD P. MULCAHY
Progress and Persistent Problems:
Sixty Years of Health Care in Appalachia

224 JOHN HENNEN
1199 Comes to Appalachia: Beginnings, 1970–1976

Section III: Politics

253 JEFFERY B. COOK
Mining Reform after Monongah:
The Conservative Response to Mine Disasters

283 MARK MYERS
Depression, Recovery, Instability: The NRA and the McDowell County, West Virginia Coal Industry, 1920–1938

305 SHIRLEY STEWART BURNS
To Dance with the Devil: The Social Impact of Mountaintop Removal Surface Coal Mining

329 PUBLICATIONS BY RONALD L. LEWIS

335 CONTRIBUTORS

339 INDEX

Preface

Louis C. Martin and Ken Fones-Wolf

IT HAS BEEN more than ten years since the seminal volume *Appalachia in the Making: The Mountain South in the Nineteenth Century* was published. That collection grew out of the flourishing of scholarly work on nineteenth century Appalachian history and inspired another decade of research on the subject. In fact, the nineteenth century has dominated the field since the 1995 appearance of *Appalachia in the Making*. In some ways Ron Eller began the discussion when the introductory chapter of his classic *Miners, Millhands, and Mountaineers: Industrialization of the Appalachian South, 1880–1930* portrayed preindustrial Appalachians as self-sufficient, egalitarian farmers. The scholarship that culminated in *Appalachia in the Making* demonstrated that Appalachians had participated in outside market economies long before the arrival of railroads, coal, and timber magnates. Furthermore, mountain communities were rarely egalitarian and were, in fact, quite diverse and thoroughly enmeshed in the controversies and conflicts of nineteenth-century America.[1] Historians continue to build on that foundation and have produced an impressive body of work exploring the impact of the Civil War on the region, further elucidating market connections and their impact on mountain society, and exploring the role of race and ethnicity among the newly discovered diverse populations of nineteenth-century Appalachia.[2]

The twentieth-century topics that once sparked such great interest in the field no longer appear to be the main focus of historical research into Appalachia, though there have been several wonderful exceptions over the past dozen years. Appalachian out-migration commanded a great deal of interest, resulting in exciting new scholarship.[3] Though not the central topic it once was, the study of coal miners and their communities generated several important new studies.[4] Likewise, the social construction of Appalachian stereotypes continued to be an important and fruitful field of inquiry.[5] Still, the topics that once seemed to demand scrutiny—industrialization, the labor movement, twentieth-century economic and political failures, and their social impact—no longer stand at the center of Appalachian historical scholarship. Indeed, there have been no recent collections of essays that survey various facets of Appalachia in the twentieth century.[6] Fortunately, the 2007 Rush Dew Holt Conference at West Virginia University offered an excellent opportunity to address this deficiency and simultaneously recognize one of the leading scholars in the field.

The History Department at West Virginia University organized the Holt Conference to honor the career of Ronald L. Lewis. In the spring of 2006, Lewis, the Stuart and Joyce Robbins Distinguished Chair in History, announced his plans to retire from West Virginia University in December 2007. A prolific author, dedicated professional, and much sought-after advisor for WVU's doctoral students, Lewis has been a key figure in the department since his arrival in Morgantown in 1985. In addition to serving two terms as department chair, mentoring more than two dozen doctoral students, and influencing countless undergraduates through his fully enrolled West Virginia and Appalachian history classes, Lewis also served as the interim director of the Regional Research Institute, on the board of the West Virginia Humanities Council, as an officer of the North American Association for the Study of Welsh History and Culture, as president of the Appalachian Studies Association, and as the founding editor of the *Journal of Appalachian Studies*. Among the honors Lewis accumulated are the Benedum Distinguished Scholar Award, the William Miernyk Award for Career Scholarly Achievement, the Eberly Family Distinguished Professor, a Fulbright-Hayes Commission Award, a National Endowment for the Humanities Fellowship, and the Governor's Award for West Virginia

History and Literature. As parting gifts to West Virginia's historical community, Lewis has helped establish both the university press and the journal, *West Virginia History: A Journal of Regional Studies* at the university to make certain that the university will build upon the legacy of regional scholarship that he worked so hard to advance.

In a tribute to Lewis's achievements, history graduate students at West Virginia University, led by Jennifer Egolf, Elizabeth Oliver Lee and Louis Martin, devoted their biennial Rush Dew Holt Conference to recognize his contributions as both scholar and mentor. The conference, "Transforming Appalachian Scholarship," met in April 2007, bringing together scholars from the fields Lewis has studied, including African-American workers in the American South, industry and labor in the Appalachian region, the industrial transformation of West Virginia, immigration and ethnic diversity within Appalachia, and West Virginia regional history. (A full bibliography of Lewis's writings is contained in the appendix.) Keynote speaker Dwight Billings, in his presentation of the History Department's Callahan Lecture, honored his colleague by surveying the evolution of Appalachian scholarship and putting the influence of Lewis on both students and colleagues into that larger context (see introduction). Two panels at the conference included such notable scholars as John Inscoe, Joe William Trotter, Bill Jones, John Hennen, Ken Sullivan, and the award-winning *Charleston Gazette* reporter Ken Ward. These panels surveyed Lewis's many contributions to the state, the region, and the scholarly community.

One highlight of the conference was that it brought together many of Lewis's former and current students whose research spanned a wide array of Appalachian history topics. While these students have examined topics ranging from the efforts to establish the state of Franklin in eighteenth-century Tennessee to the crusade against mountaintop removal over two centuries later, for this volume, we have brought together eleven essays from former students who are helping rewrite the history of twentieth-century Appalachia as well as Dwight Billings's Callahan Lecture. Together they contribute to our understanding of modern Appalachia and reflect the influence of Lewis as an advisor during his career at West Virginia University.

* * *

CULTURE, CLASS, AND POLITICS

The book is organized around the themes of culture, class, and politics in modern Appalachia, themes that have guided much of Lewis's own scholarship. The first group of essays broadens our understanding of Appalachian culture. Until the 1960s, many scholars believed that modern Appalachian culture was static and homogenous—a product of colonial settlers, isolated by the mountainous terrain, and unchanged by the forces of modernization at work in the rest of the nation. In his 1962 book *Night Comes to the Cumberlands*, a valuable and pathbreaking study in many ways, Harry Caudill wrote that by "1840 they had accomplished their work. The twig had been bent. The tree had grown. The course of the mountaineer's development was determined." Illiteracy, isolation, lawlessness, and primitive technology, Caudill continued, shaped Appalachians. "From these forces emerged the mountaineer as he is to an astonishing degree even to this day," he concluded.[7]

Henry D. Shapiro's *Appalachia on Our Mind: The Southern Mountains and Mountaineers in the American Consciousness, 1870–1920* was instrumental in unraveling many of the stereotypes about Appalachians that had dominated discussions of their history, as well as the discourse out of which the very idea of Appalachia emerged.[8] Shapiro's work, having revealed the origins of many Appalachian stereotypes, inspired scholars to do more empirical studies of Appalachian culture, which Billings discusses in greater detail in the introduction. Social scientists discovered that, in many ways, nineteenth-century Appalachians resembled rural people in other parts of the country. In *The Road to Poverty*, for example, Dwight Billings and Kathleen Blee found that Appalachian farms were just as productive as northern farms until population growth in the mountains created an agricultural crisis; not unlike the one New England experienced in an earlier era.[9]

Turning to the twentieth century, Appalachian scholars were struck by the tremendous diversity of what was previously considered to be a homogeneous population of Scots-Irish farmers and coal miners. Far from static, Appalachian culture changed so dramatically that historians realized it was no longer possible to talk of a single, essential Appalachian culture. Labor migrations from around the world brought many different immigrant and migrant groups to the region as it transitioned to an industrial economy, resulting in diverse communities with a wide array of customs, practices, and beliefs. Eastern European Jews, Italians, southern African Americans, and many other groups played central roles in the community-building process during that critical time in Appalachian history. Furthermore, they all contributed to the development of a distinctly rural-industrial culture that has yet to be fully explored.[10]

The essays of the first section of this volume continue that exploration of the diverse and complex Appalachian culture that emerged in the twentieth century. Deborah Weiner's examination of the retail economy in the coalfields during the boom period reveals the critical role that Jewish merchants played in bringing modern consumer goods to relatively isolated coal camps and county seats. She complements the recent scholarly tradition of situating Appalachia within its broader contexts both nationally and internationally. These independent merchants linked even the most remote camps to the outside merchants and provided important alternatives to the company stores that otherwise charged exorbitant prices for a limited selection. Weiner's essay also sheds light on the experience of an important part of Appalachia's middle class, a group often left out of accounts of the region's industrialization.

Next, Paul Rakes explores parallels between the cultures of mining coal and fighting wars. Both operators and miners themselves came to accept casualties as a tragic but inevitable byproduct of the coal industry. Far from being immune to the dangers of the mines, coal miners drew on personal experiences with military combat as well as the symbols and values of combat culture to steel themselves for their daily work and to try to comprehend the carnage of numerous accidents. This aspect of coal mining culture stands in stark contrast to stereotypical images of fatalistic Appalachians untouched by the currents of modern change. Instead, Rakes offers us a glimpse into the social construction of a workers' culture that served psychological and pragmatic functions, enabling miners to not only face another day underground but to take pride in what was the most dangerous of industrial occupations.

Jennifer Egolf's essay further dismantles the idea of a homogenous and static Appalachian culture. Her examination of the 1922 coal strike in northern Appalachia reveals the fissures in the communities of Somerset County, Pennsylvania, between farmers, businessmen, and native-born and immigrant coal miners. The strike occurred not long after the Red Scare at the end of World War I, and communities eyed immigrants suspiciously, concerned that they might be radical revolutionaries. While operators and newspaper editors viewed the strike as the work of "reds," union officials defended the strike on patriotic grounds. This discourse revolved around several competing notions of what it meant to be a good American. Rather than view the coal fields as a simple bifurcated battleground pitting labor against capital, Egolf illuminates the complexities and conflicts that accompanied the cultural changes of an industrializing region.

The focus by scholars on a homogenous regional culture largely obscured the history of African Americans in modern Appalachia. Pathbreaking studies of black coal miners by Ronald L. Lewis and Joe William Trotter corrected this oversight in many respects.[11] Connie Park Rice, building on these earlier studies of black coal miners and their communities, continues the inquiry into this still relatively understudied field with a portrait of Dewey W. Fox, an African-American educator in West Virginia who helped reform education in Monongalia County, especially the coal mining community of Scott's Run, during the 1920s and 1930s. Fox's experiences demonstrate how African Americans used their education and political influence to force the often uncaring white majority to concede to their demands. In doing so, they left an indelible mark on modern Appalachian culture that demands further research. Rice's research reveals the ability of Appalachians to effect change within their own communities and improve local conditions, which contradicts the long-held image of Appalachians as victims of outside forces.

The second group of essays revolves around issues of class. For generations, scholars and local color writers paradoxically portrayed Appalachians as docile, fatalistic people who could also be stubborn, violent, and unpredictable. Rather than employing static cultural traits to explain Appalachian workers' behavior, more recent scholarship has demonstrated that class has been an important consideration in virtually any analysis of life in Appalachia. During the 1970s and 1980s, Appalachian historians began to interpret the experiences of Appalachian workers, especially miners, within the contexts of internal colonialism and the larger struggles between labor and capital.[12] These studies successfully disputed the earlier notions of a culture of poverty and provided a foundation for the studies to come, but they still portrayed an essentialized Appalachian world of exploited workers rising up against outsider capitalists. By the 1990s, it was no longer plausible to write about a singular Appalachian history or such a two-dimensional struggle.

All the essays in this volume find greater complexity in the class struggles within Appalachia. Mike Workman charts the development of the Fairmont Field by a handful of powerful industrialists who were natives of West Virginia. Because of the tremendous demand for labor, these operators recruited immigrants to work in the mines. Turning previous histories upside down, Workman finds that native coal operators joined with native miners and community leaders to oppose outsider workers. Another surprising finding is that rather than fight unionization with every resource at hand, a local industrialist in the Fairmont Field instead formed a compact with the miners that paved the way for unionization of local mines during World War I.

Rebecca Bailey's study of the events leading up to the Matewan Massacre demonstrates how actions taken by politicians, coal operators, officials of the United Mine Workers of America (UMWA), and townspeople contributed to the explosion of violence on May 19, 1920, in Matewan, West Virginia. Observers at the time implied that violence was inherent to life in the mountains by noting that the Massacre occurred in the home of the infamous feud between the Hatfields and McCoys. Bailey analyzes the larger historical forces at work that came to a head in the small mining town and uncovers a far more complex set of causal factors than scholars have previously recognized. Not merely one chapter in the larger national labor movement, the shootout on the streets of Matewan resulted from distinctly local circumstances. Bailey's analysis demonstrates that a true understanding of the forces behind the massacre must consider the machinations of West Virginia politicians and the rivalries of local elite families.

The health and welfare of coal miners and their families have been subjects of numerous folk songs but relatively few historical inquiries.[13] Richard Mulcahy discusses the two-tiered system of health care delivery that left Appalachians with fewer doctors and facilities to treat them. After World War II, several individual government administrators, as well as leaders of the UMWA, waged a major battle for important reforms in health care throughout the Appalachian coalfields. Ultimately, this struggle resulted in the creation of the United Mine Workers of America Welfare & Retirement Fund as well as the construction of a chain of clinics across the region. Covering an important part of the twentieth century, Mulcahy describes the improving standard of living in the coalfields that resulted from the creation of the fund, which occurred ironically as the rest of the nation was rediscovering Appalachian poverty. Unfortunately for miners and their dependents, the fiscal crises facing the coal industry all but destroyed the system that the UMWA put into place.

By the 1970s, the health care system of Appalachia became a major employer in the region, and health care workers began to fight for a voice in the workplace and union representation from the Service Employees International Union, Local 1199. John Hennen uncovers the history of Local 1199's first campaign in Appalachia as well as the shifting tactics employed by hospital administrators. Having witnessed many positive effects of unionization in the coal industry, Appalachians working in health care were often very receptive when union organizers—some of whom were also Appalachians—arrived. Far from fatalistic and docile, these health care workers fought an uphill battle. Union organizers and activists faced professional consultants skilled in the strategies of "union avoidance," opening a new chapter in Appalachian labor history.

The final set of essays examines politics in modern Appalachia. Earlier scholarship emphasized the powerlessness of Appalachians in the face of absentee owners and outside investment capitalists. According to such interpretations, reformers often waited for the federal government to intervene while corrupt Appalachian politicians facilitated the exploitation of the region. Of course, the region cuts across several states, each of whose political history has been explored in depth, but only West Virginia is considered to be wholly within Appalachia, according to most definitions. Still, political histories of West Virginia before the 1970s focused primarily on the statehood movement, interpreting it as redress for decades of disenfranchisement. Then, in 1972, John Alexander Williams published an influential article in *West Virginia History* that argued that an entrenched elite in western Virginia used politics to consolidate their hold over land and power. They survived the statehood movement during the Civil War and regained power during the 1870s.[14] Williams's 1976 book *West Virginia and the Captains of Industry* ably demonstrates that the natural resources of the Mountain State were at the center of politics during the late nineteenth and early twentieth centuries, and the West Virginians who controlled the state government and the state's seats in the U.S. Senate eagerly facilitated their exploitation. No longer hapless victims or corrupt proxies, native politicians instead maneuvered in Wall Street circles to aid in the industrialization of West Virginia for their own enrichment.[15]

Building on Williams's work, there have been several important contributions to the historical study of Appalachian politics that rejected both the stereotypical portraits of backwards politicians as well as the internal colonialism that had influenced the way Williams framed his evidence. As Billings discusses in the introduction, John Hennen's *Americanization of West Virginia* uncovered a sophisticated and widespread campaign in the 1920s to inculcate West Virginians with values amenable to industrial capitalism and to narrow the discourse, making dissent un-American.[16] Jerry Bruce Thomas's 1998 book, *An Appalachian New Deal*, addressed a significant gap in the literature. Thomas found that while the New Deal did not result in a full recovery, some of the federal policies and programs "did much to make the Depression more tolerable."[17] Thomas's findings parallel what other historians have discovered in other parts of the nation when they examined the interplay of state and federal politics during the New Deal, belying the notion that Appalachian politicians were any more corrupt or ineffectual than their counterparts in other regions.[18]

The authors of the essays in this section expand that analysis. Jeffrey Cook and Shirley Stewart Burns argue for the agency of Appalachians as political actors, and Mark Myers explores the unintended negative consequences of federal legislation on the coalfields of McDowell County. Jeffrey Cook discusses the efforts by the conservative A. B. Fleming to create a U. S. Bureau of Mines in the wake of the Monongah disaster of 1907. Fleming's lobbying efforts initially raised concern among other coal operators, but Cook shows that profits and productivity motivated the reluctant reformer as much as concern for life and limb. Fleming was able to lobby for reform while remaining true to his belief in limited government and the free market.

Federal legislation influenced life in the coalfields, often in unanticipated ways. Mark Myers explores the ramifications of the National Industrial Recovery Act of 1933 and the National Recovery Administration (NRA) on the coal economy of McDowell County. The NRA's policies supported unionization of the coalfields, which loosened the coal operators' control over life in the coal towns and empowered the miners. The NRA's codes were a mixed blessing as higher labor costs encouraged operators to mechanize mining operations, diminishing employment opportunities in a region where there were few alternatives for steady jobs. Many reformers as late as the 1960s, including Harry Caudill, called for intervention of the federal government to, in effect, save Appalachians from themselves.[19] Myers demonstrates that mountaineers had long worked with the federal government in a variety of ways, but federal intervention did not solve the basic economic and social problems facing Appalachian society.

Many of the political changes in modern Appalachia were the result of grassroots organizing and activism. Shirley Stewart Burns analyzes the emergence of one group of grassroots activists galvanized by the deleterious effects of mountaintop removal on their environment. Writing in the tradition of the seminal volume, *Fighting Back in Appalachia,* Burns weaves together oral histories with written sources to discuss the ongoing social impact of the new mining techniques and places these events within the larger historical context. Coal companies turned to mountaintop removal because the cost of the process was lower than underground mining, but its destructive side effects made life nearly intolerable for nearby residents. Burns chronicles the efforts of these unlikely activists as they attend hearings and seek justice through state agencies and the courts. She also explores the tensions between the community residents and the workers who depend on mountaintop removal for their livelihood. Finally, she analyzes the strategies of companies to buy up whole communities and force individual homeowners off their property. The story of the fight over mountaintop removal could easily turn into another tale of victimization and exploitation, but Burns analyzes the unfolding environmental disaster from multiple angles, eschewing easy explanations.

These essays, together with Dwight Billings's Callahan lecture, give some insight into the variety of topics, approaches, and insights encouraged by the example and the mentoring of Ronald L. Lewis. This volume cannot begin to represent the full range of Ron's contributions to Appalachian history and culture. However, a shelf full of his former students' books should remind him that his legacy as a teacher and colleague is far from over.

NOTES

1   Mary Beth Pudup, Dwight B. Billings, and Altina L. Waller, eds., *Appalachia in the Making: The Mountain South in the Nineteenth Century* (Chapel Hill: University of North Carolina Press, 1995); Ronald D. Eller, *Miners, Millhands, and Mountaineers: Industrialization of the Appalachian South, 1880–1930* (Knoxville: University of Tennessee Press, 1982).

2   Wilma A. Dunaway, *The First American Frontier: Transition to Capitalism in Southern Appalachia, 1700–1860* (Chapel Hill: University of North Carolina Press, 1996); David C. Hsiung, *Two Worlds in the Tennessee Mountains: Exploring the Origins of Appalachian Stereotypes* (Lexington: University Press of Kentucky, 1997); Kenneth W. Noe and Shannon H. Wilson, eds., *The Civil War in Appalachia: Collected Essays* (Knoxville: University of Tennessee Press, 1997); Michael Puglisi, ed., *Diversity and Accommodation: Essays on the Cultural Composition of the Virginia Frontier* (Knoxville: University of Tennessee Press, 1997); Dwight B. Billings and Kathleen M. Blee, *The Road to Poverty: The Making of Wealth and Inequality in Appalachia* (New York: Cambridge University Press, 2000); Kenneth E. Koons and Warren R. Hofstra, eds., *After the Backcountry: Rural Life in the Great Valley of Virginia, 1800–1900* (Knoxville: University of Tennessee Press, 2000); Donald Edward Davis, *Where There Are Mountains: An Environmental History of the Southern Appalachians* (Athens: University of Georgia Press, 2000); Robert S. Weise, *Grasping at Independence: Debt, Male Authority, and Mineral Rights in Appalachian Kentucky, 1950–1915* (Knoxville: University of Tennessee Press, 2001); John C. Inscoe, ed., *Appalachians and Race: The Mountain South from Slavery to Segregation* (Lexington: University Press of Kentucky, 2001); and Benito J. Howell, ed., *Culture, Environment, and Conservation in the Appalachian South* (Urbana: University of Illinois Press, 2002). Certainly some of the essays that appeared in the above-cited collections focused on periods other than the nineteenth century. Furthermore, note that this brief survey does not pretend to be comprehensive; only to paint an impressionistic portrait of the published work in Appalachian history since 1995.

3   Chad Berry, *Southern Migrants, Northern Exiles* (Urbana: University of Illinois Press, 2000); Phillip J. Obermiller, Thomas E. Wagner, and E. Bruce Tucker, eds., *Appalachian Odyssey : Historical Perspectives on the Great Migration* (Westport, CN: Praeger, 2000); Carl E. Feather, *Mountain People in a Flat Land: A Popular History of Appalachian Migration to Northeast Ohio, 1940–1965* (Athens: Ohio University Press,

1998); Susan Johnson, "West Virginia Rubber Workers in Akron," in *Transnational West Virginia: Ethnic Communities and Economic Changes, 1840–1940*, ed. Ken Fones-Wolf and Ronald L. Lewis (Morgantown: West Virginia University Press, 2002); J. Trent Alexander, "'They're Never Here More Than a Year': Return Migration in the Southern Exodus, 1940–1970," *Journal of Social History* 38 (Spring 2005): 653–671.

4   Richard A. Brisbin Jr., *A Strike Like No Other Strike: Law and Resistance During the Pittston Coal Strike of 1989–1990* (Baltimore: Johns Hopkins University Press, 2002); Chad Montrie, *To Save the Land and People: A History of Opposition to Surface Coal Mining in Appalachia* (Chapel Hill: University of North Carolina Press, 2003); Alan Derickson, *Black Lung: Anatomy of a Public Health Disaster* (Ithaca, NY: Cornell University Press, 1998); Shaunna L. Scott, *Two Sides to Everything: The Cultural Construction of Class in Harlan County, Kentucky* (Albany: State University of New York Press, 1995); and Suzanne E. Tallichet, *Daughters of the Mountain: Women Coal Miners in Appalachia* (University Park: Penn State University Press, 2006).

5   Dwight Billings, Gurney Norman, and Katherine Ledford, eds., *Back Talk from Appalachia: Confronting Stereotypes* (Lexington: University Press of Kentucky, 1999).

6   There have been collections that revolved around single topics that included essays about twentieth-century Appalachia including Billings, Norman, and Ledford, eds., *Back Talk from Appalachia*; Howell, ed., *Culture, Environment, and Conservation*; Stephen L. Fisher, ed., *Fighting Back in Appalachia: Traditions of Resistance and Change* (Philadelphia: Temple University Press, 1993); and Fones-Wolf and Lewis, eds., *Transnational West Virginia: Ethnic Communities and Economic Change, 1840–1940*.

7   Harry Caudill, *Night Comes to the Cumberlands: Biography of a Depressed Area* (Boston: Little, Brown and Company, 1962; 1963), 31. Citations are to the 1963 edition.

8   Henry D. Shapiro, *Appalachia on Our Mind: The Southern Mountains and Mountaineers in the American Consciousness, 1870–1920* (Chapel Hill: University of North Carolina Press, 1978).

9   Billings and Blee, *The Road to Poverty*.

10   See especially, Fones-Wolf and Lewis, *Transnational West Virginia*.

11   Ronald L. Lewis, *Black Coal Miners in America: Race, Class, and Community Conflict, 1780–1980* (Lexington: University Press of Kentucky, 1987); Joe William Trotter, Jr., *Coal, Class, and Color: Blacks in Southern West Virginia, 1915–32* (Urbana: University of Illinois Press, 1990).

12   David Allan Corbin, *Life, Work, and Rebellion in the Coal Fields: The Southern West Virginia Miners, 1880–1922* (Urbana: University of Illinois Press, 1981); Eller, *Miners, Millhands, and Mountaineers*.

13   Exceptions include Derickson, *Black Lung*; Sandra Lee Barney, *Authorized to Heal: Gender, Class, and the Transformation of Medicine in Central Appalachia, 1880–1930* (Chapel Hill: University of North Carolina Press, 2000); Ivana Krajcinovic, *From Company Doctors to Managed Care: The United Mine Workers' Noble Experiment* (Ithaca, NY: ILR Press, 1997); and Mulcahy's own *Social Contract for the Coal Fields: The Rise and Fall of the United Mine Workers of America Welfare and Retirement Fund* (Knoxville: University of Tennessee Press, 2000)

14   John Alexander Williams, "The New Dominion and the Old: Ante-Bellum and Statehood Politics as the Background of West Virginia's 'Bourbon Democracy,'" West Virginia History 33 (1972): 317–407.

15   John Alexander Williams, *West Virginia and the Captains of Industry* (1976; repr., Morgantown: West Virginia University Press, 2003).

16   John Hennen, *The Americanization of West Virginia: Creating a Modern Industrial State, 1916–1925* (Lexington: University Press of Kentucky, 1996).

17   Jerry Bruce Thomas, *An Appalachian New Deal: West Virginia in the Great Depression* (Lexington: University Press of Kentucky, 1998), 3.

18   Thomas, *An Appalachian New Deal*, 4–5.

19   See Caudill, *Night Comes to the Cumberlands*, chapter 22, for an example.

INTRODUCTION

Writing Appalachia:

Old Ways, New Ways, and WVU Ways [1]

Dwight B. Billings

APPALACHIA has been written in diverse and contradictory ways for more than a century and a half. With the growth of a multi-disciplinary Appalachian Studies movement, especially during the last thirty years, representation of Appalachia has both changed dramatically and been highly contested. Here, I will examine old and new ways of representing Appalachia, and I will give particular attention to how historians at West Virginia University—especially Ronald Lewis and his colleagues and students—have reshaped our thinking about the region.

It is indeed a pleasure to do so at a conference honoring the contributions of Ronald Lewis. At its annual conference in 2007, the Appalachian Studies Association recognized Ron by honoring him with the Cratis Williams-James S. Brown Award for Appalachian Studies, the association's highest recognition for career-long contributions to Appalachian Studies. The award acknowledges Ron's scholarship which has done so much to advance our understanding of Appalachian history. It also recognizes his important contributions both as a past president of the association who strengthened the association's financial and organizational footing, and as founding editor of the *Journal of Appalachian Studies*. Finally, the award recognizes Ron's outstanding contributions as a teacher, the fruits of which are so evident in the

2007 Rush Holt Conference and in the numerous books and dissertations that have been published by his students at West Virginia University. This work, of course, reflects a collaborative and cumulative process of conversations that have been nurtured in and across many dissertation projects. It is this impressive body of writing that I will try to contextualize.

To do so, I want to sketch four approaches to conceptualizing Appalachia that will show where the academic thinking about the region originated and where it has been heading. One approach, cultural modernization theory, views Appalachia as a traditional or pre-modern culture. (A closely related variant views Appalachia as a region-wide culture of poverty.) A second representation, the model of Appalachia as an internal colony, helped to give birth to the Appalachian Studies movement. The work of WVU historians challenges both of these models and is, I believe, informed by two other theoretical approaches. One is world systems theory, which represents Appalachia as a peripheral region in an advanced ("core") capitalist economy. The other is a class-based model of social, political, and economic development that identifies alternative routes to modernity and shows the lasting institutional effects of what is known as "path dependency." Path dependency grows out of the comparative and historical sociology of Barrington Moore.

First I will sketch, very briefly, the model of Appalachia as a traditional subculture and the underlying assumptions about social change that were tied up with the idea of cultural modernization. Then I will describe how unexpected events in the region led to the breakdown of that approach and how, in the context of the rise of the Appalachian Studies movement, it was replaced by the representation of Appalachia as a colony.[2] From there, I will begin to characterize the work that, for convenience, I shall refer to as that of the "WVU historians"—work that challenges both models of Appalachia as a traditionalist subculture and as a colony.

Old Ways

Scholars in the 1960s sought to explain Appalachia's extremely high levels of poverty as lingering effects of geographical and cultural isolation. They viewed Appalachia as "a semi-autonomous rural social system" and believed that the continuing diffusion of values and cultural orientations from urban

America into the region's rural hinterlands would eventually bring about economic improvements in the region.[3] Their central concern was how to strengthen and speed that process. A conference organized at WVU in 1967 to prompt such change by guiding "action programs" may be taken as indicative of the approach. (Ron Lewis and I wrote about this conference in an early issue of the *Journal of Appalachian Studies*.[4]) The WVU conference, and a widely read book that grew out of it, were addressed to "professional field workers in programs of directed change," including "social workers, Extension agents, Vista Volunteers, Peace Corps members, community development experts, or field representatives of other agencies."[5] Two things are of note here. First, the conference's target audience was comprised of professional change agents, not the community activists that make up an important element in today's Appalachian Studies movement. Second, it is also significant that no historians were involved.

Scholars at the WVU conference recognized that the forces of cultural modernization—effects of improved education, transportation, and mass communications—were already breaking down traditional ways of life in Appalachia. Nonetheless, they believed that two factors—persistent isolation and cultural isolationism—were impeding social improvement. Geographically remote sections of Appalachia were imagined still to resemble a traditional folk culture while, under the duress of rapid change, some Appalachians were viewed as retreating into what WVU sociologist Richard Ball described at the conference as "an analgesic subculture"—that is, a culture of poverty that numbed the effects of economic frustration and failure.[6] Others at the conference pointed to cultural orientations such as "familism" and "personalism" that were hindering social change by insulating rural families from the mainstream cultural values of mass society and contradicting the achievement orientation that was claimed to underlay participation in the modern, urban-industrial American economy. The focus of the conference was how to change such values.

Beyond the WVU conference, such thinking had already been systematized by sociologist Thomas Ford who identified four value dimensions he took to be central to Appalachian culture: individualism and self-reliance, traditionalism, fatalism, and religious fundamentalism.[7] Ford's thinking was popularized in perhaps the best-read book ever published on Appalachia, Jack Weller's *Yesterday's People*, an account of social life in southern West Virginia.[8]

Many of us as young Appalachian scholars cut our teeth on that book. A personal anecdote. I got my start in Appalachian Studies by accident. As a high school junior in Beckley, West Virginia, and the only male that year on the school newspaper staff, I was assigned the distasteful role of sports editor. I'm a big basketball fan now, but back then I hated sports reporting. Most of all, I detested writing profiles of the always-male "sports personality of the month." After all, my articles were about the same guys that used to beat me up in junior high. Fortunately, as soon as another male joined the newspaper staff in the second semester, I was demoted from sports editor to book review editor, a new role created to ease my demotion. The first book I reviewed was *Yesterday's People*. The next year, I wrote a senior term paper on it and, later, I was told to read it when I became a college intern with the West Virginia Department of Welfare. In graduate school, I wrote my masters thesis on Weller and on Tom Ford whom I would later join as a colleague at the University of Kentucky.[9]

I want to make two points about the cultural approach to Appalachia. One is that its assumption of cultural and geographical isolation was based on bad history. As Michael Workman pointed out in his excellent WVU dissertation on the history of industrialization in the upper Monongahela Valley,[10] modernization theory's assumptions about the isolation and homogeneity of Appalachia substituted social theory for social history. I would add that it was also bad—or at least one-sided—social theory.

Modernization theory grows out of the work of Talcott Parsons who maintained that under ideal conditions social systems are integrated and directed by consensus over shared values.[11] Values tell social actors what goals to pursue and how to define social situations. Cultural values also influence the norms that regulate social practices. Finally, through childhood and adult socialization, values are internalized by individuals and function as psychological motivations. That is, values shape norms, which in turn shape motives. Hence we get the formula, stated by sociologist Rupert Vance in his preface to Weller's *Yesterday's People* that "to change the mountains is to change the mountain personality."[12]

Modernization theory thus essentialized the process of change by characterizing an idealized version of how the United States had supposedly developed—based on a consensus around middle class values—as the one sure path for other, less-developed societies (or regions) to follow in their quest for stability and prosperity. Faith in this model, however, was soon shaken. (I will take up the theme of multiple paths to modernity a little later.) The eventual breakdown of cultural modernization theory in the late 1960s and 1970s, nationally and in Appalachia, came about, in part, because of what was taking place in the streets. A highly abstract account of how social systems change, modernization theory had paid remarkably little attention to human agency. Consequently, its proponents were embarrassed by historical events in Appalachia, the nation, and throughout the world that the theory had little grounds to predict and even less power to explain, i.e., the rise of an era of social activism.

Conflicts, social movements, and power struggles in the 1960s and 1970s dramatized the potency of social activism in American life. Such events were paralleled internationally by comparable struggles in other industrial societies and by anti-colonial and anti-imperial revolts in so-called "modernizing" societies.

In Appalachia, roving pickets and work stoppages by coal miners fighting the loss of union representation; the War on Poverty, poor people's campaigns, and battles to control local community action programs; and struggles over schools and textbooks were only a few of the many grassroots conflicts that affirmed the undeniable role of activism in Appalachia. There were also widespread efforts to stop surface mining and protect local environments, to achieve equity by fairly taxing absentee land owners and levying taxes on unmined minerals, to protect local landowners (in Kentucky) from "broad-form" deeds to mineral rights that allowed mining companies to override the interests of surface owners, and to improve local health and human services. Actions spoke louder than words, testifying clearly against images and models of Appalachians as tradition-bound and fatalistic. Further, when Appalachian peoples struggled collectively at work to expand and democratize unions, to win recognition of and compensation for occupational illnesses, to make workplaces safer, or to make room for women in jobs—such as coal mining—traditionally defined for men, they challenged longstanding stereotypes of themselves as individualistic and incapable of cooperation beyond the bonds of kinship.[13]

In the context of these events, Appalachian scholars began to move beyond a narrow focus on culture to investigate political and economic power and the role of social actors and social movements in making history—issues that, if never altogether banished, had been relegated to scholarly margins. Significantly, new recognition of the importance of power and history also led to the re-conceptualization of Appalachian culture itself—a more robust understanding of culture as involving more than just values, combined with a new emphasis on social history in the context of a multidisciplinary Appalachian studies movement.

Now I want to talk about the emergence of the multidisciplinary Appalachian Studies movement. I think that at least three elements helped give it shape. First was this flourishing of social activism throughout Appalachia that stimulated what many observers have termed an "Appalachian Renaissance" among activists, scholars, educators, writers, and artists. These regional actors began to explore and deepen the understanding of Appalachia's historical and geographical diversity; its cultural resources, traditions, and creativity; and its expanding forms of social participation and engagement. Appalachian research centers and studies programs began to appear on college campuses throughout the region in the 1970s and 1980s when academic journals and multi-media centers like Appalshop in eastern Kentucky were also established. Prompted by the need for a region-wide response to the destructive floods that struck the region in the late 1970s (seemingly brought about by the intensification of surface mining), Appalachian Alliance was formed. The alliance was an advocacy group composed of membership organizations as well as individual activists and educators. Since the idea of a multidisciplinary Appalachian Studies Association was first discussed in the context of the Appalachian Alliance, it is not too far off the mark to say that grassroots activism helped give birth to Appalachian Studies. The Appalachian Studies Association was established in 1977 "to coordinate analysis of the region's problems across disciplinary lines," "to relate scholarship to regional needs and the concerns of the Appalachian people," and to "overcome the separateness of regional activists, citizens, and academics."[14]

A second influence was the Highlander Center. Located in Appalachian Tennessee and a leader in the training of labor and civil rights activists, the Highlander Center played a prominent role in redirecting the style and focus of research on Appalachia.[15] Scholars associated with the center in the 1970s and 1980s, such as John Gaventa and Helen Lewis, promoted a blend of scholarship and activism by advocating participatory action research. A prime example was the landmark study *Who Owns Appalachia?*[16] The task force, consisting of both activists and academics, trained local indigenous researchers (that is to say, citizen researchers) to investigate mineral rights, land ownership, and taxation on over 20 million acres of land in eighty counties spanning six states in the region. Its documentation of vast amounts of absentee-owned, but minimally taxed land and mineral resources spearheaded tax reform efforts in several Appalachian states and led to enduring grassroots organizations such as Kentuckians For The Commonwealth (KFTC), presently the largest progressive statewide organization in Kentucky and a group currently at the forefront of that state's battle against mountaintop removal coal mining.

Participatory research remains one of the new ways of writing Appalachia. A recent issue of the *Journal of Appalachian Studies* (Spring/Fall 2005), for instance, published three articles on environmental issues that were the result of faculty, undergraduate student, and community collaboration. Traditional academic writing as well has been reshaped by its connection to activism as the 2007 Rush Holt Conference session on "Ronald Lewis's Contributions to History and Activism" made clear. And from the other direction, activists learned that re-writing local history, conducting oral history interviews, and even staging plays and pageants that reinterpret the past are important elements of community renewal and mobilization.[17]

Finally, a third influence on Appalachian Studies was a strong current of postmodernism that also influenced parallel academic efforts that likewise grew out of social movements and grassroots activism such as Women's Studies and African-American Studies.[18] Postmodernism encouraged younger scholars to be critical of the God's-eye view of objectivistic approaches like cultural modernization theory, to be more sensitive to the power relations embedded in the production of scholarly knowledge, and—again—to be more open to dialogue and partnership between citizens and scholars. Under the influence of postmodern sensibilities, the universalism and essentialism of the 1960s modernization theory gave way to greater stress on difference and diversity in the region. Unitary notions of Appalachia and Appalachian identities have been replaced by plural and complexly constructed conceptions of the region and its peoples. This recognition of regional diversity, along with the intersection of racial, national, ethnic, and gender identities with class, is a hallmark of how Appalachian history is being written at WVU.

Postmodernism in Appalachian Studies also brought attention to representation and writing. Rather than transparent vessels that convey truths without influencing their content, representations are now understood to help shape what can been seen and accepted as true, leading to important studies of the discursive formation of the very idea of Appalachia as a unified region and the role of stereotypes in both popular and academic writings. Henry Shapiro's intellectual history of the idea of Appalachia, *Appalachia on Our Mind: The Southern Mountains and Mountaineers in the American Consciousness, 1870–1920*, examines how local color writers, missionaries, settlement house workers, folk song collectors, leaders of the handicraft revival, spokespersons for mining and railroad interests, and regional intellectuals established the myth of Appalachia as a "strange land and peculiar people."[19] Other scholars have followed Shapiro's lead by looking at the continuing development of Appalachia representations throughout subsequent decades and in other media including film and television.[20]

What these books have in common is the effort to redirect attention from "actual" mountain life to the how myths about Appalachia are constructed in discourse. Showing that assumptions about the isolation of Appalachia and the tenacity of its traditional values are just that—assumptions (i.e., assumptions that are the product of a way of writing about the place rather than the reality of the place), offered a powerful way to challenge cultural modernization theory. But so, too, did the writing of history.

One of the most important efforts on the research agenda of early Appalachian Studies was documenting the historical impact of extractive industries such as coal mining and timbering on communities, patterns of economic development, and poverty. The impulse was to show that Appalachia was poor because of the nature of its integration with—not isolation from—the wider American economy. By shifting attention from culture to coal—as Helen Lewis once put it, from fatalism to the coal industry—Appalachian scholars responded critically to the representations of Appalachian deficiency in cultural modernization and culture of poverty theories by asserting that the region's problems should be seen as the result of economic exploitation and political domination, not outmoded values, norms, and attitudes.

Landmark studies that put history squarely at the center of Appalachian Studies include Ronald Eller's *Miners, Millhands, and Mountaineers*, John Alexander Williams's *West Virginia and the Captains of Industry*, and John Gaventa's *Power and Powerlessness*.[21] Each of these studies described the political economy of Appalachia as a colony. Appalachia was characterized by a rich land but a poor people, as a national television documentary put it, because the region's natural resource wealth flowed outside the mountains like that from a colony to its imperialist ruler.

Helen Lewis and her co-authors Linda Johnson and Sue Kobak formulated a highly influential model of Appalachian colonization that emphasized four historical phases of development.[22] First, the colonizers gained entry and achieved control over Appalachia's resources. Then they solidified control by eliminating opposition and resistance. Next, they changed the values of the local social system by educating and converting the natives. Finally, they have maintained control through lasting forms of political and social domination.

Despite what I said about postmodernism earlier, it should be obvious that the turn to history did not yet signal an end to essentialism. The metaphor of Appalachia as a colony replaced that of Appalachia as a backward culture, but the mythical unity of the region and the homogeneity of its population remained largely unquestioned. Even the cultural patterns associated with Appalachia by cultural modernization theory such as familism and religious fundamentalism were left intact by proponents of the colonial model, although to be sure, they were reinterpreted as defensive reactions to colonization rather than simply the persistence of traditionalism.[23]

An even greater problem is that talk about insiders and outsiders obscured class differences within the region. The impacts of absentee ownership on communities eclipsed the analysis of class exploitation in coal mining. "Natural" wealth in the form of coal was stolen from rightful, i.e., "inside," property owners by the smooth-talking agents of "outside" corporations who bought mineral rights on the cheap.[24] Off the hook and out of theoretical sight were the capitalist labor processes that paid the wage-laborers who mined the coal too little in exchange for the value of the commodity they produced. Therefore, ownership and theft were stand-ins for class and exploitation in the early drama of Appalachian Studies. While the colonial model claimed that the wealth of Appalachia lay underground, stolen from its rightful owners, the truth of the matter is that property ownership had little to do with it other than to legitimate the transfer of wealth. The colonial model proved unable to see that Appalachian wealth was created by the underpaid labor of working people who blasted, dug, sorted, cleaned, loaded, and transported coal.[25]

Despite the glaring absence of class analysis, or perhaps for that reason, the representation of Appalachia as a colony captured the imagination of regional scholars and many activists alike. The collaborative study of absentee land ownership was its finest fruit.

New Ways

Now I want to turn to new ways of writing Appalachia. For the reader willing to read only one book on Appalachia, Ron Lewis's *Transforming the Appalachian Countryside* is the one I'd recommend.[26] It's a terrific, comprehensive look at how one industry—timbering—transformed the non-coal mining, "backcountry" counties of West Virginia in the thirty years from 1880 to 1910. At the beginning of that period, two thirds of West Virginia was still a virgin forest, an "underdeveloped wilderness" as Ron puts it. By the end of the period, also in Ron's words, "entire ecological systems had been destroyed, the backcountry had been tied to the market system, independent farmers had become wage hands, and the political system had been transformed into a mechanism for the protection of capital."[27]

Ron shows that at the beginning of the era, farmers of the West Virginia backcountry still made their living with a mix of subsistence and commercial practices. Poor transportation in the backcountry meant that food had to be grown for home consumption but the sale of livestock provided the backbone of the commercial agricultural economy. Foodstuffs were too bulky for easy transport but livestock, cattle especially, could be walked to the regional gathering places and then led to distant markets by professional drovers. This system made Appalachia the livestock raising center of the United States, but the mix of subsistence and commercial farming strategies required extensive free-range grazing that the forest commons provided. Likewise, subsistence agriculture required large amounts of land for the rotation of fields so that grounds could lie fallow and be replenished. Penetration of the backcountry by railroads, the transfer of lands to timber companies, and the massive deforestation that followed closed the door on this rural way of life. At the end of the era, only commercial farming was viable and few farmers were able to make the difficult transition to it.

Besides its impact on farming, Ron looks at how timbering also led to industrial wage labor. By 1910, thousands of West Virginia workers were employed in the forest industry cutting trees, sawing wood, transporting lumber, tanning leather, and making wood and pulp products. New towns, many of them centered around giant saw mills, were established in once remote locales and old towns took on new economic functions. Ron explores the labor processes in timbering and he looks at how local elites, eager to be the beneficiaries of heightened land values, competed with one another over where railroads would be routed and over the location or relocation of county seats. Such "county seat wars," as Ron describes them, resemble in some ways the inner-elite conflicts over the economic development of natural resources in Kentucky which were known as "feuds." (On feuds, I should point to the decisive work of Altina Waller, a former member of the WVU history department.[28])

The result of this pattern of development was, as the case of farming suggests, completely unsustainable. Once the great forests were consumed, loggers moved on to other regions. It was, as Ron says, cut out and get out. Competition and overproduction—factors similar to those that plagued the Appalachian coal industry as well—led to "monumental waste." Clear-cutting led to erosion, pollution, and floods, and millions of acres of land lay in ruin. Ron's descriptions of post-timbered West Virginia makes the aftermath of Armageddon look like Eden, images that evoke today's mountaintop removal mining, but on a much vaster scale.

Finally, Ron looks at the political order that was established in West Virginia to protect industry from the consequences of such devastation including changes in the law. Whether the cow hits the locomotive or the locomotive hits the cow, it's bad news for the cow. In West Virginia, the dead cow was the farmer's problem, not the railroad's. The old Virginia legal theory of natural rights—rights that protected agrarian property, above all slave property—were superseded in West Virginia by the theory of positive law that stresses correct procedures of enactment over long-standing rights, that is, procedural correctness over community standards of morality. Law became simply what you made it and industrial corporations were the beneficiaries, not farmers whose customary rights were abridged.

Ron uses these legal changes to gauge the building of a modern capitalist system in West Virginia. They were part of a wider political process that severed western Virginia's ties to the eastern slaveocracy of the Old Dominion. I will return to how Ron and his colleagues have dealt with state building, but first I want to comment on his departure from colonial theory.

Ron acknowledges that there is much in the story of timber to recommend the theory of colonialism. Absentee capitalists did, after all, cut and run. But their entry was hardly a sudden invasion as the theory implies. Speculators from outside the mountains had long claimed the lands of western Virginia and it took many years of litigation for local backcountry elites and the state's rising capitalists to thwart their claims. Barbara Rasmussen's important WVU dissertation, published as *Absentee Landowning and Exploitation in West Virginia from 1760 to 1920*, bears this out nicely.[29] Even more importantly, however, from the standpoint of challenging the colonial model is the fact that extremely powerful West Virginians led the process of developing the backcountry. West Virginia capitalists like Johnson Newlon Cameron, Henry Gassaway Davis, and Stephen B. Elkins, among others, built the railroads, organized the land acquisitions that made timbering possible, and represented the interests of corporate capital in state politics and Washington, D.C.[30] Of course, West Virginians were hardly alone in exploiting the backcountry. Ron analyzes corporate records to show that half of the firms involved in the timber industry were indeed absentee (that is, out of state) but 71 percent of the individuals named as officials in them were native West Virginians. West Virginia capitalists made their fortunes by developing railroads or from land sales while outside entrepreneurs built most of the large milling centers.

Rather than internal colonialism, Ron uses Immanuel Wallerstein's world systems theory to portray the backcountry timber counties as a peripheral region in a core capitalist economy.[31] The two theories are kin but distinct.

World systems theory divides the world into "core" regions where knowledge, capital, and manufacturing are centered, and "peripheral" regions where cheap (often forced) labor, primary production like farming, and natural resource extraction do the heavy lifting. ("Semi-periphery" regions do a little bit of both.) Most importantly, according to Wallerstein, economic exchanges between core and periphery regions typically reproduce or extend the inequality between zones since the services and products of the periphery are less highly rewarded that those of the core. The periphery sells cheap and pays dear. West Virginia's economic and political leaders made handsome profits for themselves by yoking the underdeveloped backcountry to "core" industrial capitalism through railroad building and timbering, but they guaranteed a road to peripheralization for the areas they opened up.

World systems theory does much to bring attention to the structural inequalities between core and periphery but it focuses too narrowly on exchange, trade, and commerce. It needs to be complemented by analyses of class processes as well.[32] Nations and regions do, sometimes, upgrade or loose their positions in the world system order, and it is class processes, combined with state actions, that help to explain these transitions. The WVU historians understand this fact, and they also know that the backcountry was not West Virginia's only story.

Michael Workman's important 1995 dissertation on industrialization and the development of mining in the Upper Monongahela Valley and Ken Fones-Wolf's recent study of the glass towns of the same region offer contrasting pictures of urban, commercial, and industrial development to that of the cut-over backcountry.[33] Both show how antebellum railroads and, later, river improvements in northern West Virginia, led to more complex class structures than those of the backcountry. A market economy, a modernizing political culture, and the better balanced growth of diversified commercial agriculture—mixed with mercantile firms, mills, foundries and machine shops, glass plants, artisan crafts, and manufacturing including the making of mining machinery—added to coal, coke, oil, and gas production to make northern West Virginia a more dynamic place than the southern West Virginia or the timber-rich backcountry. An interurban trolley system linking cities to outlying mining communities further enhanced economic integration and the growth of middle class and professional career opportunities.

According to Workman, even the development of coal mining, portrayed in essentialist terms by the colonial model as the Appalachian absentee-owned industry *par excellence*, was indigenous to the Upper Monongahela Valley. Local capitalists, led by members of the Watson family, consolidated mining operations around Fairmont and undersold their competitors in other coalfields through the operation of non-union mines. In 1903 the Watsons merged with the Consolidation Coal Company of Maryland to establish the largest coal company operating in the United States up until 1928. Consol was a big player in southern West Virginia and it partnered with eastern Kentucky's most important local capitalist, John C. C. Mayo, to help develop that region's coalfields as well.

Workman suggests that the so-called Fairmont Ring, as it was known, operated with a regional identity that helped to promote the well-being of the local economy. The Watson group's role in building the interurban transit system is a good case in point, but I'm not entirely convinced of Workman's argument. The finding that Consol—its local genealogy obscured by the infusion of Rockefeller capital—was really the creation of our long-lost hillbilly cousins from northern West Virginia does little to negate David Walls's observation that no one "would seriously suggest that the region would be better if all the coal companies were owned by the local elite of 'hillbilly millionaires.'"[34] Walls, of course, was arguing for a model of class analysis instead of the colonial model's myopic stress on absentee ownership.

A better case for the potential benefits of indigenous development, however, is made in Ken Fones-Wolf's recent study of *Glass Towns*. In this important new work, Fones-Wolf shows how the window glass, bottle glass, and tableware industries helped to chart a distinctive road to modernity in northern West Virginia that contrasted with that favored by the coal industry even though, ultimately, its promoters lacked the power of coal. He shows as well how glassmaking played out differently in towns like Moundsville, Clarksburg, and Fairmont that, in part, grew because of its influence. What might, for shorthand, be called "the glass road" involved an economic development strategy favored especially by certain state Republicans who stressed high wage manufacturing, protective tariffs, and infrastructure investments, including education, over the low-tax, low-wage policy favored by most West Virginia Democrats and extractive industries like mining. The latter strategy, I would suggest, was consistent with the postbellum Southern road to modernization. The development of a home market for local energy resources like coal and gas might, indeed, have helped offset the high costs of transporting West Virginia coal to distant markets, a factor that augmented the need to keep labor costs in mining so low. Fones-Wolf's book offers a sophisticated and complex rendering of the diverse agents and contradictory political-economic forces and policies—including those sponsored by conflicting elites and variously favored by working class and ethnic groups—that were in contention in the making of "modern" West Virginia.[35]

It is significant that all three WVU scholars—Lewis, Workman, and Fones-Wolf—each rethink the statehood movement in West Virginia. During the antebellum period, long simmering conflicts over taxation, internal improvements, and political representation had pitted elites in northwestern Virginia who favored economic and political modernization against the conservative planters of eastern Virginia who politically dominated the Old Dominion. Workman, in particular, refers to West Virginia's achievement of statehood as the "bourgeois revolution of 1863" while making reference to Barrington Moore's analysis of the Civil War as a struggle for dominance between bourgeois capitalism and slaveholding.

I find this really interesting. I lived with a cat for 20 years who I named Barrington Moore because of my admiration for both the cat and the sociologist. In contrast to cultural modernization theory that essentialized a one true path to modernity and reduced it to the adoption of a middle class value scheme, Moore identified three principle roads to modernity.[36] One was the bourgeois route of middle class revolution where the bourgeoisie of England and France overthrew old landlord regimes to bring about change. In this vein, Moore interpreted the American civil war as a similar struggle between two competing and antithetical class systems, bourgeois capitalism and slaveholding. Another route was revolution from below, the communist path of the Russian and Chinese revolutions. Most interesting to me is Moore's concept of revolution from above where the old dominant agrarian classes, the Junkers of Germany or the Meiji of Japan, brought about economic modernization without political emancipation, binding industrialization to authoritarian rule and sometimes culminating in fascism.[37] In a book titled *Planters and the Making of a New South*, I used this model to counter

C. Vann Woodward's interpretation of the New South as a society brought about by the fall of the planter aristocracy after the Civil War and the rise of a new class of entrepreneurs unconstrained by having to maintain a slaveocracy. Instead, I tried to show that the old planter class both survived the Civil War and built a new, industrial, but undemocratic, society in the Upper South. Disenfranchisement of African Americans and the passage of Jim Crow laws allowed postbellum landlords to defend the racial caste system, impose labor-repressive measures in the New South industries they built such as textiles, and to use the state—once purged of African-American and populist voters—to bring about economic modernization policies like road building and educational improvements. [38]

Michael Workman shows how the Baltimore and Ohio Railroad linked northern West Virginia to capitalism and northern markets during the antebellum period while Kenneth Noe shows that the Virginia and Tennessee Railroad linked the counties of what would become southern West Virginia to the Confederacy.[39] If northwestern Virginia indeed had its bourgeois revolution of 1863 with the founding of West Virginia, what about the southern counties of the state that favored secession and identified more with Richmond than Wheeling, but were nonetheless annexed into the new state? The WVU authors tell us that southern West Virginia Democrats achieved a partial counter-revolution in the 1870s by eliminating the Yankee township system of governance that the state founders had designed in favor of a return to the Virginia county-government system that favored rural elites. But I wonder how else the southernism of southern West Virginia may have influenced development in the counties below the Kanawha River? And what about Appalachian Virginia, Kentucky, and Tennessee? The WVU historians invite us to question whether the still-lasting effects of the colonial model as the way to write Appalachia have led us away from important questions about subregional differences in the configurations of race and class throughout Appalachia.

Sociologist Larry Griffin recently shocked an audience at a recent Appalachia Studies Conference and readers of the *Journal of Appalachian Studies* with his report that contemporary opinion polls reveal that, today, white Appalachians identify more strongly with the South and adhere even more vigorously to arguably racist ("neo-Confederate") views on race and region than do whites anywhere else in the South.[40] Recent studies of postbellum lynching, as well, suggest that apart from its mining communities, Appalachia looked a lot like the rest of the South in terms of the prevalence of racial violence.[41]

One of the new ways of writing Appalachia involves weighing the impact of slaveholding on early Appalachia.[42] But what about slavery's lasting impact on Appalachia after legal emancipation? Have we neglected to question how the making of a postbellum racial order in Appalachian states other than West Virginia has shaped the region ever since? Barrington Moore's fundamental insight was that rural class relations, including slavery and other forms of coerced labor, set certain societies—and I would add regions[43]—on distinctive paths or routes to modernity, shaping whether they chose to foster or resist industry and democracy. While I don't think we have exhausted this insight, I know of no better demonstration of such path dependency than Ron Lewis's *The Black Coal Miner in America*.[44]

Having already explored the importance of slavery in the industries of antebellum Virginia and Maryland in an earlier book,[45] Lewis next analyzed the regionally specific paths by which African Americans entered or were barred from the coal industry. He shows that in the northern coal industry of Pennsylvania, Ohio, and Illinois, coal mining jobs were defined as jobs for whites only. Memberships in the UMWA and the KKK appear to have overlapped as white miners fought to exclude blacks from the coal industry. By contrast, the states of Georgia, Alabama, and Tennessee routinely leased convict labor to mining companies. Here, since much of the prison population was black, racial oppression, coal mining employment, and economic development intersected in a distinctly southern fashion. The West Virginia coalfields followed a different route. Hoping to maximize labor control through division, West Virginia employers preferred what they referred to as a "judicious mixture" of the native white, black, and immigrant European workers. Significantly, according to Lewis, all three groups entered employment in the coalfields of central Appalachia on more or less even footing. The "ironic fruit" of this judicious mixture, as Ron writes, was "that within a generation black and white miners had not only accepted equality as a democratic ideal, at least in rough outline, but had come to identify along class lines on economic questions. This was, of course, the foundation of the interracial unity which united the southern West Virginia miners against the operators."[46]

Lewis points out that social and political segregation was written more deeply into the state constitutions of Kentucky, Tennessee, and Virginia than in West Virginia and, for this reason, African Americans preferred jobs in the West Virginia coal industry. To Lewis's work, then, must be added William Trotter's insights into the formation of black political power in the southern West Virginia coalfields, as well as Connie Rice's important study of the role of African-American lawyers in northern West Virginia, such as J. R. Clifford who used the courts to ensure that separate institutions for African Americans were equally funded and directed by African Americans.[47]

WVU Ways

As one might imagine, the theme of miners against operators runs through quite a number of the WVU history dissertations—a theme that transforms commonplace assumptions about Appalachian fatalism into questions about agency and activism. Thus Paul Rakes looks at the history of coal mining accidents in Appalachia and workers' struggles to challenge the disregard for human life in the industry.[48] Richard Mulcahy examines the UMWA health and retirement fund as a fascinating, if ultimately doomed, system of private social security that has important lessons for us today as the links between employment, benefits, health care, and welfare are increasingly severed.[49] Turning a lens on activism in southern West Virginia, Rebecca Bailey asks why Matewan? Why Mingo? In a probing analysis of the infamous Matewan massacre, Bailey looks beneath the legend of Sid Hatfield to ask why, surprisingly, law officials in this southern West Virginia county chose to side with miners rather than operators. One familiar only with John Sayles's movie *Matewan* might imagine it was solely strength of character that led Hatfield to stand up for what was right. But Bailey shows that UMWA officials selected Mingo as the site of an organizing drive in the first place because of the (correct) belief that cleavages within the divided local elite could be manipulated. Her dissertation, now published as *Matewan Before the Massacre*, offers insights into the formation of the southern West Virginia coalfields, that, especially when read against Michael Workman's study of the Upper Monongahela Valley, point to the kind of sophisticated local studies that we still have too few of in Appalachian Studies. Such studies could begin to address the questions about race and class outside of West Virginia that I raised above.

Carletta Bush's recent dissertation on the role of religion in the radical class politics and organizing efforts of miners in Harlan County, Kentucky, during the 1930s provides another exemplary study of local class activism.[50] She shows that Pentecostal holiness ministers provided much of the leadership and legitimacy for the miner's insurgency there, even during Harlan County's radical flirtation with communist-led unions. She shows that unions were so important to working-class congregants of the Church of God throughout the coalfields that that Appalachian—now world-wide—denomination had to abandon its prohibition against membership in secular organizations like unions in order to survive. Surprisingly, religion is a somewhat neglected topic in Appalachian Studies. Carletta does much to sever the presumed connection between religion and fatalism, and to probe the contributions of religion to coalfield activism.

Shirley Stewart Burns has connected this history of coalfield activism to today's struggles against mountaintop removal coal mining (MTR) in a recently published dissertation.[51] She analyses the impact of MTR on the environment, people, and communities of southern West Virginia, including its economic benefits and costs. She also examines the UMWA's role in promoting MTR; the history of legal challenges by citizens; and the relationship between the coal lobby, West Virginia's political leaders, and campaign finance.

In Harlan County, Kentucky, in 1931, Florence Reece wrote one of the most frequently sung songs of the American labor movement, "Which Side Are You On?" proclaiming that "If you go to Harlan County, There are no neutrals there, You'll either be a union man, Or a thug for J. H. Blair." At about the same time, another Harlan Countian, Sarah Gunning, wrote one of the most radical songs in the Appalachia folk song treasury that ended with the exhortation, "Let's sink this capitalist system in the darkest pits of Hell."

The strength of such opposition by miners and their wives suggests that Appalachian capitalists had much more to do than simply build railroads and coal towns to produce coal, monumental as those tasks were. They had to prevent unionization as well, since a comparative advantage of Appalachian coal lay in the ability to keep labor costs as low as possible. As if taking a lesson from the Italian theorist of revolution, Antonio Gramsci, they augmented the use of force, especially the notorious system of private mine guards, with struggles to win the hearts and minds of their subalterns, miners and the small middle class, as well. John Hennen analyzes this struggle for hegemony in his study of the Americanization of West Virginia in the period from 1916 to 1925.[52] As his subtitle ("Creating a Modern Industrial State") suggests, Hennen eschews the representations of West Virginia as a colony or a subculture of poverty like his fellow WVU historians. Instead, he represents West Virginia's economic and political development—and by extension, Appalachia's—in the broader context of the development of the corporate capitalist American state at the end of the World War I period.

Gramsci conceptualized the state as the effective combination of force and hegemony.[53] After World War I, West Virginia's capitalists sought to build a new corporate industrial state—can we say, an extension and transformation of the bourgeois revolution of 1863?—by sanctifying free enterprise and securing the loyalty of workers and citizens to corporate authority. Hennen shows how industrial ideologues elicited the support of Sunday schools and public schools, organizations such as the American Legion, and local chambers of commerce in addition to civic clubs like Rotary, Kiwanis, and women's clubs, in a wide ranging campaign to identify unionism with Bolshevism and Americanization with loyalty to the new corporate order. He demonstrates that the model for this hegemonic project was provided by the mass propaganda efforts to mobilize the state's population for participation in World War I. Significantly, Hennen shows that this post-war effort was led, in part, by WVU historians including the scholar for which this Callahan Lecture is named.

Elizabeth and Ken Fones-Wolf examine similar right-wing methods deployed by Weirton Steel in a successful effort to defeat a unionization drive among steel workers in 1950.[54] Company leaders appealed to a blend of Cold War Americanism, patriotism, welfare capitalism, and fears of foreign subversion as celebrated in community pageantry. Their work, like Hennen's, offers a chilling parallel to similar ideological maneuvers in post-911 America.

Works such as these take us far beyond the narrow identification of culture with values that Appalachian scholars inherited from modernization theory. They utilize a much richer conceptualization of culture that, in addition to values, analyzes how culture is objectified in texts and creeds; how it is produced and performed by active agents of cultural intervention in speeches, ceremonies, and public events; and, finally, how it is institutionalized through the deployment of power and organization.[55] This richer approach to culture is also evident in Sandra Barney's *Authorized to Heal*.[56]

Feminist historians like WVU's Barbara Howe, the author of pathbreaking works on women in the West Virginia economy, have been showing us for a long time how writing Appalachia changes dramatically when we employ the lens of gender analysis.[57] In *Authorized to Heal*, Barney examines the profound transformations in health care that accompanied industrialization and the transition from subsistence to market in Appalachia. She renders a non-stereotypical account of how traditional or vernacular health practices, led by midwives, were displaced by professionalized medicine in the Appalachian coalfields. In a complex account that I cannot do justice to here, she argues that maternalist reform impulses, expressed by such dissimilar agents as middle class club women and settlement house workers, got linked to the ambitions of university-trained physicians. The latter sought to secure market advantages over both traditional healers and the older, less formally trained doctors of the mountain middle class by the monopolization and legitimation of so-called scientific medicine. Club women, women reformers, and public health nurses played key roles in this crusade, but their visions of public health were eventually marginalized by the hegemony of fee-for-service providers whose authority they had helped to construct in the first place. Barney's historical narrative puts meat on the bones of Foucault's theories of the subjugation of subversive knowledges and the rise of disciplinary power, and on Jurgen Habermas's critical theory of the colonization of the communicative lifeworld by the systems media of money and bureaucratic power. For its combination of rich cultural analysis, political economy, and gender, I can think of no more fully realized contribution to Appalachian Studies than *Authorized to Heal*. It's a favorite of mine that I use repeatedly in teaching.

Finally, I want to mention one more new way of writing Appalachia that the WVU historians have pioneered and this is the analysis of transnational migrations, ethnicity, and labor struggles in the region, a signal contribution of their historical corpus. No single work does more to shatter the essentialistic images of Appalachia as a culturally homogeneous and isolated region than the volume edited by Ron Lewis and Ken Fones-Wolf titled *Transnational West Virginia*.[58] Not only does this volume make visible overlooked peoples of Appalachia—Irish Catholic railroad workers, Belgian glassmakers, Swiss farmers, Jewish merchants, German factory workers, Welsh iron workers, and black and Italian coal miners—but it documents how these ethnic groups indelibly shaped community, politics, work cultures, and activism in the Appalachian region.

I will suggest how this approach works with two examples from the volume that have since been published in book-length treatments. I have already discussed generally how Ken Fones-Wolf's analysis of glass manufacturing contributes to understanding the political economy of northern West Virginia. But more specifically, his study situates manufacturing and craft work in the West Virginia glass industry in relation to patterns of technological change, industrial and manufacturing restructuring and plant relocation, internal and transatlantic migration, ethnicity, community formation, unionization, labor force segmentation, and local politics. Further, he shows how these forces were impacted by the transnational transactions of Belgian craft workers. Not only did these labor aristocrats frequently cross national and oceanic spaces to find work, but they struggled on both sides of the Atlantic to control labor conditions by regulating the supply of labor and struggling to maintain craft work imperatives. In West Virginia, when mass production threatened craft work, they also created at least a temporary niche by creating worker-owned cooperatives to perpetuate hand blowing in the window glass industry. What is most impressive and important about this study is Fones-Wolf's skillful handling of materials in the United States and Belgium to show the transnational character of these labor struggles, especially the role of the Knights of Labor's Local 300 in organizing glass workers in both Belgium and America. With workers and ideas flowing back and forth across laboring spaces, Belgian workers pursued conservative forms of craft unionism that were suited to the conditions in Belgium while they experimented with forms of craft socialism in the very different Appalachian context. *Glass Towns* is a powerful demonstration of how the method of transnational study at once reshapes our understanding of both the local and the global.

In her important new book, *Coalfield Jews*, Deborah Weiner uses world systems theory to show that global forces and tensions of capitalist industrialization simultaneously pushed Jews from shtetl origins in Eastern Europe while pulling them to the new industrializing coalfields of Central Appalachia.[59] Here, she shows, Jews drew upon old-country, "middlemen" traditions of commerce to play a prominent role in the emergence of middle class institutions, the advance of consumer culture, and the civic, religious, and political life of the region's small towns. Many first entered the region on horseback as pack peddlers but soon became small-town retailers. More than anything else, Weiner contends that the boomtown atmosphere of rapid economic and social change in Appalachia enabled them to successfully establish a commercial niche in coalfield communities.

According to Weiner, a key to Jewish success in Appalachia came initially by securing a viable economic niche separate from the competition of company-owned stores in the "coal camps" and from already established merchants in the older county-seat towns of the region. Jews first did best in the boomtown milieu of Appalachia's small, independent towns before a second generation began to re-concentrate in county-seat towns like Beckley, West Virginia. By the 1950s, however, the mechanization of coal mining led to the massive depopulation of Appalachia's industrial villages while, at the same time, the growth of shopping malls and chain stores undercut the vitality of downtown shopping districts. As older merchants retired, and many of their children chose to leave the region for careers elsewhere, the Appalachian Jewish population began to diminish. By the 1970s, only a remnant remained, working primarily as professionals in a few of the larger coalfield communities that offered extensive medical and government services.

That two of the most important recent contributions to Appalachia Studies would be about transnational processes linking Belgian workers, Jewish merchants, and Appalachian development represents a powerful statement about how far the practice of history at WVU has moved Appalachian Studies away from the notion of Appalachia as a "semi-autonomous rural social system" that we inherited some 30 years ago.

NOTES

1   This introduction was first given as the Callahan Lecture in History at West Virginia University in conjunction with the 2007 U.S. Senator Rush Holt

Conference on "Transforming Appalachian Scholarship: A Conference Honoring the Contributions of Ronald L. Lewis," April 12–14, 2007.

2   For an elaboration of these comments, see Dwight B. Billings, "Appalachian Studies and the Sociology of Appalachia," in *The Handbook of 21st Century Sociology*, ed. Clifton Bryant and Dennis Peck (Thousand Oaks, CA: Sage Publications, 2006), 390–439 and 543–45.

3   John D. Photiadis, "Rural Southern Appalachia and Mass Society," in *Change in Rural Appalachia: Implications for Action Programs*, ed. John D. Photiadis and Harry Schwarzweller (Philadelphia: University of Pennsylvania Press, 1970), 5–22.

4   Ronald Lewis and Dwight Billings, "Appalachian Culture and Economic Development: A Retrospective View on the Theory and Literature," *Journal of Appalachian Studies* 3, no. 1 (Spring 1997): 3–42.

5   John D. Photiadis and Harry Schwarzweller, eds., *Change in Rural Appalachia: Implications for Action Programs* (Philadelphia: University of Pennsylvania Press), ii.

6   Richard Ball, "The Southern Appalachian Folk Subculture as a Tension-Reducing Way of Life" in *Change in Rural Appalachia*, ed. Photiadis and Schwarzweller, 69–79.

7   Thomas R. Ford, "The Passing of Provincialism" in *The Southern Appalachian Region: A Survey*, ed. Thomas R. Ford (Lexington: University of Kentucky Press, 1962), 9–34.

8   Jack E. Weller, *Yesterday's People: Life in Contemporary Appalachia* (Lexington: University of Kentucky Press, 1965).

9   Dwight B. Billings, "Culture and Poverty in Appalachia: A Theoretical Discussion and Empirical Analysis," *Social Forces* 53, no. 2 (December 1974): 315–323.

10   Michael E. Workman, "Political Culture and the Coal Economy in the Upper Monongahela Region, 1776–1933" (PhD diss., West Virginia University, 1995).

11   Talcott Parsons, *The Social System* (New York: Free Press of Glencoe, 1951).

12   Rupert Vance, "An Introductory Note" in *Yesterday's People*, Weller, v–ix.

13   The best overview of Appalachian activism remains Stephen L. Fisher, ed. *Fighting Back in Appalachia: Traditions of Resistance and Change* (Philadelphia: Temple University Press, 1993).

14   Quoted from official documents in Alan Banks, Dwight B. Billings, and Karen W. Tice, "Appalachian Studies, Resistance, and Postmodernism," in *Fighting Back in Appalachia*, ed. Fisher, 283.

15 John M. Glen, *Highlander: No Ordinary School* (Knoxville: University of Tennessee Press, 1996).

16 *Appalachian Land Ownership Task Force, Who Owns Appalachia? Landownership and Its Impact* (Lexington: University Press of Kentucky, 1983).

17 See Mary Ann Hinsdale, Helen M. Lewis, and S. Maxine Waller, *It Comes from the People: Community Development and Local Theology* (Philadelphia: Temple University Press, 1995).

18 See Banks, Billings, and Tice, "Appalachian Studies, Resistance, and Postmodernism."

19 Henry D. Shapiro, *Appalachia on Our Mind: The Southern Mountains and Mountaineers in the American Consciousness, 1870–1920* (Chapel Hill: University of North Carolina, 1978).

20 Allen Batteau, *The Invention of Appalachia* (Tucson: University of Arizona Press, 1982); J. W. Williamson, *Hillbillyland: What the Movies Did to the Mountains & What the Mountains Did to the Movies* (Chapel Hill: University of North Carolina Press, 1995); Anthony Harkins, *Hillbilly: A Cultural History of an American Icon* (New York: Oxford University Press, 2004).

21 Ronald Eller, *Miners, Millhands, and Mountaineers: Industrialization of the Appalachian South, 1880–1930* (Knoxville: University of Tennessee Press, 1982); John Gaventa, *Power and Powerlessness: Quiescence and Rebellion in an Appalachian Valley* (Urbana and Chicago: University of Illinois Press, 1980); and John Alexander Williams, *West Virginia and the Captains of Industry* (1976; repr., Morgantown: West Virginia University Press, 2003). (That John A. Williams and Altina Waller (below) were both members of the WVU history department when their important studies were being produced is another indication of this department's long-standing contributions to Appalachian Studies.)

22 Helen Lewis, Sue Kobak, and Linda Johnson, "Family, Religion and Colonialism in Central Appalachia" in *Colonialism in Modern America: The Appalachian Case*, ed. Helen Lewis, Linda Johnson, and Donald Askins (Boone, NC: The Appalachian Consortium Press, 1978), 113–139.

23 Lewis, Kobak, and Johnson, "Family, Religion and Colonialism in Central Appalachia."

24 See, for example, John Gaventa, "Property, Coal, and Theft" in *Colonialism in Modern America*, ed. Lewis, Johnson, and Askins, 141–59.

25 For a theoretical account of class processes as the production, distribution,

and receipt of the fruits of labor rather than property ownership, see S. Resnick and R. Wolff, *Knowledge and Class: A Marxian Critique of Political Economy* (Chicago: University of Chicago Press, 1987).

26   Ronald L. Lewis, *Transforming the Appalachian Countryside: Railroads, Deforestation, and Social Change in West Virginia, 1880–1920* (Chapel Hill: University of North Carolina Press, 1998).

27   Lewis, *Transforming the Appalachian Countryside*, 7, 9.

28   Altina L. Waller, *Feud: Hatfields, McCoys, and Social Change in Appalachia, 1860–1900* (Chapel Hill: University of North Carolina Press, 1988).

29   Barbara Rasmussen, *Absentee Landowning & Exploitation in West Virginia, 1760–1920* (Lexington: University Press of Kentucky, 1994).

30   Even though he used the model of internal colonialism that he adapted to Appalachia from his mentor C. Vann Woodward's analysis of the postbellum American South, John Alexander Williams's examination of these same figures in *West Virginia and the Captains of Industry* remains extremely valuable.

31   Wallerstein's approach was first formulated in Immanuel Wallerstein, *The Modern World-System I: Capitalist Agriculture and the Origins of the European World-Economy in the Sixteenth Century* (New York: Academic Press, 1974), and subsequent works.

32   This critique was most forcefully argued in Robert Brenner, "The Origins of Capitalist Development: A Critique of Neo-Smithian Marxism," *New Left Review* 104 (1977): 25–92. For an important discussion, see Robert Denemark and Kenneth Thomas, "The Brenner-Wallerstein Debate," *International Studies Quarterly* 32 (1988): 47–65.

33   Workman, "Political Culture and the Coal Economy"; Ken Fones-Wolf, *Glass Towns: Industry, Labor, and Political Economy in Appalachia, 1890–1930s* (Urbana and Chicago: University of Illinois Press, 2007).

34   David Walls, "Internal Colony or Internal Periphery? A Critique of Current Models and an Alternative Formulation," in *Colonialism in Modern America*, ed. Lewis, Johnson, and Askins, 340.

35   Ken Fones-Wolf, *Glass Towns*.

36   Barrington Moore, Jr., *Social Origins of Dictatorship and Democracy: Lord and Peasant in the Making of the Modern World* (Boston: Beacon Press, 1966).

37   In addition to Moore, see Ellen Kay Trimberger, *Revolution from Above: Military Bureaucrats and Development in Japan, Turkey, Egypt, and Peru* (New Brunswick, NJ: Transaction Books, 1978).

38   Dwight B. Billings, *Planters and the Making of a "New South"* (Chapel Hill: University of North Carolina Press, 1980).

39   Workman, "Political Culture and the Coal Economy"; Kenneth W. Noe, *Southwest Virginia's Railroad: Modernization and the Sectional Crisis* (Urbana and Chicago: University of Illinois Press, 1994).

40   Larry J. Griffin, "Whiteness and Southern Identity in the Mountain and Lowland South," *Journal of Appalachian Studies* 10, nos. 1–2 (Spring/Fall 2004): 7–37.

41   W. Fitzhugh Brundage, "Racial Violence, Lynchings, and Modernization in the Mountain South," in *Appalachians and Race*, ed. John C. Inscoe (Lexington: University Press of Kentucky, 2005), 302–316.

42   See, for example, John C. Inscoe, *Mountain Masters, Slavery, and the Sectional Crisis in Western North Carolina* (Knoxville: University of Tennessee Press, 1989); Dwight Billings and Kathleen Blee, *The Road to Poverty: The Making of Wealth and Hardship in Appalachia* (New York: Cambridge University Press, 2000); and Wilma A. Dunaway, *Slavery in the American Mountain South* (New York: Cambridge University Press, 2003).

43   An important application of Moore's approach to American regional development is Charles Post, "The American Road to Capitalism," *New Left Review* 133 (1982): 30–51.

44   Ronald L. Lewis, *Black Coal Miners in America: Race, Class, and Community Conflict 1780–1980* (Lexington: University Press of Kentucky, 1987).

45   Ronald L. Lewis, *Coal, Iron, and Slaves: Industrial Slavery in Maryland and Virginia, 1715–1865* (Westport, CT: Greenwood Press, 1979).

46   Lewis, *Black Coal Miners*, 164.

47   Joe William Trotter, Jr., *Coal, Class, and Color: Blacks in Southern West Virginia, 1915–32* (Urbana and Chicago: University of Illinois Press, 1990); Connie Lee Rice, "For Men and Measures: The Life and Legacy of Civil Rights Pioneer J.R. Clifford" (PhD diss., West Virginia University, 2007).

48   Paul H. Rakes, "Acceptable Casualties: Power, Culture, and History in the West Virginia Coalfields, 1900–1945" (PhD diss., West Virginia University, 2002).

49   Richard P. Mulcahy, *A Social Contract for the Coal Fields: The Rise and Fall of the United Mine Workers of America Welfare and Retirement Fund* (Knoxville: University of Tennessee Press, 2000).

50   Carletta A. Bush, "Faith, Power, and Conflict: Miner Preachers and the United Mine Workers of America in the Harlan County Mine Wars, 1931–39" (PhD diss., West Virginia University, 2006).

51   Shirley Stewart Burns, *Bringing Down the Mountains: The Impact of Mountaintop Removal on Southern West Virginia Communities* (Morgantown: West Virginia University Press, 2008).

52   John C. Hennen, *The Americanization of West Virginia: Creating a Modern Industrial State, 1916–1925* (Lexington: University Press of Kentucky, 1996).

53   See Dwight B. Billings, "Religion as Opposition: A Gramscian Analysis," *American Journal of Sociology* 96 (1990): 1–13.

54   Elizabeth Fones-Wolf and Ken Fones-Wolf, "Cold War Americanism: Business, Pageantry, and Antiunionism in Weirton, West Virginia," *Business History Review* 77 (Spring 2003): 61–91.

55   See Billings, "Appalachian Studies and the Sociology of Appalachia."

56   Sandra Barney, *Authorized to Heal: Gender, Class, and the Transformation of Medicine in Appalachia, 1880–1930* (Chapel Hill: University of North Carolina Press, 2000).

57   For example, Barbara J. Howe, "'Home Work,' and Nineteenth-Century West Virginia Women," *Upper Ohio Valley Historical Review* 18 (1989): 16–26 and "The Status of Women's Historical Research in West Virginia" in *West Virginia History: Critical Essays on the Literature* (Dubuque, IA: Kendall/Hunt Publishing Co., 1993), 149–184.

58   Ken Fones-Wolf and Ronald L. Lewis, eds., *Transnational West Virginia: Ethnic Communities and Economic Change, 1840–1940* (Morgantown: West Virginia University Press, 2002).

59   Deborah R. Weiner, *Coalfield Jews: An Appalachian History* (Urbana and Chicago: University of Illinois Press, 2006).

Section I

Culture

"Scrip Was a Way of Life":

Company Stores, Jewish Merchants, and the Coalfield Retail Economy

Deborah R. Weiner

IN THE WOODY ALLEN movie *Zelig*, the title character finds himself a bystander at an implausibly large number of important historical occasions, so the audience sees these famous events through the lens of an ordinary Jewish everyman. This essay will engage in a somewhat Zelig-like feat. Jews who settled in the coalfields seemingly had little to do with the relationship between miners and coal operators. As merchants, their activities were peripheral to an economy dominated by coal—but the coal industry certainly was not peripheral to them. My study of coalfield Jewish communities led me to examine how the coal industry affected local retailing and to investigate that famous icon of coalfield life, the company store. Like Zelig, Jews are bit players here, but their retail role sheds light on an important regional story: the impact coal companies had on their workers, on other residents of the coalfields, and on the region's economy.[1]

First, some background. The coal boom in south-central Appalachia at the turn of the twentieth century attracted entrepreneurs who sought to provide goods and services to a fast-growing population. Among them were Jewish immigrants who drew on a cultural heritage as "middlemen" in the European countryside to carve out a niche as retailers in the coal economy. Together, the new arrivals and a small pre-existing contingent of established merchants formed the nucleus of a coalfield middle class that lodged, sometimes uneasily, between the mining workforce and the coal barons and landowners who controlled the region's economic life.

Upon their arrival, the newcomers found that the very industry responsible for creating entrepreneurial opportunity had also developed a formidable system to take advantage of it. Coal company stores abounded in the region and presented a distinct challenge to local merchants. But while much has been written about company stores, their competitors have received little attention. Bringing independent retailers into the discussion offers a chance to see the company store—and the workings of the coal economy—from a new perspective. Taking off from the point of view of Jewish merchants, this essay examines the competition between company stores and independent stores in the coalfields of southern West Virginia, southwestern Virginia, and southeastern Kentucky.

The first fact of life for retailers in this single-industry region was that the health of their businesses fluctuated with the health of the coal business. When coal mining boomed, merchants flourished. During mine layoffs or strikes, local stores lost customers, and storeowners could even land in bankruptcy. As one retired Jewish merchant explained, with the air of someone merely stating the obvious, "everybody was dependent on the coal companies." So retailers rooted for the success of local coal companies, even as they competed against them for the trade of the local population.[2]

Company Stores and How They Operated

Company stores started out as a necessity. At the dawn of the coal boom, the local farming population's small size and relative self-sufficiency supported a limited number of merchants. When coal operators first imported labor into the region, they had to provide a place for their workers to obtain basic goods and services. The stores soon turned into a profitable sideline, and continued to proliferate even as independent merchants came on the scene. The power coal companies wielded over their workers, not to mention their control over the socioeconomic and political system, gave them considerable advantages over other retailers. For decades, the institution of the company store loomed over the consumer economy.[3]

Many historians have depicted the company store as a key tool used by coal operators to exploit their workers. The picture, broadly painted, goes as follows. Coal companies undertook to monopolize the trade of their employees. They compelled miners to shop at the company store by threatening to fire them or place them in the least productive areas of the mine (miners were paid by the ton) if they shopped elsewhere. They then charged "exorbitant" prices and further ensnared workers through the use of scrip: coal company currency that could be drawn in advance of payday and could be used only in the company store. When payday came around, miners could find themselves with a negative balance, forced into debt peonage—or, at best, would receive little pay since most of their earnings had already been spent. Either way, they were obliged to draw more scrip to meet their expenses, binding them further to the company store and to the company as employer.[4]

Revisionists have challenged this picture, portraying the company store as a benign institution that provided convenience, quality products, and reasonable prices to miners and their families. Scrip, they maintain, was a service that, if used judiciously, offered flexibility in family budgeting and shopping. They insist that coal operators did not force their employees to use the company store, that few miners became indebted to the company, and that company-owned stores, far from being monopolistic, faced "stiff competition" from merchants, mail-order houses, and stores owned by other local coal companies.[5]

Only by ignoring a substantial body of evidence can revisionists sustain claims that the company store operated in "a competitive market situation" and that coal operators abstained from using coercive tactics. Numerous sources, from contemporary newspapers to oral histories to the coal industry's own trade journals, document attempts by coal operators to drive out independent retailers and to pressure employees into trading at the company store.[6] However, the evils of the company store have been exaggerated by its harshest critics. It did indeed provide essential retail services in relatively remote areas, while serving as the community center and focal point for coal town life. Moreover, as the very existence of an independent retail sector suggests, its "monopoly power" was limited. Nevertheless, the company store as a dominant coalfield institution imposed an inequitable and restrictive set of conditions that miners and their families (as employees and consumers) and local merchants (as competitors) were forced to confront. The tension between its monopolistic tendencies and the response of these consumers and competitors helped shape the coalfield retail economy.

Coal companies pressured their employees to use the company store with strategies ranging from subtle to heavy-handed, with the more egregious tactics applied before the 1930s, at which point unionization reduced their control over their workers. A resident of West Virginia's New River coalfield explained that at one local mine, "before the unions got so strong, if you worked there, you spent your money there. If they caught you spending your money uptown, they'd either lay you off or put you in a water hole—you couldn't make a living." A Kentucky coal miner similarly recalled that in the 1930s, "If they caught you trading, going to Harlan and trading in the store, instead of trading in the commissary, why, they'd run you off from here, they didn't take you back."[7]

Employers freely acknowledged such policies to outsiders and amongst themselves, while vehemently denying them in government hearings. A 1915 letter in the trade journal *Coal Age* conceded that "all mine officials are expected to make their men realize, in some subtle but effective way, that the man who spends most at the company store will receive more favors than the one who buys elsewhere." In the early 1920s, a coal operator told journalist Winthrop Lane that "he had formerly discharged men 'for buying two dollars' worth of flour' somewhere else," though Lane reported that miners were no longer "compelled to trade at company stores." In the early 1930s, a company bookkeeper slyly informed researcher James Laing, "We wouldn't fire a man for not trading at the store—but he could be let out for something else."[8]

Not all coal companies attempted to coerce their employees, and those who did try did not always succeed. As revisionists point out, the tight labor market that sometimes pertained in the region curbed their ability to dictate to their workforce: if miners were dissatisfied, they could easily quit and find a job elsewhere. One operator told Lane that competition for good workers led his company to attempt to keep its men "contented and satisfied." A 1915 *Coal Age* letter writer expressed the same view more candidly: "Now about this store-profit business. . . . If I abused my workmen any more than any [other] coal operator does, my men would leave me." Some operators professed indifference as to where their workers spent their pay. A former company store manager in the New River field recalled that his firm's miners often purchased groceries at a nearby independent store. "They thought they could get 'em cheaper than they could in the company store. . . . It was all right. It didn't hurt us any."[9]

Meanwhile, some miners and their families refused to submit to pressure. A woman who grew up in a southern West Virginia coal camp in the 1920s recalled that her mother "declined to shop at the company store, and when the store manager tried to pressure her, she told him she would shop where she pleased." Similarly, an African-American coal miner's daughter interviewed for this study recalled that her family shopped at a local grocery despite the coal company's disapproval.[10]

Evidence from independent retailers suggests that the coercive policies of company stores hampered their businesses but did not necessarily prevent them from operating. A storekeeper in eastern Kentucky in the 1920s asserted that a mine superintendent tried to keep miners from trading with him: "He'd tell them it might mean their job if they didn't patronize the company store." They ignored the threat and shopped at both stores. Some former Jewish merchants interviewed for this study believed that miners were required to spend a portion, but not all, of their earnings at the company store. One man stated that miners, paid twice per month, spent their first paycheck at the company store and came to town to shop with their second paycheck.[11]

Yet, even if coercive policies were unevenly or halfheartedly applied, the relationship between coal companies and miners gave company stores an inherent edge over independents. Author John Hevener, comparing three different mines in 1930s Harlan County, noted that one did not apply any pressure at all while at another, the superintendent "bluntly told his employees: 'If you trade at Piggly Wiggly's you can get your job at Piggly Wiggly's.'" The third company "sent its miners a letter that they interpreted as a warning" to trade at the company store or be fired. Indeed, the line between coercion and aggressive marketing could be fuzzy, as the U.S. Coal Commission noted in its 1922 study. Though miners typically were "no longer . . . openly" forced to shop at the company store, "the energetic store manager . . . has access to the mine company's pay roll and can find out which families do none or but part of their buying at his store. If he solicits the trade of these families . . . he is doing no more than any wide awake merchant would do.

Yet he is exposing himself and his company to the charge of 'forcing purchasing.'" The coalfields' independent "wide awake merchants," of course, had no such access to their customers' payroll records and no such targeted opportunities to "solicit" trade.[12]

A 1913 *Coal Age* article delineated other advantages company stores enjoyed over independent retailers: they "pay no rent, do no advertising, should have no losses on bad credits and have small delivery charges, all of which cost usually from 8 to 10 percent of the gross profit." A 1915 article asserted that some operators faced "fierce" competition from independent stores, "and they frequently succeed so well . . . that they take away the bulk of the business from the local merchants, owing to the fact that they have a large trade absolutely secured from bad debts and of a volume which can be relied on."[13] In other words, only company stores could deduct employees' expenditures from their pay and impose measures to guarantee that their miners shopped in their stores.

Also, access to the financial resources of their parent corporations enabled company stores to carry a larger (and by most accounts, higher quality) stock than small family-owned businesses. Their ability to buy in bulk allowed them to undersell competitors if they so chose. At least one company applied this strategy selectively. In 1909, a general store owner in Bramwell, West Virginia, complained in a letter to the governor that the Caswell Creek Coal Company was "soliciting business from this town, and selling goods, at ruinous prices" in order "to choke off the small local merchant." The prices offered townspeople for order and delivery of goods were far lower than the prices it charged its workers in its store, the letter claimed.[14]

Their proximity to the bulk of the coalfield population gave company stores their most obvious advantage. In 1922, for example, more than 80 percent of West Virginia's miners and their families lived in company-owned towns. The Coal Commission found that in the New River district, representative of the coalfields as a whole, the company-owned store served as "the most important source of supplies." And in "newly-developing" areas where no other stores yet existed, it was "still a necessary institution." Until improvements in the road system and the spread of automobile ownership starting in the 1930s, company stores did indeed have a "virtual monopoly" (as Winthrop Lane put it) in the newer or more remote coal camps.[15]

Some coal companies were determined to protect this geographical supremacy. A 1911 U.S. Immigration Commission study observed that "in most instances the companies own large tracts of land and keep out competitors very largely." Many coal operators refused to allow independent stores on their property and also banned peddlers from their camps. An article about Lebanese merchant Asaff Rahall of Beckley, West Virginia, notes that during his peddling days in the 1910s, "the company police would try to chase him out of the houses." In the early 1920s, West Virginia's labor commissioner told Winthrop Lane that "in nearly every case hucksters and peddlers are prohibited" from coal camps.[16]

Yet this statement must have been an exaggeration, since peddlers such as Rahall managed to earn enough income to move into the less precarious role of coalfield shopkeeper. While some companies enforced strict "no trespassing" rules, others encountered limits to their ability to bar independent retailers. A Harlan County local historian writes that peddlers, mostly Jews and Assyrians, "sold in all the communities and the coal camps" in the 1920s and 1930s. "They were not welcome in some coal camps," he states, "but they did well and prospered." Rudy Eiland operated a small grocery store just outside of Logan, West Virginia, and made deliveries into nearby coal camps in the 1910s. Company officials "weren't happy about his coming," stated his son, but they could not stop him because he had purchased the land on which his store sat.[17]

In 1915, Superior Pocahontas Coal Company Superintendent George Wolfe sued a local firm, McNeal & Goodson, for trespassing when the firm tried to deliver goods to miners in one of his McDowell County, West Virginia, coal camps. To close off access to the camp, he erected a barrier across a public road used by rural residents traveling to and from the hamlet of Davy, angering villagers and farmers. An editorial in the *Bluefield Daily Telegraph* criticized Superior for attempting to prevent employees from trading with local merchants. "In this free country, a suit of this kind causes considerable wonder," the newspaper fumed. "It is just such foolish actions as this that cause agitation and labor troubles." The incident apparently pitted residents of the countryside, backed by the regional Bluefield newspaper, against the coal company and its allies in the local county seat, Welch. Wolfe insisted in a letter to his employer, coal baron Justus Collins, that "the entire population of Welch is with us." Indeed, the Welch newspaper defended the company, saying that the *Telegraph*'s editorial was "unjust, unfair, and untrue."[18]

Wolfe publicly denied the restraint-of-trade charge, maintaining that his miners could "trade where they please" and that he was merely attempting to keep his mining operations safe from intruders. But, the *Telegraph* pointed out, his protestations rang hollow since, during the legal proceedings, "the company [stated] that it is deprived of profit by the merchants and their employees using the streets and roadways." Wolfe himself explained to Collins that "net earnings on our store will fall from Eleven Thousand to Six Thousand Dollars per year" if he allowed deliveries from merchants. In a rare court defeat for local coal operators, Wolfe's suit was dismissed and his firm was ordered to pay all court costs.[19]

However, the incident does not support the conclusion that public opinion routinely prevented coal companies from getting away with such behavior. For years Wolfe had enforced a policy barring independent retailers from company property. As he noted in a 1913 letter to Collins, "About once a year I have a round with these merchants at Davy, as there seems to be always some new one that has to be broken in." Many of the Davy merchants were, in fact, Jewish immigrants, newcomers unlikely to challenge their powerful neighbor. Only after McNeal & Goodson (a non-Jewish firm) began to ignore the rule did Wolfe bring his lawsuit.[20]

By resorting to the courts rather than relying on his usual methods of intimidation, Wolfe apparently overreached his authority and raised the specter of bad publicity, as Collins pointed out when he wrote to him, "you can see what a club you have . . . placed in the hands of the enemies of the coal industry." Collins, himself a strong believer in authoritarian rule, recognized that "the Company's stores have every advantage in many ways" and thus there was no need for "coercion in any shape or form," which needlessly aroused the antagonism of miners and townspeople alike. He preferred to save heavy-handed tactics for keeping out "undesirable people . . . during strike times."[21]

Collins's instincts were correct. While public opinion may have become unsettled over lawsuits against local businesses and the manipulation of miners as consumers, coal companies could conduct their campaigns against unionism virtually unchecked. Conveniently, their no-holds-barred anti-union measures had the effect—intended or not—of restricting interaction between their retail competitors and their workers. When Jewish immigrant James Pickus, newly arrived from Russia, entered one coal camp to post signs advertising his brothers' Beckley store in the mid-1910s, he was attacked and badly beaten by a mine guard. The incident left "a real deep scar" on his forehead, according to his son. This company's officials cared little about retail competition, or so they said; Pickus had been mistaken for a union organizer.[22]

Routine inquisitions carried out by coal industry watchdogs ranged far and wide, and must have had a chilling effect on trade. A 1920 Baldwin-Felts Detective Agency memo reported that one agency spy based in Keystone, West Virginia, "has failed to locate any Union organizer or come in contact with any agitator, [although] he has investigated dozens of miners . . . picture agents, peddlers, insurance agents, newspaper collectors, and others." Local authorities, operating as an adjunct to the coal industry, joined in the repressive measures that stifled commerce as well as union activity. A traveling salesman told a reporter in 1920, "Every operation has its armed guard—usually two. I was served with a little notice by a deputy sheriff in Logan . . . to not go up Island Creek any more, that it was dangerous and I was liable to get shot. . . . The notice forbade me using the railroad going up there and back."[23]

Though efforts to monitor and regulate daily life in their towns arose from a desire to stamp out unionism, coal operators recognized the additional benefits derived from their social-control methods, and many came to view these benefits as their prerogative. Some coal towns developed into armed camps, and labor organizers were not the only undesirables deliberately barred. George Wolfe described admiringly the arrangement of a fellow coal operator in a 1915 letter to Collins. "Bradley has a five-strand barbed wire fence, ten feet high, around his property, with only one gate and at that gate he keeps someone day and night. . . . Works absolutely nonunion and controls the situation and people come for miles to work for him. His net profits on the store are 12 percent and there are no outside teams delivering anything."[24]

A compelling incentive led coal companies to stifle their retail competition and coerce miners to use company stores: an acute need to make high store profits in order to overcome losses in their coal operations. A government survey found "not a few" coal operators who stated that "while there was a loss in their coal-mining business proper this was more than counterbalanced by profits from selling merchandise and renting houses," reported *Coal Age* in 1915. One letter writer lamented, "Too often has the company store . . . been required . . . to carry the burden of other departments of a coal-mining company by making good the losses sustained in those departments." The phenomenon was widespread. For example, a *New York Times* reporter observed in 1931 that "many Harlan mines were turning their only profit at the commissary."[25]

Given the unpredictability of the national coal market, it is not surprising that many company officials viewed independent retailers as a threat. In his letters to Justus Collins, George Wolfe repeatedly and obsessively referred to the "nine competitive stores" in Davy. In his eyes, their practice of underselling the company store made them as subversive as union organizers:

> If we attempted to meet all of this competition, we would wind up selling the whole shooting match at cost. . . . The merchants of Davy have cut our Powder profits from three thousand dollars per year to eight hundred dollars and if they have their way, they will make a serious dent in our store business and profits. Superior with her thin coal and decreasing selling price and little business and serious shutdowns and increased protection from liability . . . and general increase in taxes and everything that we have to buy in the way of supplies cannot for ever stand the strain and it is for this reason that we must try and find some legitimate way to protect ourselves.[26]

Coal companies generally did a good job of using their stores to "protect" themselves from coal losses. As one government study emphasized, "the company store in all cases is decidedly a paying institution." Historian Crandall Shifflett, who tends to view company stores favorably, points out, "The fact that a company store siphoned much of the payroll of the company back into the company's coffers while at the same time making a profit was a business arrangement of manifold importance, even at low returns." The company he analyzed showed "excellent" returns of 10 to 15 percent. But profits were often much higher than that. George Wolfe, despite his complaints, netted an 80 percent profit at his Superior store. One Harlan County company store produced a 170 percent annual profit between 1934 and 1937. (The local sheriff and a county judge shared a financial interest in that particular store.) A company store operator interviewed by James Laing revealed gross profits around 50 percent.[27]

Coal Age took a defensive stance regarding the tendency for coal companies to rely on store profits, stating that it "may appear a damaging indictment against the coal operator, but surely because the mine owner has lost money in coal production is no reason why he should lose it in all his other ventures." But some of the journal's correspondents pointed out the disturbing implications. One letter writer noted that an average mine "can easily afford to lose three cents a ton in the cost of production if 40 percent of the men's wages goes into the store." (This was an achievable goal: Laing found one store manager who "boasted of the fact that he had been able to get back 46 percent of the pay roll" at his mine.) As a result, some operations became geared to selling merchandise rather than mining coal, and strove to employ good consumers rather than good miners. Not only were workers rewarded for store patronage instead of mining skill, the letter writer charged, "there are numerous examples of superintendents and foremen who hold their positions mainly because they are good trade-getters."[28]

Contributors to *Coal Age* would not blatantly admit to coercing their miners to shop in their stores, but they acknowledged that heavy store users were favored, which ultimately amounted to the same thing. One writer pointed out that miners who spent the most—and became the most indebted to the company—would naturally be placed in the most profitable sections of the mine, if only so that the company could be assured of being paid back. Indebted miners would be the first to be called for overtime and the last to be laid off. Those who did not patronize the store were inevitably shunted aside. "It may not be the intention of the company to show favoritism, [but it is only natural for it to encourage debt payment] by every means in its power," one commentator stated. Problems arose when the "improvident" miner realized that "more is to be gained by keeping in his employer's debt" than by working hard: "He counts on his indebtedness to the company as his best protection."[29]

Some *Coal Age* correspondents concluded that a dependency on store profits was unhealthy, resulting in inefficiency in mining, low morale, and increasing reliance on the store to make up for these deficiencies.[30] The result for the entire region was to prop up basically unprofitable enterprises, thus contributing to the problems of overproduction and stagnation that plagued the coal industry and caused economic distress for the population generally.

The Impact of Scrip

By issuing scrip, coal companies exacerbated the dilemmas the company store posed for independent retailers and for the economy as a whole. Like the company store itself, scrip began as a necessity. Currency fabricated by individual coal companies (in coins or paper) helped ease the shortage of cash in a region characterized at first by a lack of financial institutions. Scrip served as a practical accounting mechanism for companies involved in numerous transactions with their workers and it became standard in company stores throughout the coalfields. Although the use of scrip to pay miners soon became illegal, says historian Glenn Massay, "there was nothing to prevent employers from issuing scrip" as an *advance* on pay, "and this became a common practice." While employees could receive their wages in cash if they waited until payday, scrip advances were often unavoidable: not only did miners need to purchase job supplies such as tools and blasting powder before they could even start working, but also, until the 1920s, many were paid only once per month.[31]

Once enmeshed in the scrip system, it was difficult for miners to break away. Company officials quickly realized that scrip offered them a way to boost store purchases. As Crandall Shifflett concedes, "coal companies encouraged easy credit." Analysts differ on how much debt the average miner incurred, though they agree that immigrants tended to be less indebted, and to rely less on scrip, than native whites or blacks. This was partly because they were more likely to be single men without local families to support (and if they were sending money home, they needed to receive their pay in cash). But whether or not miners found themselves in debt on payday, the use of scrip greatly reduced the amount of cash they received. And since scrip could be redeemed at full value only at the company store, it served as a powerful disincentive for them to shop elsewhere, regardless of whether or not their employer attempted to coerce them. As one coalfield study notes, "If a family had only scrip to spend, purchasing outside the coal camp could be difficult."[32]

But that was not the only problem scrip caused local merchants. The system put pressure on the entire retail economy to operate on a credit basis. For one thing, scrip greatly contributed to the lack of cash in circulation. As one Jewish man explained, his family's Northfork store had no choice but to offer credit, because "miners had scrip in their pockets."[33]

Also, miners and their families became accustomed to relying on credit. The company stores investigated by the Coal Commission in 1922 made up to 78 percent of their sales on a non-cash basis. Independent stores had to offer credit as well, if they wanted to compete. But company stores were shielded from losses by their ability to deduct from employees' pay. In contrast, small coalfield retailers were vulnerable to bad debts, especially given the volatility of the coal economy. The Commission concluded that because they operated "on a credit basis, . . . a number of independently owned stores come into existence when the mines begin work and go out of business when the mines are not operating."[34]

For retailers, therefore, the scrip system heightened the hazards of mine layoffs or strikes. Although miners generally paid their bills as soon as they could, for small businesses trying to make it through hard times, that was often not soon enough, recall some Jewish retailers. When customers who normally settled part or all of their debt on payday suddenly found themselves unable to do so, the consequences could be "quite difficult." For as long as they could, merchants would "carry" their customers through these periods and hope that good times would soon return. "We just had to wait 'em out," one merchant explained.[35]

Aside from offering credit, some independent retailers made another accommodation to the scrip system: they accepted scrip as payment, in lieu of cash or credit. Far from being usable only at company stores, scrip in fact circulated throughout the coalfields, aided by the activities of local scrip dealers who exchanged it (at a discounted price) for cash that could then be spent in local enterprises.[36]

Peddlers, coalfield stores large and small, theaters, and other businesses took in coal company currency in their daily course of operation. Before the post–World War II era, its impact was so widespread that, according to one Jewish merchant, scrip was "a way of life in those days." An interviewee who car-peddled in the coal camps in the 1930s to 1950s confirmed, "to us, scrip was just like money." It is unknown how many retailers accepted scrip, though it would be safe to say that county seat stores were less likely to take it than stores located nearer the mines. Peddlers and shopkeepers accepted scrip at anywhere from 60 to 90 percent of its face value (most accounts suggest that the typical discount was 75 to 80 percent). Faced with the task of disposing of the currency, they either turned to scrip dealers or took it themselves to the company store. Companies maintained a variety of policies on how they redeemed their currency. Many demanded the purchase of goods, while others would exchange it for cash—but often at a discount. In this manner, company stores could gain back part of what they lost to competitors who accepted scrip.[37]

In West Virginia until the mid 1920s, the law required coal companies to redeem their scrip, in cash, at full value, to anyone who brought it to the company store and wanted to exchange it. But journalist Winthrop Lane found that penalties for non-compliance were minor and enforcement was lax, so company store operators universally disregarded the law. George Wolfe again provides supporting evidence. In a letter to Justus Collins he expressed concern that redeeming scrip in cash "would give the nine competitive stores at Davy the chance of their lives." He added, "as the matter is only a misdemeanor and at the most would not cost us over $100 we will make no change in our method of doing business." In the 1920s the state changed the law to allow companies to make their scrip non-transferable, so that they could legally refuse to take it from non-employees. This law stayed on the books until 1975. In Virginia, coal operators had successfully lobbied for a similar law in 1919.[38]

Merchants forced to deal with scrip tried to profit off the system in various creative ways. Some used company stores as a sort of wholesaler: they would accept scrip at a significant discount (say, 65 percent), then use it to purchase large quantities of cigarettes, candy, or other goods at the company store. They would then mark up these goods and sell them in their own stores. Not surprisingly, this practice caused some coal companies to retaliate by refusing to sell goods to non-employees using scrip. Despite these complications, to small entrepreneurs accustomed to offering credit, scrip at least did not present the danger of uncollectible debts. Some peddlers in the coal camps were even willing to accept scrip at full value from their customers, and then sell it at a loss to scrip dealers—the mark-up on their goods was high enough for them to make a profit anyway.[39]

Since miners who drew scrip were at an automatic disadvantage if they wanted to spend it outside the company store, the system provided opportunities for scrip dealers and retailers to exploit them. In fact, scrip and credit transactions invited abuse by merchants as well as coal companies. Like today's overly-aggressive credit card companies, some retailers worked to convince people to buy things they could not afford. But unlike huge corporations, small businesses found that such tactics could easily backfire. One Jewish man who peddled in the southern West Virginia coal camps observed, "It was easy to sell to [miners] . . . but sometimes you'd have trouble collecting." His method of collection shows how retailers could become involved in the larger system of exploitation: he used the enforcement powers of local justices of the peace, who kept for themselves a portion of the monies they collected on his behalf. "It was nothing for me to have two or three justices of the peace to turn over my accounts to," he noted. "Of course, sometimes we had trouble collecting from the justice of the peace!"[40]

Conversely, scrip also offered possibilities for savvy miners to manipulate the system. It was not uncommon, for example, for miners' families to *purchase* scrip from scrip dealers—at marked-down prices—and use it in local company stores, thus in effect receiving a discount on the goods they bought. Using scrip prudently could work to their advantage in somewhat the same way that today's consumers can benefit from careful credit card use. Crandall Shifflett suggests that "miners brought to their dealings with the company store the same realism and practicality they exhibited elsewhere, and in doing so, they limited the hegemony of the store and the operators over their lives." If they had not, the independent retail sector would have been considerably smaller than it was. Nevertheless, seen within the context of the dominant position coal companies held over their employees and retail competitors, scrip overwhelmingly worked to benefit the region's reigning power, at the expense of the rest of the population.[41]

Competing with the Company Store

While some mining officials saw local retailers as interlopers who interfered with their profits, others saw them as junior partners in the region's development. In fact, most of the independent towns that sprang up in the region did so because land companies and/or their coal subsidiaries sold chunks of property to real estate developers (or created real estate companies of their own) with the goal of encouraging the creation of business and residential districts. In Pocahontas, Virginia, the coal company sold its land in the top half of town and kept the bottom half. The company's power plant supplied power to local businesses, and its store, located on the main downtown street, engaged in friendly competition with the stores ranged alongside it (many owned by Jews). In McDowell County, West Virginia, the massive U.S. Coal & Coke Company looked benevolently on small retailers, perhaps seeing them as part of an environment conducive to keeping a productive labor force. In the New River Company town of Scarbro, West Virginia, in the 1920s, miners lived at the top of the hill while at the bottom could be found the company store, post office, and privately-owned homes and stores, including those of a few Jewish families. To these companies, the benefits of town life outweighed the negative impact of competition—which they probably saw as negligible in any case: adding up all the advantages, the 1922 Coal Commission concluded that "in every respect less effort is required on the part of the miner's family to buy at the company store than elsewhere."[42]

The company store posed so many difficulties for independent retailers that interviewees for this study differed on whether or not their families' stores even attempted to solicit the trade of coal miners. Some who grew up in county seat towns maintained that because of coercive coal company policies, scrip, and the remoteness of many coal camps, for many years their customers consisted of the region's non-mining population—a narrow slice of the overall demographic. They contended that miners did not become a major part of the customer base until unionization, the gradual elimination of scrip, and improved transportation loosened the grip of coal companies. However, other interviewees stated that merchants had always vied with company stores for the trade of miners' families. Certainly businesses based in smaller towns, closer to the coal mines, had no choice but to compete directly with company stores.[43]

Aside from devising schemes to maneuver around the scrip system, independent retailers found other ways to compete. The 1911 Immigration Commission noted that foreign-born miners preferred to buy supplies from "grocery stores and markets conducted by members of their own race, which spring up in nearly all communities settled by immigrants." Even though Jews did not share the same ethnicity, they did share languages and customs with the region's many Eastern European miners. Hungarian-born Rudy Eiland's ability to speak several languages helped him compete with company stores around Logan. Immigrants felt more comfortable trading with him since "he could talk to them, they couldn't," his son observed. Many Jews spoke more than one Eastern European language; some stores displayed signs such as "Polish spoken here," while others sold steamship tickets on the side in a bid to encourage the trade of immigrant miners.[44]

Max Roston's old country style helped him gain customers in the coal town of Wilcoe, West Virginia. The local population "could have found a greater selection at the company store, but we had the lower prices, and our customers knew that if they made a large purchase they could haggle with Uncle Max," his nephew writes. "Bargaining was part of the fun." This popular old country custom was definitely not an amenity offered by the company store. Jewish merchants sometimes helped non-English speaking miners in encounters with American bureaucracy; for example, by acting as an interpreter in court. This role hearkened back to Eastern Europe, where part of the relationship Jews had with their often-illiterate peasant neighbors included helping them deal with officialdom.[45]

As Roston's nephew suggests, underselling was another strategy independent retailers pursued. In fact, the high prices charged by company stores offered merchants their primary leverage in attracting trade away from coal companies. One Jewish retailer observed that the Island Creek Coal Company had operated thirty-five stores around Logan, but since "their prices were higher," independent stores could compete. As another merchant put it, "Their charges was a little bit exorbitant. . . . There was no bargains in the coal company store." Much anecdotal evidence supports the view that coal companies exploited their workers by overcharging them. Revisionists have argued that company stores did not inflate prices, citing the 1922 Coal Commission study, which found company store prices not significantly higher than those at other stores. Nevertheless, the Commission noted that "there is a universal belief among miners' families that prices are higher in company stores," a complaint that remained constant through the decades. Accompanying this belief, it should be stated, was the common perception that goods in company stores were of better quality than in independent stores.[46]

No doubt, considerable variation existed in coal companies' pricing policies. The 1911 Immigration Commission report concluded that smaller coal companies in isolated areas "take advantage of the situation and charge extortionate prices, but the larger companies seldom exact more than a reasonable profit." (A mitigating circumstance may have been the higher cost of transporting goods to remote locations.) A decade after the Coal Commission study, James Laing found a large gap between independent and company store prices. He attributed the discrepancy partly to the introduction of chain stores and a general lowering of prices in independent stores.[47]

One significant generalization can be made: independent retailers, by their very existence, curbed the worst abuses of company stores. The Immigration Commission in 1911, the Coal Commission in 1922, and James Laing in 1931 all found that company store prices tended to be lower in areas where local merchants vied for trade. The Coal Commission further reported that "quality of merchandise, upkeep of merchandise and store, and business methods were better where company stores and independent stores were in direct competition." The anecdotal evidence concurs: an eastern Kentucky storeowner noted that even though prices in the nearest company store were generally higher than at his own store, the company would occasionally engage in price wars with him, to the benefit of consumers. The sole interviewee for this study who insisted that company store prices were comparable to other store prices lived in Pocahontas, Virginia, where independent competition thrived.[48]

George Wolfe's frustration with the dampening effect local merchants had on company store prices led him to offer a revealing analysis of his competitors and how they managed to survive in a difficult business environment. "This is the way these outside merchants figure," he explained to Justus Collins. If they can earn even a modest amount, "they claim that they are getting rich." As he saw it, their willingness to accept low profits gave *them* an unfair advantage, not the coal company. "We have nine competitive stores in Davy and each one of them make a leader out of something and sell it at cost," he complained. "Most of the merchants in business here own their own property and are satisfied with very small profits, as their store provides them with a job and their families with eatables, so they are satisfied. Then some of them do not mind failing and beating their creditors."⁴⁹

While Wolfe is obviously not an unbiased source, his comments provide a glimpse into the marginal nature of independent coalfield retailing that rings true in many respects. The Coal Commission offered a similar assessment in discussing one reason why company store prices were often higher than their competitors: coal companies needed to pay salaries and show a profit to shareholders, "whereas many of the independent stores are operated as family concerns with no salaries to pay and only the family living to get." Given the pitfalls surrounding local enterprise, it is not surprising that those independent concerns tended to be small, dependent on family labor, and low in capital outlay and profitability.[50]

Interestingly, the observations of Wolfe and the Coal Commission echo the literature on small Jewish businesses that describes their longstanding tendency, arising out of the exigencies of European society, to carve out a precarious place on the edges of the economy. The Jewish merchants in Davy at the time Wolfe wrote (the mid 1910s) included general store owner Sam Masil, clothing store owners Meyer and Israel Sneider, and the three Michaelson brothers, who owned a tailoring shop. The town also had two Jewish saloon owners, while several other Jewish merchants came and went during the course of the decade. A few remained in Davy into the 1920s (Meyer Sneider's widow Lillian even outlasted Wolfe, operating the family store into the 1950s) while others left to pursue opportunities elsewhere in the coalfields. Most, however, vanished from the local historical record.[51]

Davy's Jewish contingent offers a contrasting image to the more stable and prosperous merchant-based Jewish communities in coalfield county seat towns, but in many ways they were two sides of the same coin. From the standpoint of the region's retail economy, the problem with coal company stores was that they promoted the marginality of all small businesses—whether they intended to or not. They did not have to be guilty of the most egregious practices in their power in order to have a negative impact. Scrip, for example, boosted them over their competitors whether meant as a convenience or a snare. Though several factors restrained their monopolistic tendencies, their overall effect was to subdue commerce and narrow its scope. Their impact was greatest, of course, in towns located closest to the mines.

Company stores inhibited town development by providing ample incentive for merchants to relocate at their earliest opportunity. But even the county seats contended with the loss of a major portion of the region's potential customer base, as some Jewish merchants noted.

For decades independent stores served as a secondary shopping site for miners' families, after the company store. It is impossible to know what the economy would have looked like had that not been the case. (And stores were not the only businesses affected, since coal companies also operated theaters, barber shops, saloons, pool rooms, and other enterprises.) Moreover, by taking for themselves a large share of the retail trade, coal companies drained financial resources from the economy. Their profits, unlike the profits of local merchants, traveled straight out of the region. Scholars have shown how corporate control of the region's land and resources stunted diversification, as economic activity focused almost exclusively on the extraction of coal for outside markets. As shown here, during periods when mining coal proved unprofitable, miners' consumer dollars were relied upon to maintain the status quo. Company stores therefore constituted an integral part of a single-industry order that proved unable to sustain a healthy economy.[52]

But Jewish merchants were not inclined to take a look at the larger picture. Because they became absorbed into a middle class that supported the coal industry and rooted for its success, they accepted this situation without question.[53] They simply went about the business of establishing their own niche in the economy, attempting to transform themselves from marginal shopkeepers in coal towns such as Davy into county seat business leaders. Many of them succeeded in doing so—but that's a whole other story.

NOTES

1   This essay is part of a larger study that delves into the economic, social, and cultural history of Jews in the Appalachian coalfields. See Deborah R. Weiner, *Coalfield Jews: An Appalachian History* (Urbana: University of Illinois Press, 2006).

2   Edward Eiland, interview with author, Logan, WV, May 19, 1998. On the boom and bust nature of the coal industry, see Richard M. Simon, "The Development of Underdevelopment: The Coal Industry and its Effect on the West Virginia Economy, 1880–1930" (PhD diss., University of Pittsburgh, 1978); Jerry Bruce Thomas, "Coal

Country: The Rise of the Southern Smokeless Coal Industry and its Effect on Area Development, 1872–1910" (PhD diss., University of North Carolina, 1971).

3   On the establishment and operation of company stores, see Crandall A. Shifflett, *Life, Work, and Culture in Company Towns of Southern Appalachia, 1880–1960* (Knoxville: University of Tennessee Press, 1991); Mack Gillenwater, "Cultural and Historical Geography of Mining Settlements in the Pocahontas Coal Field of Southern West Virginia, 1880–1930" (PhD diss., University of Tennessee, 1972); Charles Kenneth Sullivan, "Coal Men and Coal Towns: Development of the Smokeless Coalfields of Southern West Virginia, 1873–1923" (PhD diss., University of Pittsburgh, 1979).

4   David Alan Corbin, *Life, Work, and Rebellion in the Coal Fields: The Southern West Virginia Miners, 1880–1922* (Urbana: University of Illinois Press, 1981), 10; Ronald D Eller, *Miners, Millhands, and Mountaineers: Industrialization of the Appalachian South, 1880–1930* (Knoxville: University of Tennessee Press, 1982); John W. Hevener, *Which Side Are You On? The Harlan County Coal Miners, 1931–1939* (Urbana: University of Illinois Press, 1978).

5   Price Fishback, "Did Coal Miners 'Owe Their Souls to the Company Store'? Theory and Evidence from the Early 1900s," *Journal of Economic History* 46, no. 4 (December 1986): 1011–1029; Shifflett, *Life, Work, and Culture*, 183–184.

6   The revisionist view relies primarily on analysis of a single coal company, a hefty dose of neoclassic economic theory, and two governmental reports: U.S. Immigration Commission, *Immigrants in Industries*, Vol. 7 (Washington, DC: GPO, 1911), and the 1922 *Report of the United States Coal Commission* (Washington, DC: GPO, 1925). The two reports, however, are equivocal, presenting evidence on both sides. The following discussion cites these and other sources on tactics company stores used to attempt to monopolize trade.

7   Wallace Bennett, interview with Paul Niden, Oak Hill, WV, October 1, 1980, transcript, West Virginia and Regional History Collection, West Virginia University, Morgantown, WV, hereafter WVRHC; Alessandro Portelli, *The Death of Luigi Trastulli and Other Stories: Form and Meaning in Oral History* (Albany: SUNY Press, 1991), 206. In *Life, Work, and Rebellion*, Corbin cites similar statements from letters to the *United Mine Workers Journal*.

8   "Discussion by Readers," *Coal Age* 8 (1915): 895; Winthrop D. Lane, *Civil War in West Virginia: A Story of Industrial Conflict in the Coal Mines* (New York: Arno Press, 1921), 28; James T. Laing, "Negro Miner in West Virginia" (PhD diss., Ohio

State University, 1933), 311. Miners and operators argued the point in government hearings. See *West Virginia Coal Fields: Hearings before the Committee on Education and Labor, United States Senate, 67th Congress, 1st Session* (Washington, DC: GPO, 1921), 76–79, 245–247, 286–287, 472, 502–503, 918.

9   Shifflett, *Life, Work, and Culture*, 189; Lane, *Civil War*, 35; "Discussion by Readers," 813; E. H. Phipps, interview with Paul Niden, Beckley, WV, August 16, 1980, transcript, WVRHC.

10   Janet W. Greene, "Strategies for Survival: Women's Work in the Southern West Virginia Coal Camps," *West Virginia History* 49 (1990): 51; Frances Monroe, interview with author, Bluefield, WV, April 27, 1998.

11   Laurel Shackelford and Bill Weinberg, eds., *Our Appalachia* (New York: Hill and Wang, 1977), 225; various interviews (see Weiner, *Coalfield Jews*, 227–228).

12   Hevener, *Which Side Are You On?*, 22; *Report of the U.S. Coal Commission*, 1462.

13   B. F. Roden, "The Commissary: Its Indispensability," *Coal Age* 4 (1913): 240–241; "Company and Other Stores," *Coal Age* 8 (1915): 716. See also *Immigrants in Industries*, Vol. 7, 201.

14   Letter from W. J. Francis to Governor William E. Glasscock (Box 3, Glasscock Papers, WVRHC).

15   *Report of the U.S. Coal Commission*, 1462–1463, 1466, 1514; Lane, *Civil War*, 28.

16   *Immigrants in Industries*, Vol. 7, 201; Yvonne Snyder Farley, "One of the Faithful: Asaff Rahall, Church Founder," *Goldenseal* 8, no. 2 (Summer 1992): 52–53; Lane, *Civil War*, 29.

17   William D. Forester, *Before We Forget: Harlan County, 1920 through 1930* (n.p.: William D. Forester, 1983), 25; Eiland interview.

18   "Company Store Seeks To Force Employees To Buy" and "Coercion Still Continues," *Bluefield Daily Telegraph*, December 22, 1915, and "Injunction Is Applied For by Coal Company," *Bluefield Daily Telegraph*, December 24, 1915; George Wolfe-Justus Collins correspondence, December 25 and 28, 1915 (Justus Collins Papers, WVRHC); *McDowell Recorder*, December 24, 1915.

19   "Injunction Is Applied For by Coal Company," *Bluefield Daily Telegraph*, December 24, 1915; Wolfe-Collins correspondence, December 25 and 28, 1915; Circuit Court records, McDowell County Courthouse, Welch, WV.

20   Wolfe-Collins correspondence, 1913 to 1915, Justus Collins Papers. The ethnic origin of Davy merchants was determined from: Thomas C. Hatcher and Geneva Steele, eds., *The Heritage of McDowell County, West Virginia, 1858–1995* (War, WV: McDowell County Historical Society, 1995); Abraham I. Shinedling, *West Virginia Jewry: Origins and History, 1850–1958* (Philadelphia: Maurice Jacobs, Inc., 1963), 1275; U.S. Bureau of the Census, McDowell County, WV, Manuscript Census, 1910, 1920.

21   Wolfe-Collins correspondence, 1913 to 1915, Justus Collins Papers.

22   Manuel Pickus, interview with author, Charleston, WV, May 18, 1998.

23   Jack M. Jones, *Early Coal Mining in Pocahontas, Virginia* (Lynchburg, VA: Jack M. Jones, 1983), 135; Arthur Gleason, "Company-Owned Americans," *The Nation* (1920).

24   Wolfe-Collins correspondence, December 28, 1915, Justus Collins Papers.

25   "Company and Other Stores," 716; "Discussion by Readers," 812; *New York Times* quoted in Hevener, *Which Side Are You On?*, 22. See also Laing, "Negro Miner," 308; Roden, "The Commissary," 240; and "Store Checks vs. Thrift," *Coal Age* 8 (1915): 619.

26   Wolfe-Collins correspondence, December 25 and 28, 1915, Justus Collins Papers.

27   *Immigrants in Industries*, Vol. 7, 204; Shifflett, *Life Work, and Culture*, 188; Hevener, *Which Side Are You On?*, 22. On company store profits see also William Tams, interview with Richard M. Hadsell, Tams, WV, June 9, 1966, transcript, WVRHC; "Negro Miner," 308.

28   "Company and Other Stores," 716; "Discussions by Readers," 812–813; Laing, "Negro Miner," 308.

29   "Company and Other Stores," 716; "Discussions by Readers," 812-813.

30   The solution was simple, according to one letter writer: the company store and the coal business should be run as two separate operations, with no administrative link between the two.

31   Fishback, "Did Coal Miners 'Owe Their Souls?,'" 1022; *Immigrants in Industries*, Vol. 7, 201; Glenn Massay, "Coal Consolidation: Profile of the Fairmont Field of Northern West Virginia, 1852–1903" (PhD diss., West Virginia University, 1970).

32   Shifflett, *Life, Work, and Culture*, 182; Fishback, "Did Coal Miners 'Owe Their Souls?'"; Tams interview, 37; Lane, *Civil War*, 25–26; Laing, "Negro Miner,"

312; *Immigrants in Industries*, Vol. 7, 202; *Report of the U.S. Coal Commission*, 1463; Greene, "Strategies for Survival," 49–51 (quote).

33   Milton Koslow, interview with author, Charleston, WV, May 13, 1998.

34   *Report of the U.S. Coal Commission*, 1462–1463, 1522.

35   Weiner, *Coalfield Jews*, 61. See also Bennett interview.

36   Phipps interview; Shifflett, *Life, Work, and Culture*, 184; *Immigrants in Industries*, Vol. 7, 202; Irving Alexander, "Wilcoe: The People of a Coal Town," *Goldenseal* 16 (Spring 1990): 29; Greene, "Strategies for Survival," 51. None of the sources mentioned Jewish scrip dealers. Laing offers the interesting observation that "the company doctor not infrequently adds scrip discounting to his professional services. . . . [It is] surprisingly lucrative." ("Negro Miner," 282).

37   Phipps interview; Isadore Gorsetman, interview with author, Charleston, WV, May 13, 1998; Sam and Harvey Weiner, interview with author, Logan, WV, November 8, 1996; Weiner, *Coalfield Jews*, 61.

38   Lane, *Civil War*, 28; Wolfe-Collins correspondence, March 23 and 26, 1915, Justus Collins Papers; West Virginia Code, 1891, 1923, 1927, 1973; Acts of the West Virginia Legislature, 1925, 1927, 1975. On Virginia, see Shifflett, 184.

39   *Immigrants in Industries*, Vol. 7, 202; Roden, "The Commissary," 241; Shifflett, *Life, Work, and Culture*, 184; Ira Sopher, interview with author, Beckley, WV, May 31, 1996; Weiner, Gorsetman interviews.

40   Gorsetman interview.

41   Shifflett, *Life, Work, and Culture*, 189; Weiner, *Coalfield Jews*, 62.

42   "Most Stupendous Land Deal Ever Pulled Off in Keystone," *McDowell Times*, January 12, 1917; Houston Kermit Hunter, "The Story of McDowell County," *West Virginia Review* 17 (April, 1940): 165–169; "Northfork-Clark Is 1200 Town," *McDowell Recorder*, June 30, 1922; Jones, *Early Coal Mining*; Tazewell County Deed Indexes (Virginia State Library, Richmond, Virginia); Alexander, "Wilcoe," 28–33; *Report of the U.S. Coal Commission*, 1462. In small ways, direct economic relationships developed between coal companies and merchants. One Jewish man recalled that company stores in Mingo County sold tickets to his father's Williamson movie theater, keeping 10 percent of the amount collected. Local tailors came into company stores to fit customers. In later years, some Jewish merchants even had concessions to run departments within company stores. Various interviews, Weiner, *Coalfield Jews*, 227–228.

43   Various interviews, Weiner, *Coalfield Jews*, 227–228.

44   *Immigrants in Industries*, Vol. 7, 213; Eiland interview.

45   Alexander, "Wilcoe," 30; Sylvan Bank, phone interview with author, March 4, 1998; Weiner interview.

46   Shackleford and Weinberg, *Our Appalachia*, 225; various interviews, see Weiner, *Coalfield Jews*, 227–228; Herman Monk, interview with Paul Niden, Beckley, WV, October 1, 1980, transcript, WVRHC; *Report of the U.S. Coal Commission*, 1461.

47   *Immigrants in Industries*, Vol. 7, 212, 204; Laing, "Negro Miner," 309–310. See also Sullivan, "Coal Men and Coal Towns," 198–199.

48   *Report of the U.S. Coal Commission*, 1460, 1463; *Immigrants in Industries*, Vol. 7, 201; Laing, "Negro Miner," 308; Shackleford and Weinberg, *Our Appalachia*, 225; Edna Drosick, interview with author, Pocahontas, VA, April 26, 1998.

49   Wolfe-Collins correspondence, December 25, 1915, Justus Collins Papers.

50   *Report of the U.S. Coal Commission*, 1461.

51   U.S. Bureau of the Census, McDowell County, WV, Manuscript Census, 1910, 1920; McDowell County naturalization records, McDowell County Courthouse; Hatcher and Steele, *Heritage of McDowell County*, 49; Shinedling, *West Virginia Jewry*, 1275. On the marginality of small Jewish businesses see for example Walter P. Zenner, *Minorities in the Middle: A Cross-Cultural Analysis* (Albany: SUNY Press, 1991); I. M. Rubinow, "Economic Condition of the Jews in Russia," *Bulletin of the Bureau of Labor* 72 (September 1907): 487–583.

52   Thomas, "Coal Country"; Eller, *Miners, Millhands, and Mountaineers*. On coal company ownership of entertainment and service businesses, see Shifflett, *Life, Work, and Culture*; Sullivan, "Coal Men and Coal Towns."

53   In contrast to the quiescence exhibited by coalfield merchants both Jewish and non-Jewish, merchants in Steelton, Pennsylvania, protested the local steel companies' use of scrip with petitions and other actions, and won some concessions. The difference attests to the nature of the hegemonic status that coal operators managed to attain in the southern coalfields. John Bodnar, *Immigration and Industrialization: Ethnicity in an American Mill Town* (Pittsburgh: University of Pittsburgh Press, 1977), 10.

A Combat Scenario:

Early Coal Mining and the Culture of Danger

Paul Rakes

ON JANUARY 29, 1907, coal production at the Stuart Mine in Fayette County, West Virginia, came to a halt after a loaded car on the cage shifted position,[1] tearing out supporting timbers and metal guides near the shaft bottom. Soon after, coal loaders at the various working areas of the mine found themselves without empty mine cars to load, and the men began making their way to the shaft bottom. Trapper boys,[2] youths usually from eleven to fourteen years old, noted the exodus of many of the miners and joined in the procession. Situated in various areas around the bottom, the men waited and watched as workers above them attempted to repair the damages to the shaft structures.[3]

Possibly, even probably, a critical ventilation trap door was left ajar during the movement of so many workers from the mine. In such case, a halt or reduction of airflow to the working faces would have occurred. Richard Lee, a black miner from Alabama with twenty-four years experience, had not joined the exodus having decided to prepare his working area for loading on the following day.[4] Lee drilled the coal face in preparation for blasting and, after a short absence from his working face, he returned to pack the holes with black powder. Unknown to Lee, methane, the nemesis of coal miners, had dispersed into the area and mixed with the surrounding atmosphere producing a gaseous combination known as firedamp. As Lee turned the corner into his working room, the open flame of his headlamp ignited the mixture. The ignition killed Lee instantly and caused the suspension of fine particles of coal left behind by the mining-process.[5]

Once suspended in the atmosphere, fine coal dust is explosive, and the ignition that began in Lee's area increased in fury as it pulled available oxygen to its center, raising more particles into the air. Speeding along the mine's entries, the intensity of the blast had reached its apex when it approached the more than seventy men waiting at the shaft bottom. Two groups of men had narrowly escaped minutes before when the cage had become operational again, but a majority of the work force remained in the path of the explosion. Engulfed by the blast, men, mules, and equipment were hurled against the walls of the entry and many of their bodies piled on top of one another. The force of the explosion projected mules along the mine's entries and ripped the harnesses from their bodies, depositing them several feet away. Blasting powder carried by several workers exploded from the intense heat, dismembering numerous bodies. Much of the force turned up into the shaft and damaged the steel guides and timbering that had just been repaired, but a substantial portion of the explosion continued toward the other end of the mine.[6]

Although the explosion had lost some of its force by passing up the shaft, it had not completed its course. State law required two mine openings in any mine that employed more than twenty workers, but Stuart had only one. Reaching the other end of the mine, the explosion had no way to exit to the outside and simply rebounded, heading back toward the shaft. Any of the men who had miraculously survived the first deluge, found themselves caught a second time as the blast returned and traveled up the shaft or dispersed throughout the areas where the fuel had been used up.[7] Indeed, some men did survive even the second experience, but had practically no chance of remaining alive. Coal mine explosions consume available oxygen and produce an atmosphere of lethal gases known as afterdamp. Anyone left alive simply had no air to breathe. One miner, caught in the afterdamp some distance from the shaft bottom, realized the inevitability of his death and kneeled against the rib to pray,[8] using his finger to draw a cross in the coal dust.[9]

Descriptions of the carnage of tangled men, mules, and equipment at the Stuart shaft bottom and at other mine explosions resemble reminiscences of wartime battlefield scenes. Mining engineer W. P. Tams recalled the sight as "pitiful," with over fifty men lying dead. Tams noted that "my transit man for the Stuart and Parral mines, Jesse Arthur, was lying on a mine locomotive with the top of his head blown off. One of the men was sitting with his back against the coal rib, his lunch bucket between his legs, and a piece of bread in his mouth, held by his hand." At the Monongah, West Virginia, disaster in December of 1907, Consolidated Coal Company official Frank Haas witnessed similar scenes during the recovery operation: "The heat left by the explosion was intense, and the stench from the decaying bodies of men and the carcasses of horses and mules, with the necessarily poor ventilation, made the work at times almost unbearable."[10]

Such battlefield-type descriptions accurately portrayed the reality of coal mining at the turn of the twentieth century. The industry experienced unprecedented growth between 1890 and 1910, but coal mining casualties increased proportionately with that expansion. In 1847, the nation produced 4,865,522 tons of coal and suffered only seven reported deaths, while in 1907, 3,242 men died in the extraction of over 400 million tons. Although the horrendous death toll of over 3,000 in 1907 caused some public outcry, for the most part, these dead men represented the inevitable: inherent hazards existed in coal mining, and as in combat, men die.[11]

Indeed, the increasing death rate at the turn of the twentieth century primarily served to solidify the public and, arguably, the miners' perceptions that mining was a naturally risky occupation. High casualties during the early expansion era heightened the subculture of danger among miners and equally influenced public acceptance of such unnecessary deaths. Although technological changes altered the underground environment over the years and reform rhetoric surfaced after catastrophes, coal mining's intrinsic dangers provided the industry an official and personal defense of its abysmal safety record throughout much of the twentieth century.[12]

Certainly, miners died in amazing numbers during the industry's years of expansion. Between 1890 and 1907, at least 26,434 fatalities occurred in the nation's coal mines. Of course, these figures do not include the numerous injured. Journalist Malcolm Ross noted that each year the number of men with wooden legs or empty sleeves grew. Statistics indicate that, annually, 22 percent of Pennsylvania's miners were killed or crippled for life while another 28 percent were either crippled or seriously injured with the prospect of eventually returning to work. The figures suggest that Pennsylvania's miners had a less-than-even chance of surviving for twelve years, and that they could expect to be seriously crippled or injured in six. While these figures seem incomprehensible for the simple pursuit of making a living, West Virginia miners endured even higher fatality and injury rates in 1907—double that of the Pennsylvania numbers.[13]

Much of the increase in deaths resulted from the changing conditions of mining. Coal fueled the industrial era and entrepreneurs rapidly sought to make profits from the increasing market. Industrialists could open mines with small investment and several companies began with no more than a few thousand dollars in capital. Enjoying high demand for their product, these enterprises quickly expanded and increased both the number and size of their mines. Taking advantage of new machines to undercut the coal and huge ventilation fans to force air into the mines, the individual operations tunneled underground for miles. Bound to traditional thinking that applied to small undertakings of the past, the budding professions of mine engineers and mechanics did not yet fully understand the forces unleashed by such operations. Like the military professionals who failed to comprehend the killing power of new technologies before World War I, mine operators seemed unable, or unwilling, to fully appreciate that mining with few casualties required the careful maintenance of a chain of mutual dependencies.[14]

Although technology enabled mines to extend further underground and transport larger amounts of coal to the surface, the actual process of extracting coal from the face remained labor intensive and required hundreds of men. Each member of the various working units depended upon the alertness of other workers: the fire boss to note any presence of methane before work commenced, the trapper boys to see that ventilation doors remained closed, company men to maintain open air courses, and several other details that demanded close attention. Like a battle plan that requires various units to perform certain actions, and falls apart whenever all the criteria are not met, the act of coal mining could result in disaster without constant vigilance to these details. Pointing to the increased killing power of coal mines in 1908, Thomas Keighley of the Oliver and Snyder Steel Company warned that traditional methods clouded the judgment of mine owners and that they would have to cease engaging in certain practices "we have been in the habit . . . which until late years were taken for granted to be right."[15]

By the turn of the century, miners worked deeper below the earth where geological forces increased the potential of large releases of methane and larger mines increased the statistical possibility of tapping into reservoirs of the gas. Such potential danger required that miners no longer intentionally ignite what they perceived as small pockets of gas in order to burn them away, and operators needed to provide more ventilation than ever before.

Despite the rhetoric concerning "mutual dependency" and safe operation of coal mines, operators continued to oppose and openly defy mine-safety laws, occasionally exhibiting more concern for production and equipment loss than casualties.[16] Following the extensive loss of life from explosions in 1907, one mine official noted that operators could make mistakes in ventilation—the usual cause of disasters—and recover from them, but could not undo the errors associated with the improper extraction of coal which left several tons unclaimed. Perhaps the best example of the priorities of some operators came from the Winding Gulf Colliery Company General Manager George Wolfe. Writing to Company President Justus Collins, Wolfe observed that two men, against company policy, had made use of an underground locomotive to ride to their work area. Wolfe noted that, "one of the men was torn all to pieces, but no harm was done to the motor."[17]

Over time, the concept of callous coal operators and/or companies certainly melded into the consciousness of mining labor culture. No doubt, some individual operators, such as Justus Collins, adequately fit the stereotype. Later, the large companies that replaced the pioneer managers, simply by the impersonal structure of the corporate system, added to the callous reputation. Yet, it is equally probable that some of the stories providing evidence were either embellished or taken out of context. One example centers around the popular perception that operators placed more value on their mules than on their workers.[18] An often repeated story—usually applied to a variety of disasters—suggests that when the 1907 Monongah explosion claimed at least 361 lives, one government official queried as to how many mules the company lost.[19] News of the frightful loss of life in the disaster spread quickly. Anyone arriving at the site, where a multitude of anxious family members and sightseers waited on the surface, would already have had some idea of the magnitude of the human destruction. And, it is just as probable that the official had already discussed the number of human casualties with company leadership. In an era when horses and mules remained an integral part of everyday life, it seems only natural for someone to ask about the loss of these living creatures. The comment may have been insensitive to those who overheard it, but it is doubtful that it specifically demonstrated a belief that a mule had more value than a worker.[20]

Despite the popularly perceived general callousness of mining entrepreneurs, both the American public and the coal miners themselves continued to accept the high death toll and the serious injuries that totaled at least two for each fatality.[21] Past historians have noted the high death toll or pointed out the lack of human concern by operators, but the underground combat reality of the miner has received scant attention. In *Coal-Mining in the Progressive Period: The Political Economy of Reform*, William Graebner suggests that the miners themselves resisted progressive reforms designed to produce safer coal mines, but offers little insight into the world of these reluctant reformers. Historians tend to concentrate on the early independence of workers and the concept of miners' freedom as an explanation for many of the events that occurred in the coalfields. Two notable exceptions are Anthony Wallace's *St. Clair: A Nineteenth-Century Coal Town's Experience with a Disaster Prone Industry* and Donald Miller and Richard E. Sharpless's, *The Kingdom of Coal: Work, Enterprise, and Ethnic Communities in the Mine Fields*. Wallace's chapter "Politics of Safety" and Miller and Sharpless's "Working in the Black Hell" provide some insight into the world of the miner by describing what life was like underground, but apply no historical analysis to their discussions. Indeed, if politics played an important role in mine safety and men did work in a "black hell," then some self-perception of their role must have enabled miners to accept the validity of the high death and injury rate. Combat soldiers acquiesce to the possibility of death in combat and it is conceivable that miners perceived their job as inherently hazardous and consented to the risks involved. Certainly, public perceptions of the trade at the turn of the twentieth century included the idea that the hazards of the occupation meant every man who entered a coal mine would sooner or later meet his fate. Wives of miners also accepted this potential reality and many developed a ritual of always kissing their departing husbands goodbye for fear that they may not see them alive again.[22]

Serving on the Mining "Front Line"

Born in the Maryland coalfields shortly after the turn of the century, Raymond Densmore associated the words "coal mining" with the memory of a passing hearse. Living at Carlos Junction on Georges Creek near the Consolidation Coal Company's No. 9 mine, Densmore remembered the passage of the horse-drawn vehicle carrying an injured or dead miner as a frequent occurrence. As part of the general mining community, Densmore had exposure to other aspects of the day to day lives of miners, but the terrible aftermath associated with serving on the mining "front line" left the dominant impression. Undoubtedly, his mental image attests to the dangers of coal mining as well as the integration of the risks into the occupational culture.[23]

Like the stark environment of a battlefield, the underground world that produced the casualties witnessed by Densmore represented something of an alien landscape for human habitation. When a coal loader entered his workplace, known as a "room," he found himself in an area that resembled a long, narrow tunnel. The inside end, the actual coal face, receded a few feet each day as the miner blasted and extracted the coal. Depth of the seam determined the height of the room, but often a miner could not stand upright. Width of the area varied from twenty to thirty feet, and, because connecting tunnels or "breakthroughs" were driven between advancing entries, the individual tunnel usually extended little more than one hundred feet and existed as part of a veritable maze of hallways. Breakthroughs provided pathways for ventilation and access to other rooms, and railway track ran into each working place to transport coal cars. Excluding the scant visibility produced by the light of his open-flamed headlamp, the miner moved about in semi-darkness and sometimes developed "miner's eye," an inability to steadily focus his vision on an object.[24]

Although machines to undercut the coal seam existed, many turn-of-the-century mines relied on manual labor for this process. Miners often worked two to a room and the men lay on their sides on the floor (known as the bottom) and used a pick to cut a trench three to four feet deep under the seam. This usually took two to three hours of manual exertion followed by the physical demands of drilling holes in the coal face with a breast auger.[25] Placement of the holes required skill and adequate knowledge of the particular coal seam. Well placed, the subsequent blast would dislodge a depth of three or four feet of coal. Poorly placed, the explosion broke-away little or no coal, dropped the material in a solid block onto the undercut, or blew away the safety props and scattered the material all about the room. Worst of all, handling the extremely flammable black powder could be dangerous, and a lack of vigilance might produce a catastrophe. Following the blasting, miners loaded the coal into available cars, set safety props (known as timbers), and extended the railway track further into their room for the next cycle.[26]

Throughout this process, the miner exposed himself to constant danger. While undercutting the seam, the weight of the coal above occasionally forced the material to drop prematurely, occasionally trapping or even decapitating the worker. Because early mines relied on one current of air, and blasting took place at all times of the day, the atmosphere resembled that of the powder-smoke filled nineteenth-century battlefield. Workers inhaling the acrid smoke often complained of chronic headaches. The roof (known as the top) had to be constantly scrutinized for signs of dripping (the falling of small flakes of rock that indicate an impending fall) and even the walls of the surrounding pillars could roll over on the worker (referred to as a rib roll). In some mines, the stresses of geology threatened to "squeeze out" an entire pillar, which exploded in a shower of coal and rock into the nearby tunnels. Known as a "bump,"[27] few miners lived through the larger ones to tell of such an occurrence. If the miner proved careless during blasting by tamping (the process of placing material—sometimes coal dust—inside the drilled holes to transfer the force of the blast into the coal seam) or used too much explosive, the flames from the black powder could expand into the working area, suspend and ignite surrounding particles of coal, and begin the process of a full scale mine explosion.[28]

Indeed, dealing with mine explosives often resembled the handling of nineteenth-century artillery shells. The first combat casualty of the Civil War resulted not from the intense bombardment of Fort Sumter, but from the explosion of a powder magazine during the surrender of the post. Many miners died in similar occurrences. A few examples represent numerous such incidents. Having three separate fuses to light, one miner died when the first explosion occurred before he could attend to the other two fuses. Another worker, after lighting a fuse, decided that the shot had not detonated as soon as expected and returned to ascertain the problem, only to be killed by the exploding powder as he approached the coal face. While pushing dynamite into an overhead area to blast away rock,[29] one miner struck the ignition cap of the charge with his tamping stick, causing an explosion that dislodged slate onto his working partner.[30]

While blasting and coal mine explosions produced the most publicized and dramatic deaths, far more miners died from roof falls. Some officials estimated that at least eight out of ten miners lost their lives to this cause.[31] Noises and a dripping top generally supplied some indication of impending danger, but often death came with no warning at all. As with combat soldiers who fear the presence of enemy snipers, coal miners dreaded the presence of kettle bottoms. Miners, through experience and occupational communication, came to regard the phenomenon of kettle bottoms as synonymous with a silent killer from which there was little defense.[32] One skilled Welsh miner described the danger:

> This clay roof is full of kettle bottoms as they are called here (bells at home) and we have also an abundance of the fossil remains of huge trees in this roof. It is a most dangerous roof and we have to watch it for our lives. Those bells are often eight feet in diameter and don't give the slightest warning by [sic] simply drop out without any warning whatever.[33]

Even progressive and conscientious mine inspectors agreed that roof falls and kettle bottoms represented a risk of the trade. When one miner was killed by a kettle bottom in 1903, West Virginia's Chief Mine Inspector noted the adequate safety-preparation of the worker's room and observed that "it was an entirely unforeseen accident."[34] Another inspector informed the chief that the Elkhorn coalfield of southern West Virginia contained the "dangerous proposition" of kettle bottoms, and that even the most skilled miners could not detect their location. He reported that, "The miners or laborers in any mine may be testing the roof, and it may seem absolutely solid [but] it falls, which has been the case repeatedly in this district."[35] Mine roofs often proved unpredictable, and, if miners were like soldiers in combat with nature, the potential of death from this cause resembled something like the enemy sniper's bullet. Even when taking a break, miners needed to be constantly alert to the sights and sounds of their surroundings. Withdrawing from the face itself did not afford protection from an unpredictable top, as demonstrated by the 1908 roof fall death of James Williams while he sat eating his lunch in the Chatham Shaft No. 1 Mine in West Virginia. Nor was a large roof fall, with its usual rumbling warnings, required to produce severe injury or death: the density of kettle bottoms meant they could inflict injuries that belied their size or the sharp edges of broken slate could produce fatal lacerations.[36]

Indeed, a government pamphlet warned miners that "there are so many ways YOU can be killed that it is highly important for YOU to be forever alert" (emphases in original).[37] The many ways to be killed included the simple act of riding the cage to the shaft bottom. Although rare, a hoisting cable could break or an inexperienced operator could mistakenly drop the cage to the bottom at coal-hoisting speed. As a result, some mine directors insisted that no two family members ride the same trip to the shaft bottom.[38]

Although often forgotten as a cause of serious mining injuries, mules and horses also added to the "many ways" miners could die on the job. Members of the mining community know well that, later in the century, mine locomotive haulage caused many fatalities and injuries, but nostalgic views of earlier operations dependent on animal power must not obscure the actual danger of working with mules and horses in close quarters. These beasts of burden, both from temperament and environment, increased risks in the mines similar to those of the nineteenth-century battlefield. One Civil War veteran remembered the danger soldiers faced from their proximity to horses and mules, particularly members of artillery units. At Gettysburg, Robert Stiles recalled seeing men trapped among the harnesses and bodies of wounded horses. In agony, the horses lashed out with their hoofs, crushing the heads of the men among them.[39] Drivers of mules in the coal mines often experienced similar situations. The ribs of the coal mine and the low roof kept drivers in close proximity to their animals; any mishap could unleash the fury of a frightened mule directly onto anyone nearby. Occasionally, the simple, cantankerous behavior of a mule would kill a driver; eighteen-year-old Massie Barker died when a mule kicked him over the heart. In other cases, mules became unwilling participants in the mishaps associated with coal mining. In 1900, a rock fall in the Sugar Creek Mine of Fayette County, West Virginia, trapped both a mule and a miner. Caught under the fall, the animal crushed the miner beneath him. Other mishaps reflected the individual nature of deep mining. Operators usually brought mules to the surface at least once a year for a few days of freedom and fresh grass. In 1908, an underground stable-boss at the Stuart Mine loaded an animal onto the cage for transfer to the outside. At some point during the ride, the mule crowded his handler off the cage and forced a fatal fall to the shaft bottom.[40]

Stiles also recorded the attachment between military men and animals that evolved from exposure to danger; that same affection developed in the underground world. A young driver in Pennsylvania, emotionally attached to one of his mine's mules, hurried toward the stable when he learned that a massive rock fall had blocked ventilation to, and escape from, underground. Popular tales suggest that rescue workers discovered the dead boy lying beside his favorite mule with arms wrapped around the animal's neck.[41]

Civil War artillery horses exhibited their mortal fear of combat by trembling and squatting close to the ground. Perhaps the behavior of mining mules adequately represents the comparative dangers of coal operations. The underground world exposed the mule to a radically changed environment. The animal had to accustom himself to "unnatural" sounds, dark working and living quarters, and an entirely new laboring environment. Besides the obvious exposure to the threat of slate falls, mules occasionally endured an ignition of methane that burned their hair and skin. Animals brought to the surface after long periods underground often trembled and became delirious with joy when first reexposed to sunlight, fresh air, and grass. Maintaining a memory of the dreaded environment underground, mules would often refuse to reenter the mine and violently resist any efforts of the drivers to entice them back down.[42]

While the drummer boys of nineteenth-century armies found themselves caught in the danger of combat, drivers, trapper boys, and spraggers (known as "runners" in the anthracite fields) found similar risks in coal mining. Usually, young boys in their mid- to late teens served as mule drivers. In order to survive, these boys had to possess both quickness and dexterity. Riding the narrow chain or leather strap between the mule and first coal car, the driver placed one hand on the animal's rump and the other on the bumper of the front car. If the driver lost his balance, he could be crushed by the advancing coal car. When reaching the top of a steep incline, he had to rapidly disconnect the mule and turn to the side while in motion, allowing gravity to pull the trip of cars along down the hill. Any lack of concentration could result in severed fingers or being crushed between the cars and the rib. As the cars turned down the incline, spragger boys or runners became involved. Usually nimble boys of quick hands and feet, runners had to place sprags (sticks of wood or steel) into the spokes of the moving cars. Kneeling beside a stack of sprags, the lad placed one in each wheel as the cars sped past, locking the wheel whenever the spoke brought the sprag into contact with the bottom of the car. As the speed of the cars increased, the runner usually found himself running alongside, dodging low roofs, narrow places, and jumping over obstacles as he attempted to place the last of the sprags and keep the trip from speeding out of control. Occasionally, if the driver failed to unhook in time, the runner, driver, and mule were dragged along in a tangled mass of wood and metal. Although the work of the trapper boys required little physical exertion, the youths sat alone in the darkness on a crude bench beside the assigned door, built to direct airflow for ventilation purposes. Consequently, trappers sometimes fell asleep and, on occasion, became casualties when a trip of cars crashed into the closed doorways.[43]

While both adult and adolescent miners remained alert to the everyday dangers of mining, methane represented the most-feared hazard. Undetectable except by testing with a safety lamp, the gas usually appeared without warning. Whenever the proper lethal mixture occurred, any spark or the open flame of a miner's headlamp caused an ignition, engulfing the worker.[44] If the other men in the mine were lucky, the concussion and heat would fail to suspend the surrounding coal particles and start a dust explosion. The danger from methane proved so great that coal mine operations developed a system that basically supplied a scout to seek out any presence of the gas. As mines probed deeper into the earth, they liberated more marsh gas. In the same way that military units sent advance personnel to ascertain the presence of the enemy, operators sent a scout in ahead of the rest of the labor force. In the earliest days of mining, whenever the scout located the presence of methane, he covered himself with a wet burlap sack and attempted to ignite and burn the gas from any working areas. Known as a fire boss, the individual's task became easier with the advent of a safety lamp, a device which could detect a mixture of methane in the atmosphere. Fire bosses attempted to force out any volatile combinations of a methane-charged atmosphere by increasing ventilation to the area, but, despite the risks involved, early twentieth-century miners occasionally undertook the dangerous practice of placing an open flame on the end of a long stick and "flashing" or igniting the gas to burn it away. Obviously, several miners received mortal burns from such procedures.[45]

Miners who were seriously injured underground by the many mishaps received inferior medical attention. Actually, the treatment of most injured coal miners in the early 1900s resembled that of mid-nineteenth-century soldiers when first wounded on the battlefield. Generally, men underground in adjacent rooms came to the aid of the injured miner, but after extracting him from whatever situation caused the mishap, fellow workers could do little more than throw coats over the victim and assist efforts to transfer him outside. Often, severely hurt miners were simply taken to their homes where wives attended to the injuries. One mine official noted that a casualty at his mine lay in the blacksmith shop for two hours before receiving any treatment. He added, not surprisingly, that the man died.[46]

The numerous casualties resulting from the everyday aspects of coal mining encouraged one physician to insist that the industry should develop a first-aid corps similar to that of military units. Attempting to organize a mining first-aid unit at a Pennsylvania coal mine, Dr. J. M. Shields pointed out the success of such military medical groups in the Turko-Russian War of 1878 and the Spanish-American War of 1898. Whenever British industrial first-aid units performed exemplary service during the Boer War in 1900, Shields became more determined and intensified his efforts. Although two Pennsylvania mining companies assisted Shields, most operators labeled the doctor as something of a "crank" on the subject of medical preparation for mining emergencies.[47]

However, the frequent casualties in coal mining eventually convinced operators to accept the wisdom of Dr. Shields and by 1909 at least one West Virginia company established something of a nearby "field" hospital. The United States Steel Corporation provided an emergency facility with two physicians and an operating table for the express purpose of providing immediate care to the injured before moving them to the State Hospital at Welch. Such an undertaking represented the atypical, and West Virginia's mine officials asked the legislature to enact laws that would require mines to provide the simple apparatus of a stretcher and waterproof blankets for the care of underground casualties.[48]

Although miners realized the high probability of inadequate medical treatment of injuries, bravery in dangerous situations remained a requisite for all individuals. In the early 1900s, society and the subculture of underground work perceived mining as a masculine endeavor, and the stoic male in the face of danger comprised a desired, and generally required, positive attribute. Combat soldiers generally function in a setting of mutual dependency and, similarly, coal miners needed fellow workers that could be "counted on" in emergencies.[49]

An informal but sacred code produced a common understanding among workers that, in a disaster, the miners on the surface would work ceaselessly until recovering the last body. Cultural mandates required workers to further risk their lives whenever fellow miners were trapped or endangered. Writing in 1942, Edward Wieck summed up the miner's code by quoting from novelist Karel Capek's *The First Rescue Party*: "All miners know what it means. All respond to the call for volunteers to go below after an explosion, regardless of the risking their own lives."[50]

A majority of miners always responded with bravery and were often acknowledged with glorious praise. In 1909, when twelve men died as they attempted to aid 257 trapped miners in Cherry, Illinois, officials pointed to the "thrilling rescues by a heroic band of men who finally perished in an act of supreme sacrifice and heroism."[51] Even operators could be caught up in the demands for gallantry. When former coal operator Selwyn Taylor died while attempting rescue operations at a Pennsylvania mine following an explosion, a fellow operator observed that his death attested to his courage. Taylor had entered the mine "because the grief of the stricken widows and orphans, many of whom were aliens and not of his class, appealed to his great and manly heart."[52]

If the tributes to fallen mining heroes resembled acknowledgment of battlefield sacrifice, many officials suggested that casualties could be reduced if the mining occupation itself embraced a system of military discipline. Speaking to a group of coal operators who obviously agreed with the principle, a member of the United States Geological Service observed that "there is no industrial occupation in which the establishment of discipline of a military character is so necessary as in coal mining. Any infraction of orders by a miner, a laborer, a door-boy, a pit-boss, a driver, an engineer—by anyone employed in and about a coal mine, may endanger the lives of hundreds."[53] John Laing, West Virginia's chief inspector in 1909, agreed with the premise and pointed out that the "strength of a chain is its weakest link, so is every man's life who enters a mine under [dangerous conditions], and is at the mercy of the most reckless miner."[54]

Perhaps the behavior of the least free of America's coal miners best captures the actual hazards of the occupation. Convict labor, forced into the mines of Alabama, reasoned much in the manner as the Union Civil War troops described by Gerald Linderman in *Embattled Courage*. Faced with the prospect of reenlisting to receive a furlough home, soldiers reasoned that they would probably not survive the upcoming campaign before their current enlistment ran out. Under such circumstances, the risk of surviving their enlistment period was not worth never seeing one's family again and, consequently, soldiers reenlisted in order to go home once more. Slaves and, later, African Americans unjustly convicted of crimes and brought to the coalfields, dreadfully feared the treatment they would receive if caught trying to escape. However, the terrors of the coal mine proved more intimidating and these reluctant miners used every opportunity to try and escape—the realities of mining made it worth the risk.[55]

The perils of coal mining encouraged miners to compose songs related to the dangers of the occupation. Like Civil War soldiers who sang about their combat world in refrains such as "Tenting Tonight on the Old Campground," or "Just Before the Battle Mother," miners expressed their fears in such tunes as "The Old Miner's Refrain," "The Miner's Life," or "Down In A Coal Mine."[56] Such songs often spoke of bravery in times of crisis. One example is a stanza from a composition concerning the Avondale, Pennsylvania, mine disaster of 1869 and the attempts at rescue:

Two Welshman brave, without dismay,
And courage without fail,
Went down the shaft without delay
In the mines of Avondale.[57]

Although the miner did not fight against a human enemy, but rather against the four ancients of earth, wind, fire, and water, his occupation required courage and fortitude and led composers to compare the mining world with that of the battlefield. Using such terms as "fearless" and "battle," one miner poet compared military and mining occupations:

> The Sailor's life has its round of charms,
> The storm and chase with their wild alarms;
> And he sings as the boatswain pipes to arms,
> "Ho! ho! for the sea with its mystic charms."
>
> The soldier fights for a scanty hire,
> Yet pants for the strife and the battle's fire;
> As he to the rampart's heights aspires,
> Sing—"ho! for the strife and battle's fire."
>
> The Miner pants for no gory goal;
> In vain to him may the battle roll;
> Yet his manly heart and his fearless soul
> Sing—"ho! success to the gleaming coal."[58]

Stephanie Elise Booth suggests that the general public has failed to realize that it took skill and courage to be a good coal miner. Historically, popular perceptions contained an element of contempt for miners because it seemed obvious that "anyone with any sense would not undertake such mindless work underground."[59] Yet, the songs and poems of coal mining folklore suggest that the miners perceived an element of courage and bravery necessary for their pursuit of a dangerous occupation.

Adjusting to the Carnage

> Yet the grimy, coughing coal miner is inured to the dangers of his calling. He is stoic about the deaths of friends and relatives and accepts as "part of the game" the high possibility that he may be crushed, burned, suffocated, or drowned in his own Cimmerian tomb.[60]

Self-control and emotional restraint are central to one's behavior in all societal situations. The miners who continued working under the auspices of death and injury as a natural part of the job generally attempted to exhibit the restraint society desired of the early twentieth-century male. Courage, endurance, and toughness represented a miner's perception of masculinity, and the occupation's cultural mandates required a miner to maintain fortitude and calmness even in life and death situations. Like combat units, such behavior became binding upon the entire male work force. Journalist Malcolm Ross noted a masculine pride in miners and observed that coal miners perceived themselves as brave fellows set apart from those on the surface. Ross suggested the presence of a certain swagger by miners. During his youth, Raymond Densmore noted the same behavior in young coal miners leaving the pits: "They had a jauntiness and swagger about their walk that belied their age. . . ."[61]

Cultural demands that coal miners risk their lives to save comrades trapped or killed in a disaster represented the solidarity of a strongly knit group. Any miner who openly expressed fear in dangerous situations made it difficult for others to respond and added to the unpredictability of an already hazardous situation. Sociologist Rex Lucas observed that "it was most important in the mines as in combat that each man was able to maintain his own self conception as a male."[62] Indeed, scholars have discovered that military units exposed to the highest casualty rates and faced with the most difficult combat situations maintain a group attitude that prohibits overt expressions of anxiety and fear.[63]

Yet, these demands leave no outlet for the release of tension and anxiety. Combat troops experience intense emotional strains from battle experiences and the American military implemented methods of aiding soldiers in learning to cope with the fear and anxiety.[64] Yet, other than one's reputation within his working subculture, the miner who witnessed death and injury and faced daily dangers had no formally established coping methods. A coal miner's only release from such emotional strain was absenteeism. The practice of "laying-off" plagued coal operations from its earliest years and industrialists, unaware of the real cause, blamed the practice on laziness and shiftlessness.[65]

Physicians recognized the effects of protracted tension among coal miners as early as 1869. One doctor observed that many miners suffered from a neurological syndrome that included dyspepsia, tremors, vertigo, and palpitation without any indication of organic disease. Although mid-nineteenth-century physicians had no term for the psychosomatic response, what they witnessed eventually became known as combat fatigue. Later studies convinced psychiatrists of the realities of stress in coal mining. In 1957, physicians at a Bluefield, West Virginia, hospital noted that coal miners not only "experienced dramatically tense situations, but also worked under protracted tension because as long as he is on the job there is the potential threat that he may be the next casualty. He is like the soldier under fire thinking: 'Maybe the next bullet has my name on it.'"[66]

While it took several years for medical personnel to recognize the importance of the psychological factors, one subtler occupational injury manifested itself as early as 1856. Physicians noted the softening of lung tissue in many miners and attributed it to the breathing of large quantities of coal dust. Autopsies demonstrated that the lungs of miners often turned black and were little more than carbon themselves. Consequently, mine doctors and mine inspectors pointed out the affected workers died from coal miner's consumption by age thirty-five.[67] Falling prey to this silent killer, these men resembled World War I veterans who would suffer years after the war from exposure to mustard gas.

One miner from the 1870s drew an analogy of coal mining to combat: John Maguire asserted that a coal miner "is a soldier every day that he works, while the man in the army is only taking risks when he goes to the front, which is only occasionally." Recalling his life in the mines, Maguire suggested that a "miner may expect, in the natural course of events, to ultimately meet his death in the performance of his duties."[68]

While many people accepted coal mining deaths as the natural result of a hazardous occupation, the drama of disasters eventually forced government officials to react. After some legislative arguing in 1883, West Virginia established a Department of Mines to monitor the coalfields and see that various operations provided adequate ventilation and drainage. However, the agency had little authority and, as the industry expanded, the death rate continued to rise.[69] Much to the chagrin of operators, various states attempted to halt the employment of young boys in the coal mines. One operator complained that such legislation hindered "budding manhood." Without the income of these youths, operators contended, many widows and smaller siblings would go hungry and the young boys would roam the streets and "go to the devil as fast as [his] legs can carry him."[70] Industrialists, however, conveniently failed to note that many of the widows probably resulted from coal mine deaths.

Although Pennsylvania set the minimum age for coal miners at sixteen, West Virginia permitted boys of fourteen to work underground and several youths died in the numerous explosions that rocked the Mountain State in 1906–1907. One government official, West Virginia Chief Inspector John W. Paul, refused to accept the inevitability of these disasters. Many coal operators and some coal mine inspectors suggested that explosions resulted from the lack of skill among the increasing number of immigrant workers, but Paul intended to see that industrialists conformed to statutory laws.[71]

Appointed by Governor Atkinson in 1897, Paul enjoyed powerful political connections and an impressive background in coal mining. After acquiring a degree in civil engineering from West Virginia University in 1894, Paul held a position as mine engineer in the New River coalfields. His expertise encouraged Fairmont Coal Company executives to have Paul recommend engineering changes at the Monongah mine and he so impressed officials that he received employment as their mining engineer. Paul furthered his mining education at the School of Mines at Columbus College in New York and later took a position with the politically influential Davis Coal & Coke Company.[72] Affiliated with influential political-businessmen such as Henry Davis and Stephen Elkins and well-versed in mining technology, Paul seemed a perfect choice for chief of the Department of Mines. However, within the context of his era, Paul proved to be a conscientious mine inspector and soon caused problems for entrepreneurs such as A. B. Fleming of the Fairmont Coal Company and Samuel Dixon of the New River Company.

English immigrant Samuel Dixon, who built the coal dynasty of the New River Company in West Virginia's Fayette and Raleigh counties, violated mining laws by working more than twenty men inside his Parral mine with only one opening. Driving entries to connect Parral and Stuart mines, Dixon apparently hoped to join the two mines—providing the necessary openings—without interference from state officials. On February 8, an explosion occurred in the Parral mine killing twenty-three men. When a Fayette County coroner's jury absolved Dixon's company of any responsibility, Paul pointed out to a grand jury that mine officials openly violated statutes by permitting thirty-five men to work underground. Jurors agreed with Paul's assessment and issued two indictments of manslaughter against Samuel Dixon and his son Fred, who served as superintendent of the mine.[73]

Local gossip suggested that Paul's challenge of politically-connected Dixon constituted an exercise in futility, but the chief inspector refused to relent. With an impending trial, Dixon continued to work a large number of men in both the Stuart and Parral mines. Whenever rumors suggested the potential visit of a mine inspector, the company directors reduced the working force to twenty. On January 29, 1907, the explosion at the Stuart mine, described in the introduction to this chapter, occurred. Undoubtedly, Paul pointed to his earlier warning and West Virginia legislators quickly formed a commission to investigate the Stuart disaster as well as other explosions throughout the state.[74]

Certainly, the available financial resources of the Dixon regime had influenced the staffing of Paul's own department. Commenting on the Stuart explosion, one company official observed that the corporation employed the best safety man in the state to examine the mine, at a salary in excess of that paid by the state. "We have Mr. Pickney, recently a state mine inspector."[75]

Although Dixon could use higher salaries to provide an appearance that knowledgeable men maintained safety standards at the mines, the legislators could not escape the reality that explosions occurred all around them as they conducted their hearings. Before the commission could even begin, an explosion at Thomas, West Virginia, killed twenty-five workers and another Dixon mine experienced a blast that resulted in sixteen fatalities on May 1. Explosions in other states occurred during the hearings, and on December 6, 1907, a disaster at Monongah, West Virginia, killed at least 361 men.[76]

The sanguinary loss of life aroused public demands for stricter coal mine regulation and, coupled with Paul's conscientiousness, forced government officials to continue at least an appearance of dealing with the dangerous reality of underground mining. A. B. Fleming of the Fairmont Coal Company complained to Governor William Dawson about the increase of government and public pressure. Dawson assured Fleming and other operators that he would oppose any radical legislation until the memory of Monongah had faded. Paul's legal pressure on Dixon caused some consternation among operators and they began covert efforts to remove the chief inspector. Whenever federal officials, concerned about the bloody events in the nation's coal mines, formulated plans to create a bureau of mines, West Virginia politicians and operators used the opportunity to have Paul promoted to the new agency. In his place, Dawson appointed John R. Laing, a former and future coal operator who maintained more of a commitment to the industrialists than to the oath of his job.[77]

Conclusion

When the Stuart explosion occurred in 1907, miners quickly assumed the role of rescuers. Despite the smoke escaping from the shaft, men assembled at the mine site to attempt a rescue if possible. Superintendent W. G. Colburn and General Manager Fred Dixon even had themselves lowered into the shaft by an improvised bucket, but the noxious gases and smoke prevented their descent. Despite repeated attempts to reach the shaft bottom, it was not until the night following the explosion that rescuers made their way into the mine. Although some hope remained for survivors, the single opening providing access to the pit restricted rescue efforts, hindered the ability of workers to remove noxious gases from the mine, and limited the potential of any survivors remaining alive in the isolated workings.[78] Stuart's dead miners represented the first in a series of fatalities that would plague both West Virginia and other states throughout 1907.

Industrialists not only successfully resisted the 1907 movement for progressive coal mining laws in West Virginia, they also worked diligently to ensure that the efforts of a Federal Bureau of Mines, established in 1910, centered on research rather than enforcement interference with individual mining operations.[79] Coal operators used political influence to keep progressive legislation at bay until public reactions to mining disasters subsided and the popular perceptions of miners returned. Whatever the number of coal miners upset by the events, they undoubtedly maintained the combat mentality associated with their craft and continued to embrace perceptions of theirs as a manly and dangerous occupation.

Certainly, the solidarity created by the tensions of the underground world played a role in the 1912–1921 coal mine wars of southern West Virginia. The social bonding that contributed to the racially and ethnically mixed assembly that marched to Blair Mountain in August of 1921 probably owed much to the camaraderie created by the shared dangers of the coal mine.

Acceptance of inevitable coal mine deaths and explosions continued. Although United Mine Worker's President John Mitchell suggested to the West Virginia Legislature in 1907 that adherence to mining laws would reduce the death rate by half, the *Fayette Journal* insisted that such events would occur as long as people mined coal. After waiting for the public memory of Monongah to fade, the legislative commission investigating coal mine explosions concluded that men become acquainted with danger and become careless. Significantly, the legislators observed that "it is true that do what we may we can only minimize these fearful accidents, as explosions will continue as long as mining is practiced."[80] The concept of miners as acceptable casualties in war with Mother Nature had received further official sanction.

Those who did accept the inevitability of coal mining injuries and death pointed to an obvious parallel. Miners produced the fuel that benefited the life of every American and, since the government supplied pensions for the national service of Mexican and Civil War veterans, it logically followed that the same should be done for workers in the coal industry. Others observed that miner hospitals and government inspectors cost the taxpayer a substantial amount each year and that taxation of the coal industry should pay for the expenses.[81] Despite the obvious rationale behind these suggestions and the acceptable casualty analogy, neither of these proposals found fruition. Indeed, despite an increase of practical and scientific knowledge evolving from Federal Bureau of Mines research, coal miners continued to be unnecessarily crippled or killed well into the 1960s. Major disasters such as Mather, Pennsylvania, that killed 195 miners and Centralia, Illinois, with 111 dead briefly raised public concerns, but no substantial reform efforts developed. Probably even more significant, the general acceptance of coal mining as inherently dangerous meant that the public, the industry, and the miners themselves failed to seriously challenge the much higher death and injury toll related to everyday accidents.

Disasters occurred throughout the 1960s, but the smaller work forces of mechanized mining meant that most disasters resulted in many fewer casualties.[82] Since technological advancement focused primarily on production, not safety, the new behemoth machinery often created new hazards or exacerbated old ones. Faster moving equipment in close quarters, heavy metal loading booms swinging back and forth, more rapid exposure of methane-liberating virgin coal faces, and the sparks often generated by the cutting bits of continuous mining machines all contributed to the continuation of a subculture of danger in an environment of risk. Just as tanks and other equipment had changed the face of warfare, mobile mining machinery had altered the appearance of the "battlefield," but the killing power of that environment remained relatively unabated. Accidents in fully mechanized mining continued to emotionally distress the mining communities themselves and reinforced the concept of a dangerous occupation, but they usually failed to gain much attention from the national public. Yet, the reform-oriented era of the 1960s provided the social and political atmosphere necessary to challenge coal's safety record; the only thing needed was the proper catalyst.

In 1968, events at Leivasy and Farmington, West Virginia, once again focused national attention on the coal industry. On May 6, a continuous mining machine at Saxsewell No. 8 on Hominy Creek near Leivasy cut into an abandoned mine and an inundation of water drowned or fatally injured four miners and trapped twenty-one others. The ubiquitous nature of the television industry meant that much of the nation witnessed the dramatic ten-day rescue efforts and learned that, in an age of space travel technology, the accident simply resulted from faulty surveying and inaccurate mine maps. Six months later, a massive explosion at Consolidation Coal Company's No. 9 mine in Farmington, West Virginia, produced seventy-eight fatalities. A national audience once againwitnessed the emotional devastation of family members and smoke roaring from the mine portal. Most importantly, the same audience heard company and even union officials reiterate the hazardous occupation scenario. Governor Hulett C. Smith agreed by observing, "We must recognize that this is a hazardous business and what has occurred here is one of the hazards of being a miner."[83]

For decades, explanations similar to Smith's had sufficed, but by 1968 the inherently-dangerous-job defense had lost its validity. Miners rebelled against union leadership and grassroots movements demanded political changes capable of making mining safer. Congress passed the Federal Coal Mine Health and Safety Act of 1969 and began policing the mines to force miners and companies to utilize the latest research and obey safety laws. The era of explaining coal's abysmal record via the combat scenario formulated in earlier years had closed.

A COMBAT SCENARIO

Perhaps recent events at the mine disasters in Sago, West Virginia, and Crandall Canyon, Utah, demonstrate that the old dangerous occupation argument has not resurfaced. These two disasters raised questions regarding the possible relaxation of safety guidelines and enforcement procedures, but neither the mining community nor recognized authorities publicly vocalized the old risky-occupation rationale. These events suggest that, for now, the inherently dangerous argument remains safely buried in the past.

NOTES

1   "Stuart" was the early name of present-day "Lochgelly." "Cage" is mining terminology for a type of elevator. Open on two sides, one cage is usually attached to two ends of the hoisting cables and when one is at the tipping site at the tipple, the other is at the loading site on the shaft bottom. The cage is raised and lowered at rapid speed—one might say breakneck speed—whenever holding loaded and empty coal cars, and at a reduced rate with workers aboard.

2   Turn of the century coal mines—and operations into the 1960s—often depended on trap doors across the trackways to direct ventilation into the various working areas of the mine. Young boys received the assignment of opening and closing these doors whenever loaded and empty coal cars passed through. Although a rudimentary task, trap doors were critical to mining safety: any door left ajar short-circuited the airflow and, besides the possibility of discomfort to miners in the faces, any methane accumulations were no longer swept away or diluted by the movement of fresh ventilation.

3   Shirley Donnelly, *Notable Mine Disasters of Fayette County, West Virginia* (Fayetteville, WV: Fayette County Historical Society, 1951), 10–11; Hiram B. Humphrey, *Historical Summary of Coal-Mine Explosions, 1810–1958*, Bulletins 586–591, U.S. Bureau of Mines (Washington, D.C.: GPO, 1960), 27; James W. Paul, Chief Inspector, *West Virginia Department of Mines Annual Report and Directory of Mines*, (Charleston, 1907), 486.

4   For a discussion of the experience in mining acquired by blacks in the South, see Ronald L. Lewis, *Black Coal Miners in America* (Lexington: University Press of Kentucky, 1987), 3–75.

5   Paul, *WV Dept. of Mines Annual Report, 1907*, 487–489; Donnelly, *Notable Disasters in Fayette County*, 11; W. P. Tams, *The Smokeless Coal Fields of West Virginia*,

A Brief History (Morgantown: West Virginia University Library, 1963), 49. Tams, a salaried employee of Sam Dixon, the owner of the Stuart Mine, insisted that Lee's carelessness with black powder caused the explosion. Dixon also attempted to blame Lee for the disaster. The physical evidence collected by James W. Paul, West Virginia's senior mine inspector, suggests that Lee was not personally at fault.

6   Paul, *WV Dept. of Mines Annual Report, 1907*, 487–490; Map of Stuart Shaft Mine, January 1907, Plate II, in Paul, *WV Dept. of Mines Annual Report, 1907*; Tams, *The Smokeless Coal Fields of West Virginia*, 49; "Testimony of Samuel Dixon," 13 February 1907, in *Report of Hearings Before The Joint Select Committee Of The Legislature Of West Virginia Appointed under Substitute for House Concurrent Resolution No. 5 and House Joint Resolution No. 19 To Investigate the Cause of Mine Explosions Within the State and to Recommend Remedial Legislation Relating Thereto* (Charleston: Tribune Publishing Company, 1909), 26–30: hereafter cited as *Legislative Hearings*. The author is indebted to Safety Instructor Allen Van Horn of the West Virginia Department of Miner's Health, Safety, and Training for explaining the physical behaviors of a coal-dust explosion. Experts in 1906 disagreed about the explosiveness of coal dust and engaged in heated debates over the issue. See Henry M. Payne, "Coal Dust As A Factor In Mine Explosions," *Proceedings of the Coal Mining Institute of America for 1908* (Greensburg, PA: C.M. Henry & Co., 1908), 47–72: hereafter cited as *CMIA* with the corresponding year.

7   It should be kept in mind that a coal mine explosion happens in milliseconds and that the rebounding Stuart blast occurred in much less time than is required to read the above description.

8   In mining terminology, "rib" refers to the walls of an entry. These walls of the tunnel are actually the sides of the blocks of coal, or pillars, left as the twenty to twenty-four foot labyrinth of entries are driven forward. Thus, the miner in his working "place" has coal to his front, known as the face, and coal on each side, known as the rib.

9   Tams, *The Smokeless Coal Fields of West Virginia*, 49; J. J. Forbes and G. W. Grove, *Mine Gases and Methods for Detecting Them* (Washington, D.C.: GPO, 1954), 17; *WV Dept. of Mines Annual Report, 1907*, 486; "Testimony of Samuel Dixon," in *Legislative Hearings*, 27.

10   Tams, *The Smokeless Fields of West Virginia*, 49; Frank Haas, *The Explosion At Monongah Number Six and Monongah Number Eight Mines of the Fairmont Coal Company*,

December 6, 1907, 1908, Mine Safety and Health Administration's Mine Academy Library, Beckley, WV, 10. "Parral" later became known as "Summerlee."

11   R. B. Woodworth, "Steel, the Logical Successor to Wood," in *Proceedings of the Coal Mining Institute of America for 1907* (Greensburg, PA: C. M. Henry & Co., 1908), 262; Carlton Jackson, *The Dreadful Month* (Bowling Green: Bowling Green State University Popular Press, 1982), 3; John Bartlow Martin, "Life and Death in Coaltown," *New York Times Magazine*, January 13, 1952, 34–35; John Steven Fitzpatrick, "Underground Mining: A Case Study of an Occupational Subculture of Danger," (PhD diss., Ohio State University, 1974), 17. Although improved over 1847 records, statistics of 1907 relating to mining deaths and injuries often were incomplete.

12   For a discussion of technological advancements and mining death rates, see Paul H. Rakes, "West Virginia Coal Mine Fatalities: The Subculture of Danger and a Statistical Overview of the Pre-Enforcement Era," *West Virginia History: A Journal of Regional Studies* 2 (Spring 2008): 1–26.

13   Jackson, *The Dreadful Month*, 3; Malcolm Ross, *Machine Age In The Hills* (New York: Macmillan Company, 1933), 81; Anthony F. C. Wallace, *St. Clair: A Nineteenth-Century Coal Town's Experience with a Disaster Prone Industry* (New York: Alfred A. Knopf, 1987), 253.

14   For studies concerning the impact of technology on American Civil War and World War I battlefields, see Grady McWhiney and Perry D. Jamieson, *Attack and Die: Civil War Military Tactics and the Southern Heritage* (Tuscaloosa, AL: University of Alabama Press, 1982) and Trevor Wilson, *The Myriad Faces of War* (London: Oxford, 1985).

15   Tams, *The Smokeless Coal Fields of West Virginia*, 24; Walter R. Thurmond, *The Logan Coal Field of West Virginia, A Brief History* (Morgantown: West Virginia University Library, 1964), 39; Harry Atherton, "Present Mining Methods in the Bituminous Regions of Pennsylvania From the Standpoint of the Fire Boss," in *CMIA, 1908*, 270; Thomas C. Adams, "Recent Mine Explosions and Their Lessons," in *CMIA, 1908*, 40; quotation from Thomas Keighley, "Present Mining Methods of the Bituminous Regions of Pennsylvania From the Standpoint of the Mine Owner and Manager," *CMIA, 1908*, 261.

16   Samuel Dixon openly defied the law at the Stuart mine by operating with only one opening. Directed by state authorities to have no more than twenty min-

ers underground at one time, Dixon company officials had already received a Grand Jury indictment for this practice at another mine shortly before the Stuart explosion occurred. See, *WV Dept. of Mines Annual Report, 1908*, 486.

17   George Wolfe to Justus Collins, 6 August 1918, Justus Collins Papers, West Virginia and Regional History Collection, West Virginia University Library.

18   The author often heard, both before entering the mines and while working underground, the comment that coal operators historically placed more value on a mule than a man. Usually, the statement surfaced during periods of either individual or workforce dissatisfaction and served to exemplify the perceived insensitive nature of management.

19   Former MSHA director Davitt McAteer argues that the Monongah death toll may have exceeded 550. See Davitt McAteer, *Monongah: The Tragic Story of the 1907 Monongah Mine Disaster, The Worst Industrial Accident in US History* (Morgantown: West Virginia University Press, 2007), 218–241.

20   Jackson, *The Dreadful Month*, 2; Atherton, "Present Mining Methods From the Standpoint of the Fire Boss," *CMIA, 1908*, 270; Davitt McAteer, *Monongah, 1907* [video recording] (Washington, D.C.: Occupational Health and Safety Law Center, 1986).

21   Jackson, *Dreadful Month*, 1.

22   B. F. Jones, "Address By President B. F. Jones Before the Coal Mining Institute of America in Summer Meeting at Greensburg, Pennsylvania," in *CMIA, 1908*, 18; Elsie Poole, interview by author, Oak Hill, WV, October 7 and 11, 1991, tape recording, author's collection; Billie F. Martin, interview by author, Oak Hill, WV, October 10, 1991, tape recording, author's collection.

23   Raymond E. Densmore, *The Coal Miner of Appalachia* (Parsons, WV: McClain Printing Company, 1977), 1.

24   Densmore, *Coal Miner of Appalachia*, 4–5; Dorothy Schwieder, *Black Diamonds: Life and Work in Iowa's Coal Mining Communities* (Ames: Iowa State University Press, 1983), 35.

25   The author has drawn from personal experience with a breast auger— admittedly one with a dull bit—in coal mining to suggest the physical demands. Even with two "hardened" workers, the breast auger requires intense efforts at pushing forward and muscular twisting of the apparatus at the same time. With a dull bit, it can be an exhausting experience.

26   Schwieder, *Black Diamonds*, 43; Tams, *The Smokeless Coal Fields of West Virginia*, 36; *Mines and Minerals* 26 (formerly *The Colliery Engineer and Metal Miner*) (August, 1905): 23.

27   Unfortunately, in 2007 the dangers of a "bump" became evident with the catastrophe at Utah's Crandall Canyon Mine.

28   *WV Dept. of Mines Annual Report, 1908*, 268–269.

29   Although miners used dynamite in blasting rock, it was considered unsafe for extracting coal. Of course, some miners either ignored the restrictions on the use of the more powerful explosive or, in some cases, could not read English and simply did not realize their error. See Paul H. Rakes, "Technology in Transition: The Dilemmas of Early Twentieth-Century Coal Mining," *Journal of Appalachian Studies* 5 (Spring 1999): 42–43.

30   *WV Dept. of Mines Annual Report, 1908*, 268–269. Growing up in a coal community, the author heard many tales regarding the dangers of blasting coal and remembers distinctly the "bluish-tinted face" of one of the retired miners who had coal embedded under the skin by a mishap.

31   James H. Thompson, "Significant Trends in the West Virginia Coal Industry, 1900–1957," in *West Virginia University Business and Economic Studies* 6 (May 1958), 39; Charles Ashworth, "Fatal Accidents and Their Causes From 1897 to 1938," in *WV Dept. of Mines Annual Report, 1986*, 16; Thomas A. Mather, "Present Mining Methods of the Bituminous Regions of Pennsylvania from the Standpoint of the Mine Foreman," in *CMIA, 1908*, 264; West Virginia Department of Mines, *Accidents to Coal Loaders (by Hand Into Cars): Their Cause and Prevention* (Charleston, WV: Jarrett Printing Company, 1949), 15.

32   Kettle bottoms are fossilized remains of trees that remain unattached to the larger rock strata. Varying in size from approximately six inches to four or more feet, they often resemble the shape of old iron cooking kettles. Even the most experienced miner cannot note the presence of a particular kettle bottom in the roof. Working around older miners during his first years underground, the author often heard stories of particular mines or working sections plagued by kettle bottoms. A fear of them was ingrained into the thinking of experienced miners.

33   John R. Williams to William Thomas, 10 November 1895, in Alan Conway, ed., *The Welsh In America: Letters from the Immigrants* (Minneapolis: University of Minnesota Press, 1961), 205.

34   *WV Dept. of Mines Annual Report, 1903*, 148.

35   William Nicholson to John Laing, 3 July 1909, in *WV Dept. of Mines Annual Report, 1909*, 528–529.

36   *WV Dept. of Mines Annual Report, 1908*, 263.

37   West Virginia Department of Mines, *Mine Haulage Accidents (Their Cause and Prevention)* (Charleston, WV: Rose City Press, 1951), 25.

38   Schwieder, *Black Diamonds*, 47.

39   Robert Stiles, *Four Years Under Marse Robert* (New York: Neale Publishing Company, 1903), 217–218.

40   *WV Dept. of Mines Annual Reports, 1901, 1903, and 1909*, 134, 74, and 288, respectively.

41   Stiles, *Four Years*, 234–235; George Korson, *Minstrels of the Mine Patch* (Hatboro, PA: Folklore Associates, 1964), 102.

42   Stiles, *Four Years Under Marse Robert*, 234; Donald L. Miller and Richard E. Sharpless, *The Kingdom of Coal: Work, Enterprise, and Ethnic Communities in the Mine Fields* (Philadelphia: University of Pennsylvania Press, 1985), 102; Travelers Insurance Company, *Coal Mining Hazards* (Hartford, Connecticut: Travelers Insurance Company, 1916), 19–20.

43   Schwieder, *Black Diamonds*, 51–52; Miller and Sharpless, *Kingdom of Coal*, 99–100; George Korson, *Coal Dust on the Fiddle* (Hatboro, PA: Folklore Associates, Inc., 1965), 146–147.

44   The explosive mixture of methane to air is between five and fifteen percent. Any higher concentration of gas is not an explosion risk but affects adequate breathing.

45   Densmore, *The Coal Miner of Appalachia*, 5; Testimony of W. G. Colburn, superintendent of Stuart Mine, 11 February 1907 and testimony of Fred Dixon, Jr., superintendent of Parral Mine, 13 February 1907 in *Legislative Hearings*, 14–25.

46   Densmore, *The Coal Miner of Appalachia*, 1; Horace G. Painter, "First Aid To The Injured," *Proceedings of the Coal Mining Institute of America for 1906* (Pittsburgh: C. W. Smith Company, 1906), 66–70.

47   J. M. Shields, "The Organization of First Aid Corps," *CMIA, 1908*, 245–246.

48   William Nicholson to John Laing, 3 July 1909, in *WV Dept. of Mines Annual Report, 1909*, 531, 602.

49   Rex Archibald Lucas, "Social Behavior Under Conditions of Extreme Stress: A Study of Miners Entrapped by a Coal Mine Disaster," (PhD diss., Columbia University, 1967), 31–35, 69; Samuel A. Stouffer, et al., *The American Soldier: Combat and Its Aftermath*, vol. 2, (Princeton: Princeton University Press), 131–134, 136–137.

50   Lucas, "Social Behavior Under Conditions of Extreme Stress," 8–9; Edward A. Wieck, *Preventing Fatal Explosions in Coal Mines: A Study of Recent Major Disasters in the United States as Accompaniments of Technological Change* (New York: Russell Sage Foundation, 1942), 5.

51   Sherman Kingsley, superintendent of United Charities of Chicago, Bureau of Labor Statistics, *Report on the Cherry Mine Disaster* (Springfield: Illinois State Printers, 1910).

52   Fred Keighley, "Presidential Address," *CMIA, Winter Meeting, 1906*, 140.

53   E. W. Parker, "Notes on the Present Condition of the Coal Mining Industry," *Proceedings of the West Virginia Coal Mining Institute, 1908* (Fairmont, WV: Free Press Printing Company, 1908), 25–26; hereafter cited as *WVCMI*.

54   *WV Dept. of Mines Annual Report, 1909*, 618.

55   Gerald F. Linderman, *Embattled Courage: The Experience of Combat in the American Civil War* (New York: The Free Press, 1987), 261–265; Lewis, *Black Coal Miners in America*, 31.

56   George Korson, *Minstrels of the Mine Patch*, (Hatboro, PA: Folklore Associates, Inc., 1964), 273–278.

57   Korson, *Minstrels of the Mine Patch*, 181.

58   Wallace, *St. Clair*, 233. Wallace is quoting mid-nineteenth-century mining poet Samuel Harris Daddow, but does not supply the title of this poem.

59   Stephanie Elise Booth, "The American Coal Mining Novel: A Century of Development," *Illinois Historical Journal* 81 (Summer 1988): 140.

60   "Too Late for 78," *Time* December 6, 1968, 32.

61   Lucas, "Social Behavior Under Conditions of Extreme Stress," 31–32; S. A. Stouffer, *The American Soldier: Combat and its Aftermath*, vol. 2 (Princeton: Princeton University Press, 1949), 131; Ross, *Machine Age in the Hills*, 78; Densmore, *The Coal Miner of Appalachia*, 3; One scholar has identified miner codes and mores as components of a "Subculture of Danger." See John Steven Fitzpatrick, "Underground Mining: A Case Study of an Occupational Subculture of Danger," (PhD diss., Ohio State University, 1974).

62   Lucas, "Social Behavior Under Conditions of Extreme Stress," 37.

63   Stouffer, *The American Soldier*, vol. 2, 206.

64   Stouffer, *The American Soldier*, vol. 2, 192–193.

65   Lucas, "Social Behavior Under Conditions of Extreme Stress," 44–46; Ron D. Eller, *Miners, Millhands and Mountaineers: Industrialization of the Appalachian South, 1880–1930* (Knoxville: University of Tennessee Press, 1982), 166–167. As late as 1974, one Pittston Coal Company official complained to the author that the company's mines had to operate at twenty-five percent over strength to offset chronic absenteeism.

66   Wallace, *St. Clair*, 256; quotation from David M. Wayne, M.D., Lewis W. Field, and Reed T. Ewing, "Observations On The Relation of Psychosocial Factors to Psychiatric Illness Among Coal-Miners," *International Journal of Social Psychiatry* 3 (Autumn, 1957): 134.

67   Wallace, *St. Clair*, 256–257.

68   John Maguire, "Reminiscences of John Maguire After Fifty Years of Mining," *Publications of the Historical Society of Schuylkill County*, 4 (1912): 332.

69   Oscar Veazey, *Special Report to Governor Jacob B. Jackson* (Wheeling, WV: Charles H. Taney, 1883), 3; *Journal of the Senate of the State of West Virginia for the Seventeenth Session, 1883* (Wheeling, WV: Charles H. Taney, 1883), 268 and Senate Bill No. 53; *Journal of the House of Delegates of the State of West Virginia for the Seventeenth Session, 1883* (Wheeling, WV: Charles H. Taney, 1883), 79, 252–253, and House Bill No. 96.

70   Fred Keighley, "President's Address to Summer Meeting, 1903," *Proceedings of the Coal Mining Institute of America, 1903–05* (Pittsburgh: Moore & White Printers, 1906), 12.

71   Keighley, "President's Address to Summer Meeting, 1903," 10–11; William Nicholson to John Laing, 3 July 1909 and D. R. Phillips to John Laing, 8 January 1909, in *WV Dept. of Mines Annual Report, 1909*, 528 and 624, respectively; Humphrey, *Historical Summary of Coal-Mine Explosions in the United States*, 22.

72   Testimony of James W. Paul, January 23, 1908, in *Legislative Hearings*, 491.

73   Charles Kenneth Sullivan, *Coal Men and Coal Towns: Development of the Smokeless Coalfields of Southern West Virginia* (New York: Garland Publishing, Inc., 1989), 217–219; *WV Dept. of Mines Annual Report, 1906*, 203–205.

74   *Fayetteville (WV) Fayette Journal*, January 31, 1907; *WV Dept. of Mines Annual Report, 1908*, 486; *Legislative Hearings*, 6.

75   *Fayette Journal*, February 7, 1907.

76   Humphrey, *Historical Summary of Coal-Mine Explosions in the United States*, 22; *Legislative Hearings*, 80.

77   John A. Williams, *West Virginia and the Captains of Industry* (Morgantown: West Virginia University Library, 1976), 251; Humphrey, *Historical Summary of Coal-Mine Explosions in the United States*, 34–35.

78   Testimonies of W. G. Colburn, 11 February 1907, and Fred Dixon, Jr. and Albert Terry, 13 February 1907 in *Legislative Hearings*, 14–15, 20–21, and 47, respectively; *Fayette Journal*, January 24, 1907.

79   McAteer, *Monongah, 1907* [video recording].

80   *Fayette Journal*, February 7, 1907; *Beckley (WV) Raleigh Register*, February 21, 1907; *Legislative Hearings*, 636.

81   Kingsley, *Cherry Mine Disaster*, 90; *Fayette Journal*, February 14, 1907.

82   United States Department of Labor, *Historical Summary of Mine Disasters in the United States*, vol. 2, Coal Mines 1959–1998, Section 4 (Washington, D.C.: Mine Safety and Health Administration, 1998), 1–51.

83   Department of Labor, *Coal Mines 1959–1998*, Section 4, 42–45, 51–55; Tony Boyle quoted in Lewis, *Black Coal Miners*, 183.

Radical Challenge and Conservative Triumph:

The Struggle to Define American Identity in the Somerset County Coal Strike, 1922–1923

Jennifer Egolf

ON APRIL 1, 1922, the United Mine Workers of America (UMWA) issued a call to all union miners to drop their tools and join their fellow miners in a national strike. The UMWA also strongly encouraged non-union miners in the country to join their unionized brothers. Unorganized miners in western Pennsylvania districts heeded the UMWA's call and began striking during the month of April. As a result of the union's organizing success in Somerset County, Pennsylvania, which was a part of District 2 and under John Brophy's leadership, the region became a "battleground" for seventeen months. Somerset County's union and non-union miners united, each driven by their own motives. Union miners demanded a new contract when their 1920 agreement expired on March 31, 1922. For non-union miners, the strike was both a sympathetic one and what Heber Blankenhorn called "a strike for union."[1] These miners were determined to remain out of work until the operators addressed their grievances and recognized their union. Indeed, the "strike for union" in Somerset County, Pennsylvania, provides a microcosmic lens through which to examine the larger struggle between workers and capitalists during the 1920s.

Historians have written extensively about coal strikes in central and southern Appalachia, often focusing on the miners' low wages and long, dangerous hours in the mines; the company town and the employers' dominance over miners' lives; and the evictions, court injunctions, and private and public police forces that gave the employers a decided advantage during walkouts. Most recent scholarship balances the exploitation and domination of miners by outside forces (core-periphery model) with workers' agency.[2] Studies of the northern Appalachian coalfields often follow a similar pattern of demonstrating both exploitation and agency, but historians rarely connect these workers' experiences to the larger Appalachian discourse.[3] This study of the 1922–1923 coal strike in Somerset County attempts to demonstrate the complex human relationships in northern Appalachian communities during a period of economic decline. The focus is on the rhetorical battle that operators, the press, and farmers waged against the UMWA and striking miners. As each side argued their case for or against the strike in the court of public opinion, they identified what America represented to them, and although each side viewed the other as its nemesis, their often similar conclusions contributed to America's turn to the right in the 1920s.

National and Local Background for the Strike

The strike officially began on April 1, 1922, after the expiration of the union's contract with the Central Competitive Field, which included operators in western Pennsylvania, Ohio, Illinois, and Indiana. During that month, more than 600,000 union and non-union miners in these states walked off the job. By June, a coal shortage resulted, prompting the federal government to proffer its strike solution in July. (See Chart 1, p. 93.) When UMWA President John L. Lewis rejected the government's proposal, the strike continued into early August. Finally, Lewis negotiated with Pennsylvania and Ohio operators in the Central Competitive Field. By August 15, Lewis won contracts from operators employing approximately 100,000 miners in seven states. The Cleveland Agreement maintained the 1920 wage scale for miners through March 31, 1923. The official strike having ended, non-union miners in Somerset, Fayette, and Westmoreland Counties in western Pennsylvania failed to return to work until the agreement included them.[4] With the financial assistance of the national and district unions, some non-union miners, who had been evicted from their homes, continued their "strike for union" by living in union-rented chicken coops, barns, and tents.

CULTURE, CLASS, AND POLITICS

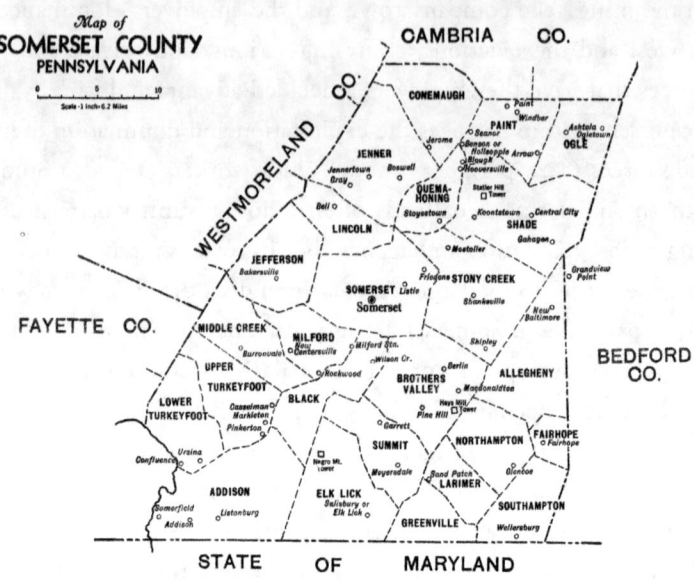

Map 1. Somerset County, Pennsylvania. (Somerset Historical Center, Somerset, Pennsylvania.)

Non-union miners had the national's monetary and moral support until January 1923, when Lewis called off the coke strike in Fayette County and discontinued relief to the strikers in Somerset County. John Brophy, president of District 2, continued to finance the strike until August, when he called twenty-two local leaders from Somerset County to a meeting in Johnstown, Cambria County, where they agreed to end the strike after seventeen months of struggle.[5]

National, political, and economic events influenced local attitudes in Somerset County during the early 1920s. The decade has been popularly termed the "Jazz Age" and the "Prosperity Decade" because many Americans improved their purchasing power and enthusiastically pursued leisure activities; however, many conservative features predominated. Political leaders underscored the pervasive economic conservatism when they lowered the budget throughout the 1920s; therefore, Americans witnessed lower taxes and an end to the government's liberal wartime expenditures.[6] Republican Warren G. Harding's "return to normalcy" speeches won the electorate's hearts and minds, especially after the tumultuous war abroad and post-war labor and anti-communist struggles at home. In Harding's new world, he envisioned industrial peace, a hands-off government, and economic prosperity, ideas that resonated with a primarily conservative population in Somerset County.[7] Throughout the decade, optimistic consumers spent liberally and sometimes beyond their means; their focus on purchasing power further contributed to 1920s economic conservatism since Americans became dependent on and avidly protected their "high" standard of living. Clinging to the "Horatio Alger" myth of success, some Americans viewed those who did not succeed as defective.[8]

When the United States experienced a recession in 1921–1922, during Harding's short presidency, coal mining, which was a major industry in Somerset County, suffered a downturn. Prior to the recession in 1917, the UMWA and coal operators in the Central Competitive Field signed the Washington Agreement under government supervision and agreed to arbitration and fixed prices and wages under the Fuel Administration until 1920 or the end of the war. The war ended in 1918, but delays in signing the peace treaty kept the agreement in force until 1920. However, because miners' wages remained fixed in spite of rising inflation, the UMWA called a national strike in 1919, a year of widespread labor unrest across the nation. Intervening in the strike, President Woodrow Wilson appointed a special arbitration commission, and union miners in the Central Competitive Field finally gained a substantial pay increase in 1920. The wage scale agreement was in effect only until 1922; therefore, the expiration of the contract on March 31, 1922, was the impetus for the 1922 strike.[9]

Reactions in Somerset County to the 1922 strike also reflected national cultural shifts in the early twenties. A "nervous generation" that feared losing its prosperity exhibited cultural conservatism that included intense nativism.[10] Particularly, anti-foreign sentiment achieved nationalistic proportions in the post-war United States when fewer opportunities to temper nativism existed. Many businessmen discontinued educational Americanization programs because of the impact of the war, the Russian Revolution, and labor strife in 1919, all of which intensified their fears of domestic Bolshevism and radicalism.[11] Although some of these relatively "benign" educational programs continued, the increasing fears of foreigners and radicals gave rise to advocates of "One Hundred Per Cent Americanism," who created a wave of hysteria against southern and Eastern European immigrants and called for immigration restriction. The public's fear of communism contributed to the rise of patriotic expressions that associated loyalty with conformity.[12]

Indeed, reactions to the coal strike in 1922–1923 exemplified many Americans' desire to achieve a cultural consensus in postwar America that they failed to secure during the progressive era. Consensus was the primary objective of progressives, who sought either to bring immigrants into cultural conformity or to protect white, Anglo-Saxon, Protestant culture by advocating for and winning immigration restrictions. National and state governments and industrial capitalists assisted their efforts to generate a mass citizenry, essentially homogeneous in its cultural and political attitudes.[13] Progressives and politicians had formidable allies in their efforts to "Americanize" immigrants.[14] In the 1920s, cultural conservative forces in America progressed closer to securing that consensus with the passage of the National Origins Act of 1924, which established immigration quotas for foreign countries. Because the protracted battle for union recognition in Somerset County involved competing definitions of democracy, the strike showcases the larger struggle to define America and a uniform American identity in the postwar, post-progressive period and further explains the triumph of conservatism in the postwar decade.

Local developments also shaped the county's response to the strike. In the 1920s, Somerset County was still primarily a rural county in northern Appalachia with agriculture and mining as the primary industries. With a population of 82,112 in 1920, a total of 3,630 owned or rented farms. Of the 661,760 acres of land in the county, 444,148 acres were farmlands.[15] The majority of farmers were only partially enmeshed in a market culture. In 1929, Pennsylvania State College surveyed agriculture in each Pennsylvania county and characterized most farms in Somerset County as one of four types: general farms, in which no single crop accounted for 40 percent or more of the farmer's income; dairy farms, in which dairy products constituted at least 40 percent of the farmer's income; self-sufficing farms, in which farm families consumed at least 50 percent of their products; and abnormal farms, which were part-time farms where the owner worked more than 150 days away from the farm.[16] Although one-fourth of the county's families resided on farms, mining became a critical secondary industry in the region, especially in the northern sections of the county. Most likely, many "abnormal farmers" or their family members worked in the mines sometime during the year.

County	Tons of Coal Production by Year			
	1921	1922	1923	1924
Cambria	15,713,340	12,383,440	19,000,381	15,954,486
Fayette	19,184,691	17,359,375	32,130,185	25,427,469
Somerset	9,141,045	7,487,409	8,517,542	8,526,928
Westmoreland	17,999,981	20,315,538	25,533,735	17,915,888

Chart 1. Effects of Strike on Coal Production by County.
In addition to Somerset County, I chose three counties that border it to offer a regional comparison. Of the four counties, Cambria County had a significant number of unionized mines. Fayette County had none in 1922; Westmoreland had a small pocket in the northern section, bordering union stronghold, Allegheny County; and Somerset had only a small unionized area in the southern fields. (Pennsylvania Department of Mines, *Report of the Department of Mines of Pennsylvania* [Harrisburg, PA, 1922–1925], annual reports, 1921–1924.)

On the eve of the strike's commencement, Somerset County was rich with coal reserves. In relationship to other western Pennsylvania counties, Somerset was a mid-level producer. Its annual coal production between 1919 and 1926 averaged over nine million tons. (See Chart 1.) The size and productivity of the mines varied depending on geographical location within the county. In general, fields employing the most miners were in the north, and mines in the central and southern sections employed fewer men per mine.[17] The miners in the north, therefore, produced the majority of the county's coal. In addition to mine size and productivity, ethnic factors distinguished the county's mining regions from each other. A larger foreign-born mining population worked in the northern fields than in the central and southern coal communities. The southern fields were overwhelmingly American-born in character.[18] These larger mines that employed the majority of foreign-

CULTURE, CLASS, AND POLITICS

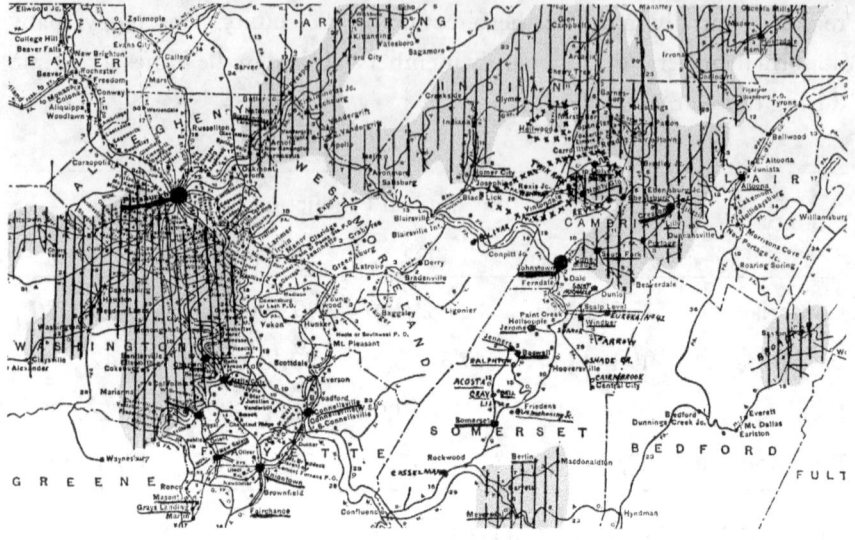

Map 2. Unions in Somerset and Surrounding Counties.
The shaded areas with vertical lines indicate unionization. Indiana, Cambria, Allegheny, and Washington Counties had significant union presence. Bedford, Fayette, Westmoreland, and Somerset Counties were predominately non-union. (Blankenhorn, *Strike For Union*.)

born miners were more likely to have links to big corporations. Indeed, the two largest mining operations in the county were Berwind White Coal Mining Company, which New York capitalists owned, and Consolidation Coal Company, in which oil magnate John D. Rockefeller held interests.[19] Although most of the mines operated as non-union, a small union pocket existed in the southern section of the county near Garrett and played a crucial role in organizing the non-union fields in the county before and during the strike.[20] (See Map 2.)

Conservative Response of Operators, the Press, and Farmers

When the UMWA called a national strike on April 1, 1922, organized bituminous and anthracite miners walked out in overwhelming numbers, and operators and the press responded.[21] The union's primary reason for calling the strike was the operators' failure to honor the contract signed in 1920, which obligated them to negotiate a new wage scale agreement by March

31, 1922. In the Central Competitive Field, southern Ohio and Pennsylvania operators refused to meet with the UMWA because they wanted to negotiate separately with the miners in their region and break the union's control over negotiations.[22]

When UMWA District 2 President John Brophy enlisted non-union miners' support in the national strike, the operators viewed this as more threatening than Lewis's call to union men. Unorganized miners had different reasons for striking than their union counterparts. The non-union miners' major complaints included arbitrary pay cuts (for example, Berwind White miners received a 40 percent wage cut just before the strike began), unfair weights (company officials did not weigh the full amount of the coal and non-union miners had no checkweighmen to ensure honesty), no payment for deadwork (the removal of rock, slate, or other obstruction to get to the coal), and lack of grievance committees.[23] John Brophy was aware that miners in his district were organizing prior to the strike and knew that winning the national strike involved limiting coal production in the non-union strongholds. His strike call in Somerset County succeeded.

Miners apparently took many operators by surprise when the strike calls were a success. Pennsylvania State Police Superintendent Lynn G. Adams sent inquiries to all mine operators in Somerset County in March 1922 in order to gauge their preparedness for the impending coal strike. The responses to these reports indicated that many operators, especially those operating non-union mines, believed that they would not have to use imported labor and that their work force would remain in the mines in spite of the union leaders' appeals to union and non-union miners. In fact, the Fair Oak Coal Company's operator in Confluence wrote confidently to Adams about his workers, "Our men are quite anxious to work and will not go out on a strike."[24] Similar to Fair Oak Coal's reply, W. E. Ambrose, managing director of Tri-State Consolidated Coal Companies, which operated in Listonburg, emphatically wrote, "We have never lost employees through a strike."[25] The non-union operator of the United Smokeless Coal Company in Humbert, echoed the bold statements of his cohorts, writing to Adams, "We will have no strike and will keep on operating full capacity."[26] Even the Berwind White Coal Mining Company, one of the largest coal operators in the county, expected to operate during the strike without the use of imported workers.²⁷ Many operators shared this positive outlook but still prepared for the worst.

When union and non-union men walked out, the operators realized their worst fears and sought injunctions against the UMWA and its representatives. The injunction petitions of small and large operators highlighted both the operators' astonishment at the walkouts and their anger toward "outside radicals." In the operators' bills of complaint to the Somerset County Court, the operators emphasized the services that they provided for their workers, including comfortable houses, medical care, and entertainment and recreation halls. F. R. Lyons, Consolidation Coal Company's vice president, stated that "cordial relations existed between the complainant and its employees" until the UMWA came into town.²⁸ John Lochrie of Lochrie Price Coal Company echoed this statement in his complaint, stating that his company won the "confidence and loyalty of its employees" and that only "good will and friendship" existed between employers and employees.²⁹ Because they viewed their relationship with the workers positively, the operators blamed outsiders for the walkout. Lyons, Lochrie, and other operators stated that when their employees did not join the UMWA strike, defendants John Brophy, James Mark, and other UMWA representatives "entered into an unlawful combination and conspiracy with one another" with the intention of destroying the good relations between the operators and their workers. All of these bills of complaint noted that the primary defendants were not residents of the county.³⁰

Both the police reports and bills of complaint indicated that unionized miners in Garrett and Macdonaldton, Somerset County, conspired with these "outsiders" to destroy the cordial relations between workers and owners. In fact, aware that violence had taken place near union fields prior to the beginning of the strike, state police solicited from the operators the names of "known radicals" who were organizing or residing near their mines. Detailed lists of the "troublemakers" came from those mine operators located near Macdonaldton and Garrett. The Tri-State Colliers Company, which provided a detailed list of "known radicals," characterized the men as "radical socialist," "agitator," "pugilist," "strong union man," or heavy drinker. The Brothers Valley Coal Company indicated in its list of "known radicals" the nationality of each.³¹ Thus, operators knew the local organizers and union members and were also cognizant of violent incidents perpetrated by these alleged radicals. Lyons claimed that unionized miners residing in the county "dynamited and burned buildings and destroyed valuable property, and committed assaults upon individuals."[32] He and other operators claimed that the purpose of this violence was to instill "fear and terror in the minds of the complainant's employees and others." Not only did these outsiders and local radicals use violent acts to threaten miners, but they also distributed "scurrilous, libelous, and defamatory literature" to turn workers against their employers.[33]

The conservative press in Somerset County proved to be a staunch ally of the operators and their claims. The *Boswell News, Somerset County Star, Somerset Herald*, and *Somerset Standard* detailed the media's perspective on the strike, which clearly sided with the operators. First, the press agreed with Lyons and Lochrie about the excellent working and living environment provided by the coal operators. Thus, the consensus of the press was that the workers, both foreign and native born wanted to work in the mines and did not want to strike. Second, the editors of these county newspapers contributed to the operators' argument that outsiders created the chaos in the coalfields by taking advantage of the naïve and un-Americanized immigrant workforce. Finally, the press bolstered the operators' view that the UMWA and its representatives threatened law and order and added that the union's demands and tactics reeked of communism. Hence, in their defense of the operators, the press proffered its pro-capitalist, anti-union message to its readers.

Newspaper editors filled their pages with glorious reports about operators bringing prosperity to the region and the miners' desire to stay on the job. One article in the *Somerset Herald* reported that Somerset County had "steady work for the miner and prosperity in all lines of business." The article continued to exalt the operators by crediting them with helping the miners to enjoy such luxuries as automobiles and fine clothes.[34] The *Somerset Standard* also heralded the operators for providing good wages that led to a contented workforce and labor peace.[35] Thus, the newspaper editors agreed with operators that neither the coal owners nor the majority of workers were to blame for the initial walkouts.

Instead, the press validated the operators' belief that the miners decided to strike because of the radical influences from both outside and inside the county. Only the *Boswell News* deviated slightly from the operators' rhetoric by blaming radicals on both sides of the negotiating table for not reaching an "amicable agreement."[36] Editors Robert Scull of the *Somerset Herald* and J. A. Lambert and F. M. Forney of the *Somerset Standard* placed the blame for non-union miners striking solely on the union representatives inside and outside the county. The *Somerset Herald* claimed that non-union miners would not strike without outside agitators pushing them to quit work. In fact, *Herald* editors claimed that these outsiders "came into this county in the manner of 'the wolf on the fold' and before the officers [of various operators] . . . knew where they were at [sic] all of their workmen were persuaded to join in the countrywide strike of unionized miners."[37] Both the *Herald* and *Standard* editors reported that the local miners had no grievances of their own against the operators, who paid high wages, and, therefore, they struck only because the UMWA encouraged them. The *Standard* echoed Consolidation Coal's F. R. Lyons's anti-union statements in an article that stated, "Somerset County is, and always has been a non-union field, and the only trouble there has ever been has been in the districts where radical union men . . . spread their damaging propaganda. . . ."[38] Thus, the *Standard* editors supported Lyons's claim that peaceful relations existed between miners and operators in non-union fields before the UMWA entered the scene. The *Herald* also asserted the claim that the agitators entered from outside the county, declaring "nothing . . . developed to show that the organizers . . . swarming up and down the county were invited by any local body of mine workmen."[39]

According to the press, the union's mere presence in the county was not sufficient to compel the miners to strike, but rather union tactics provided the stimulus for them to walk out. Supporting the operators' claims of terror in the coalfields, the *Standard* reported that men went on strike because they feared union agitators, whom the editors declared were primarily "outsiders" from other counties. The editors suggested that if the union men left the region, many non-union strikers would return to work immediately.[40]

The operators and the press not only believed that the union used terror and fear to force the miners to walk out, but they also suggested that some of the non-union miners were more easily persuaded to strike because of their national origins and their ignorance of American values. The *Herald* writers expressed their disappointment with the foreign-born miners who failed to embrace the operators' "Americanization" programs, which included "teaching their employees and their children how to become genuine citizens of the greatest democracy."⁴¹ In addition to Americanization's failure, the press highlighted the nativist, American-born miners' resentment toward the foreign-born strikers. One American-born miner from Gray, who was angry because "foreign-born miners" called him a scab, told reporters, "Many of those fellows boast of the fact that they are unnaturalized and some of them declare they will never become citizens of the United States, and yet they seem to think that they will be protected by American laws when they endeavor to restrain Americans from working." As a veteran of "The Great War," he also questioned the reasons that he fought to protect the freedom of these immigrant strikers, who now showed him nothing but disrespect. He pleaded to reporters, "When the war came on I was called to go overseas and fight to save countries these fellows came from, and because of that service must I and thousands of other young miners put up with their jeers and taunts . . . ?"⁴²

Indeed, the press suggested that the failure of Americanization left the immigrants vulnerable to the propaganda of the UMWA, created the disrespectful attitude that the foreign-born strikers demonstrated toward their replacement workers, and also contributed to nothing but hardship for miners and their families. The *Herald* editors expressed some sympathy for the non-union immigrant strikers after the UMWA signed the Cleveland Agreement in August of 1922, which ended the national strike before winning recognition for the newly organized fields. As the press reported, these naïve victims of the union, which included not only the miners, but also their wives and children who lived in tents, continued to blindly follow the union. They were "still under the spell."⁴³ These gullible miners spent months out of employment, making many sacrifices in the summer of 1922, "while organizers [cavorted] around the county in automobiles."⁴⁴ The feeling of pity did not last.

Some members of the press adopted a very negative, nativist tone when portraying foreign-born miners, suggesting that many immigrants could never assimilate to American life. Their violent behavior and disrespectful attitude toward working men proved this. After the Cleveland Agreement's signing in August of 1922, the *Somerset County Star* expressed its disgust with the non-union miners' decision to continue strike activities and violence. Then, when "radicals" bombed a bridge in Jerome in late 1922, the editor placed the blame on a lawless foreign-born population that refused to renounce its allegiance to "home country" and that failed to support policies that helped America to grow and prosper. The editor feared that if the immigrants continued to denounce Americanization efforts (i.e. renouncing loyalty to their European homelands), the United States would "be but little better than the countries of Europe" and these immigrants would be responsible for destroying "the very thing they most desired."[45] Thus, in order for these immigrant miners to secure freedom and prosperity in America, they had to abide by and respect the laws of the land.

The press concurred with and added to the operators' sentiments that the UMWA and its supporters violated the American principle of law and order by committing violent acts that threatened rural-industrial communities in Somerset County. The *Boswell News* reported on the violence and lawlessness of the union radicals during the last stages of the "strike for union" in 1923. In that year, one article reported the murder of an Italian strikebreaker at Acosta, who had left the county because he received threatening letters ordering him to quit his job. When he returned to the county, he was shot. Additional articles focused on the dynamiting of mine tipples in Windber, which caused $3,000 in property damage; strike-related violence in Jenners, Hollsopple, Jerome, and Gray; and the arrests of Acosta strikebreakers, their wives, and children.[46] Responding to the violence, the *Boswell News* editor, J. C. Oswalt, promoted both law and order and the "American way" writing that the "guilty parties should be brought to a speedy justice, for American principles are so much in contrast to that sort of crime as to make every American citizen rise with sufficient indignation to compel action in such cases." He characterized the violence as "entirely un-American" and warned that it would "not be tolerated in this country." Oswalt condemned any deviation from appropriate "American" behavior, especially when committed by any foreign resident or radical agitator. Indeed, the strikers' attacks on the operators' property represented more than a threat to law and order; the violence challenged the definition of "American."[47]

In making their case against the union, the press—like the operators—focused on both the violence of the UMWA supporters and the radical literature that sought to turn the workers against the owners and the United States government. Newspaper accounts characterized the union propaganda as "radical 'Red' literature" with the purpose of supplanting the American flag with a "red one."[48] The press agreed with the operators that the UMWA conspired to disrupt labor peace, but the newspaper editors expanded the rhetoric to include an anti-communist tone. Newspaper reports claimed that the radical unionists' anti-American conspiracy included not only stopping miners from working but also forcing the government to nationalize the mines, a communist-inspired plan.[49] Other members of the press suspected the UMWA of embracing communists in its ranks who directly threatened the democratic form of government at the core of America's economic and political values. Union representatives were "strangers engaged in breaking down the old established community spirit" in order to "install their own theories" and guilty of "exerting to upset labor conditions and social customs [in] Somerset County by substituting a totally different scheme of government for the one under which the county has grown and prospered for years."[50] Another article labeled the union leaders "Parlor-Bolshevists."[51] The carefully worded language corroborated the operators' conspiracy theory, while at the same time adding its own anti-communist, nativist dimension.

The press appealed to the community to support the operators, explaining the ways the walkout hurt everyone. When the strike settlement loomed, one article in the *Standard* suggested that the strike intentionally threatened the winter fuel needs of Americans in order to force the government to step in and conclude the strike in the union's favor. The author stated, "It would be a sorry day, indeed, should our government ever feel obliged to sacrifice the rights and welfare of 115,000,000 of its citizens [to an] insignificant radical minority."[52] The *Herald* editor also portrayed the union radicals negatively in order to win local support for the operators. Articles claimed the strike hurt local business, especially farmers.[53] In addition, press reports claimed that local farmers living near mines feared the violence and turned their homes into "arsenals." Refuting this portrayal, Heber Blankenhorn recorded incidents when operators tried to intimidate farmers by calling in their mortgages to prevent them from actively supporting the strikers. The *Herald* blamed the farmers' fear on union organizers, when it just as likely resulted from employers' actions.[54]

Whether or not the press directly influenced the beliefs of area farmers, who constituted a significant portion of the county's population in the 1920s, the Grange, an economically and culturally conservative organization for farmers and rural Americans, supported the side of law and order and industrial peace for economic and political reasons. The Cambria County Pomona Grange's resolution regarding the strike presented a neutral tone, calling for miners and operators to "take effective steps toward a settlement of the strike and resumption of production."[55] The Somerset County Pomona Grange also proffered a similar cooperative solution at a meeting in October 1922 where they urged labor leaders and operators to "adjust their differences" and end workers' idleness.[56] The Grange's apparent neutrality resulted from their connections to the mining industry in Somerset and Cambria County. Some farmers worked or had children employed in the mines and attended the same schools and churches as the miners and local operators.

In spite of their neutral tone, the economic concerns of the farmers bolstered the arguments that the operators and the press made against the strike. In a press release addressed to the union, the operators' association, the federal government, the State Grange, and the newspapers, the Cambria County Pomona Grange argued that the suspension of mining in the county and the nation diminished the market for farmers' goods. The statement concluded, "The power of the people to purchase agricultural products will soon be seriously curtailed throughout the whole country unless the coal strike is ended, and farmers . . . will in turn be unable to buy in normal volume the products of other industries."[57] In this statement, the Grange echoed the words of the operators, who feared both a downturn in their coal supply and the potential abrogation of contracts with buyers. Farmers also supported the press's claim that while coal brought prosperity to the county, a strike created hardship and despair for everyone.

In the farmers' viewpoint, because the strike was a social as well as an economic threat to the region and nation, the Grange reinforced the operators' and press' claims about the immigrants' naïveté and ignorance of American values. The Somerset County Pomona Grange expressed its negative opinion of radical unions and immigrants in the early 1920s. In June 1920, the Pomona members discussed a proposal by labor groups to join together to form a new political party but refused to participate because Grange farmers believed that many labor organizations opposed the American form of government. At the same time that they attacked radical unions, many Pomona members remained strong proponents of immigration restriction and Americanization, suggesting a possible link between their anti-union and nativist beliefs. At a meeting in

January 1921, members proposed that the government use "stringent measures" against immigrants, whom they believed were "detrimental" to America's welfare because they added to the "numbers already unemployed." Five months after the "strike for union" ended, the Pomona members met and again offered a scathing characterization of "ignorant" immigrants, adopting a resolution advocating "that a more rigid test as to the principles for which [the] flag stands be given those who seek to become American citizens through the course of naturalization."[58]

Miners and Unions Respond to Attacks

Although the case against the miners' strike and the union proved to be convincing to many in the county and the nation, the miners and their union argued persuasively in favor of the strike and unionization. First, miners and the UMWA contradicted the claims of the operators and the press regarding working and living conditions. Second, the miners and union countered the contention that the UMWA outsiders caused the strike and that naïve immigrants blindly followed the union representatives to the picket lines. Finally, the UMWA defended itself against the operators' and the press's claims that it was an un-American, communist institution, and in its defense, the union offered its definition of a good American.

The unions and miners challenged the operators' and press's claims that the coal industry provided generously for its workers. Letters from miners contradict these claims. After the UMWA began its strike in April, a non-union miner from Kelso, James Read, wrote to John Brophy expressing the desire of Kelso miners to walk out in sympathy with their union brothers. In his letter, he complained about the bad working conditions and unfair pay scales.[59] John Brophy also spelled out the grievances of many miners in the strike notice, which delineated the fears that non-union miners felt because of the mine boss, spies and spotters, gunmen, coal and iron police, anti-union civil authorities, the blacklist, and evictions. In addition, the non-union miners had to endure low wages determined by the boss, no pay for deadwork, unfair weights, and the tyranny of the company store. The strike call also expressed the miners' fear "to meet and discuss their problems as free Americans" because of the overwhelming supervision of the operators and their allies.[60]

Corroborating Brophy's claims against the bosses, Albert Armstrong, president of Gray's local, explained to Heber Blankenhorn about the conditions at the Consolidation Coal Company's mines, "Under non-union conditions for one thing the boss has it all to say as to what conditions you will work under and what pay you will get. That might be alright if people would be satisfied with conditions, but people are not."[61] He also discussed with Blankenhorn his bad experiences with Consolidation's company stores, blacklists, and policy for deadwork. When mines shut down, the company stores forced miners to pay their balances in cash, which they usually did not have in company towns with scrip. In addition, he noted that when workers refused to go into the mine because of a pay grievance, Consolidation usually blacklisted the leader of the walkout. According to Armstrong, deadwork was a chief complaint among non-union miners who sometimes worked twelve hours just getting to the location where the coal was and received no money for their time.[62] Thus, he directly contradicted the statements of the operators and the press about coal bringing opportunity and prosperity.

Indeed, many miners did not enjoy stellar working or living conditions, and as is clear from the letters of Read and Armstrong, many non-union miners invited the union into the county to organize. Thus, the miners and the union disputed the contention of the operators and the press that the outsiders from the UMWA entered the county uninvited to disrupt labor peace and terrorize the workers and operators. In fact, Read wrote to John Brophy specifically to invite organizers to a meeting that the miners at Kelso, Lochrie, and Cook mines held in April of 1922. In addition, the miner's letter scolded the workers who stayed on the job and did not support their union brothers because in continuing to work the miners, they were "hurting their own children and ours as well by accepting such jobs to keep wages down." He did not want these non-union miners to help "take bread out of other [union men's] children's mouths."[63] According to Heber Blankenhorn, who worked at the Bureau of Industrial Research in New York and followed the strike from the beginning, the miners in Somerset County had heard little about the national strike before they joined the struggle. They knew only that the UMWA was willing to support them in their own struggle for unionization. Thus, Blankenhorn confirmed that miners in Somerset solicited the help of organizers as soon as they knew that they had the UMWA's support.⁶⁴ Indeed, according to these accounts, the UMWA and its representatives did not "invade" the county and frighten the workers into striking.

Not only did union supporters argue against a UMWA invasion of the county, but they also questioned the statements of the press that the immigrant workers were the UMWA's naïve victims. In fact, in one instance, the *United Mine Workers Journal* (*UMWJ*) claimed that the operators and not union organizers manipulated newcomers by promising them a land of riches and delivering only the precarious position of scab. Addressing the foreign-born strikebreaker, one journal article stated, "Jewish agencies on both sides of the Atlantic . . . get you over by telling you that money grows on trees over in America" for the purpose of strikebreaking.⁶⁵ Basically, the operators lied to both the immigrants and to the community in order to gain support for their side because in reality, wages were low and conditions deplorable.

Disputing the press's claim that immigrant miners could not assimilate, the union supported Americanization programs to help newcomers adjust to American life. The union encouraged the miners to become citizens and respect the laws of the land. One *UMWJ* article claimed, ". . . you can see the necessity of becoming citizens of this, your adopted land, and the United Mine Workers of America have made and still are making every effort to make good law-abiding citizens of you right here." Indeed, the UMWA portrayed itself as the "true" immigrant workers' representative and as more sincerely concerned about the miners' livelihood. In fact, the union, according to the article "is a haven for all mine workers, irrespective of creed, color, or nationality."⁶⁶ Not only did the union see potential in the immigrant workers and imports to be good citizens and union members, but leaders portrayed the union as a peacekeeping American institution rather than an un-American, lawbreaking group of radicals, as the press and operators posited.

A bigger challenge for the UMWA than defending either the immigrants, the motives for the strike, or the organizers' presence in the region, was countering the press's claim that the union represented an un-American and perhaps communist institution. In order to present an American face to the public, the union's journal and leaders highlighted the organization's Christian values, its respect for law and order, and most importantly, its abhorrence of all things "red." In its propagation of these principles, the union proffered its own definition of "American." In some respects, the union's definition was almost as conservative as that of the operators and the press.

In defense of the union, the *UMWJ* clarified its principles as being in line with the American traditions of Christianity, the Constitution, and law and order. One article stated, "The United Mine Workers of America is an organization with Christian principles; its constitution is based on the rights embodied in the Constitution of the United States."[67] Following Christian and Constitutional tenets, the union denounced renegade strikers who dynamited bridges. Distancing itself from violent offenders, the journal emphatically stated, "The United Mine Workers of America is a thoroughly American institution. It stands for and supports to the fullest extent every principle of law and order . . . It has a record of patriotism second to no other organization in America."[68] In their defense of the UMWA, these articles unintentionally aligned union notions of Americanism within the narrow confines proposed by the operators and the press.

Defending itself from newspaper attacks included championing capitalism and denouncing communism and any acts that threatened the American form of government. Advancing its patriotic argument, one *UMWJ* article declared, "The United Mine Workers' organization does not wave the red flag. Its flag is the stars and stripes of this great country, and its members have proven this during the recent war for democracy."[69] The anti-communist stance in this article was not surprising since John L. Lewis, President of the UMWA, vehemently opposed communism and wholeheartedly supported capitalism, which he believed "spawned two correlative institutions, the corporation and the trade union." He not only credited capitalism for his union's existence, but he also believed that both the capitalist and the trade unionist shared a similar desire for profit and prosperity.[70] In defending God, country, and capitalism, the *UMWJ* presented a portrait of the non-union, renegade striker that mirrored the press's attitude toward the "Bolshevik" union leaders. When non-union miners committed acts of "abhorrent" violence during their "strike for union" in Somerset County, the UMWA withdrew its support, calling the perpetrators "communists and reds" whose intent was "the destruction of the United Mine Workers of America and the overthrow of the existing order of government in this country."[71] The union contended that the red agitators did not represent the UMWA, a patriotic, pro-capitalist organization.

Not only did the UMWA blame renegade miners for disrupting law and order, but the union also believed that the operators threatened American principles by reneging on their promise to negotiate with the miners. Because President Wilson's arbitration commission had negotiated the 1920 contract between miners and operators, and because the operators violated it by refusing to renegotiate in 1922, the operators flagrantly "flout[ed] the command of the United States government." Thus, not only did the UMWA absolve itself of responsibility for threatening the peace, but it placed the blame on the operators who had accused the union of wrongdoing. Placing itself on the lawful side of the struggle, *UMWJ* stated that the operators' "attitude . . . will not meet with the approval of the American people."[72] The union insisted that even President Harding supported the unions' call for a wage scale conference and sent his secretary of labor to meet with the operators to convince them to negotiate with the union in order to protect the American people's interests.[73] Clearly, the UMWA was not to blame for the breakdown of law and order because the capitalists committed the first violation.

In its defensive posturing, the union representatives and supporters generated a definition of "American" that closely resembled that of the operators and the press. The ideal American defended the principles of democracy and capitalism. Using the language of war and battle, the union and its supporters constructed an ideal American, who fights for political and economic freedom. The battleground of this "war" in 1922–1923 was not in a European setting but rather in the farmlands and mining towns of Somerset County.

When defining the ideal America and their place in it, striking miners frequently referred to the Great War, its primary mission to save democracy, and the workers' contributions to the war effort. In his inaugural address, President Harding promised to care for the returning soldiers who fought in "The Great War." He said, "Let me speak to the maimed and wounded soldiers who are present today, and through them convey to their comrades the gratitude of the Republic for their sacrifices in its defense. A generous country will never forget the services you rendered."[74] Somerset miners took Harding's message to heart and believed that as "soldiers" on the homefront, who ensured a steady production of coal for the war effort, they too had rights as "veterans" and as patriotic Americans to decent working and living conditions. During the "strike for union," miners living in the company towns of Consolidation Coal Company wrote, "We are fighting for the right to live as free American citizens and to live our lives in our own way, under the laws of the United States of America, not under the laws of a privately owned and tyrannical coal company."[75] In the aftermath of the war "to make the world safe for democracy," miners, many of whom fought in the war, linked the oppression in countries they fought to free, to their own. An article in the *UMWJ* continued the patriotic appeal, "Its members have proven [allegiance to the United States] during the recent war for democracy for which they are still patiently waiting."[76] These miners wanted American democracy to include them and their families.

To many union leaders, inclusion in the postwar American democracy meant a decent standard of living. Thomas Stiles, editor of District 2's *Penn Central News*, wrote that inclusion in a postwar democracy also required purchasing power. He believed that the high school helped to organize Somerset because "boys and girls were learning to ask their dads why they couldn't have good furniture and decent clothes and pianos, same as other people had."[77] Thus, miners, who had worked at home and abroad for the war effort, could not afford the basic luxuries that symbolized the American standard of living. In a press release, John Brophy expanded on the issue of quality of life for "veteran" miners, condemning the government for failing to honor the promises that it made to miners during the war in exchange for their cooperation and labor peace. In his formal appeal to the government for an investigation of the coal industry, Brophy wrote, "Through no fault of their own, miners in this district who, during the war, strained every muscle at the government's call to 'produce, produce,' now find themselves with nothing to live on through the winter. Thousands of women and children lack the common necessities, let alone any approach to an American standard of living."[78] Brophy wanted the government to acknowledge the miners' service and also to provide them the resources to participate in the "American dream" and become ideal Americans.

Representing themselves as ideal Americans, miners not only used their participation in the war effort to justify the strike but also compared their struggle to war in order to gain sympathy. George Gregory, a local UMWA leader, wrote to U.S. Treasury Secretary Andrew Mellon in 1922, comparing the union's current industrial struggle to a war. In his letter, he portrayed whole families as the victims of this battle. He expressed his outrage at operators and his concern for evicted families, who also suffered physical abuse from company guards. He related a personal story of his family's eviction, which directly contributed to his innocent child's death in a tent colony. In his letter, he asked Mellon, "[D]o you approve of our operators warring on women and children with evictions and guards because we joined the union?"[79]

Indeed, local miners were not alone in their view that this strike was a battle with casualties similar to the world war. The miners' efforts to gain public support and relief succeeded when the American Friends Service Committee (AFSC) agreed to raise funds for strikers' families. The organization adopted a similar war-victims analogy to justify their contributions to Somerset County miners. In July 1922, John Brophy received a letter from AFSC secretary, Wilbur Thomas, which stated, "The American Friends Service Committee is extending the scope of the relief work hitherto carried on among European victims of the International War to include American victims of industrial strife." The connection to the "war for democracy," in the viewpoint of the Friends, involved the existence of innocent war victims in both Europe and in their own country. The letter continued, "The same principle which guided us in carrying relief to German and Russian children impels us to enter this new field, namely, that Christians cannot sit by and see children stunted in mind and body on the account of the differences of opinion that exist between nations or between social groups." Although the Friends offered no support for the strikers' goals, it provided miners with money to feed and clothe their families.[80] Indeed the war analogy helped the UMWA to win at least some public support and aid.

Conclusion
In the aftermath of World War I, the Red Scare, and the strikes of 1919, President Harding called for a "resumption of our onward, normal way." Harding sought no drastic alteration in the economic system and wanted owners and workers to strive for labor peace. He also touted the United States as "an inspiring example of freedom and civilization to all mankind."[81] Although Harding was president only a short time, he paved the way for the conservative Republican administrations of the 1920s that kept industrial peace by sid-

ing with corporate America. Politicians, however, relied on their constituents to win elections and, therefore, were not the only Americans determining the nation's course in the 1920s. Businessmen and workers in the postwar period also helped to shape the path of industrial relations in the 1920s and to lead the country in a conservative direction throughout the decade.

Although the "strike for union" in Somerset County in 1922–1923 was a local event, the operators' and miners' battle to win public support by defining their ideal America represented the larger struggle for American identity in the tumultuous postwar period. The operators defined the ideal America as one in which capitalism flourished, unimpeded by government regulation and labor union demands. Any challenges to the "laissez faire" capitalist model represented Bolshevism or un-American ideas. In order to preserve capitalism, Americans had to maintain and enforce law and order, which, in the operators' opinion, the union encouraged strikers to violate.

Many UMWA leaders, especially John L. Lewis, also touted the virtues of private property and law and order to its members, especially when some miners went astray and committed violent acts. However, these leaders also believed that the operators were the real threats to American democracy. Although Lewis praised capitalists' achievements, he and other UMWA representatives believed that non-union operators, who "cloak[ed] themselves in the guise of American freedom," actually violated "institutions which [were] most dear to . . . American citizens." John Brophy concurred, arguing, "Their system destroys American community life. It lessens self respect in individuals; usually it is the means of destroying the spirit of democratic institutions. It is not compatible with the fundamentals of our government."[82] Miners defended America's democratic traditions during the Great War both at home and abroad, but after the war, democracy lost much of its meaning as a result of declining wages and worsening conditions. Thus, miners fought for their most sacred democratic rights, including the right to organize, just as they had waged war to preserve and extend democracy in Europe.

At the conclusion of the seventeen month-long strike, miners lost their battle for union recognition, and the open shop mines persisted. The United Mine Workers of America declined significantly throughout the 1920s. Union membership was over half a million in 1920, but by 1929 it had fallen to under one hundred thousand as operators moved production to the non-union fields in southern Appalachia. Progressive leader John Brophy's ideas for reform, such as nationalization of the mines and worker's education, faded from the UMWA's platform in the early 1920s. His effort to depose an autocratic Lewis for the union presidency and to replace Lewis's business unionism model failed in 1926. Lewis lost influence with the operators and government officials and had no reasonable alternative but to accept the operators' will as the coal industry declined during the late 1920s. Operators who had signed the Jacksonville Agreement in 1924, which continued the 1920 wage scale for three more years, violated it by cutting wages before the ink was dry. The final blow to Lewis's strength was the union's failure to renew an agreement with the Central Competitive Field in 1927.[83] Indeed, the operators' vision of postwar America triumphed after the 1922–1923 strike. Not until the New Deal legislation of the 1930s did the miners' version of America resurface to offer a radical challenge to the conservative victors.

NOTES

1   Heber Blankenhorn, *The Strike for Union: A Study of the Non-union Questions in Coal and the Problems of a Democratic Movement* (New York: Bureau of Industrial Research, 1924; New York: Arno Press, Inc., 1969). He chronicled the strike in response to the United States Coal Commission's request for any records unveiling the civil liberties abuses in the coalfields. Citations are to the Arno edition.

2   See David Alan Corbin, *Life, Work, and Rebellion in the Coal Fields: The Southern West Virginia Miners, 1880–1922* (Urbana: University of Illinois Press, 1981); Ronald L. Lewis, *Black Coal Miners in America: Race, Class, and Community Conflict, 1780–1980* (Lexington: University of Kentucky Press, 1987); Brian Kelley, *Race, Class, and Power in the Alabama Coalfields, 1908–1921* (Urbana: University of Illinois Press, 2001); Crandall A. Shifflett, *Coal Towns: Life, Work, and Culture in Company Towns of Southern Appalachia, 1880–1960* (Knoxville: University of Tennessee Press, 1991); and Joe William Trotter, *Coal, Class, and Color: Blacks in Southern West Virginia, 1915–32* (Urbana: University of Illinois Press, 1990).

3   See Mildred Allen Beik, *The Miners of Windber: The Struggles of New Immigrants for Unionization, 1890s–1930s* (University Park: Pennsylvania State

University Press, 1998); Muriel Sheppard, *Cloud by Day: The Story of Coal and Coke and People* (Uniontown, Pennsylvania: Heritage, 1975; Chapel Hill: University of North Carolina Press, 1947); Kenneth Warren, *Wealth, Waste, and Alienation: Growth and Decline in the Connellsville Coke Industry* (Pittsburgh: University of Pittsburgh Press, 2001); and Barry P. Michrina, *Pennsylvania Mining Families: The Search for Dignity in the Coalfields* (Lexington: University of Kentucky Press, 1993). Elizabeth Chiang does attempt to incorporate her study of the causes and effects of deindustrialization in the northern Appalachian county of Indiana, Pennsylvania, into the Appalachian historiography by using an "internal colonization" model that compares the railroad and coal mining industries' exploitation of communities in Indiana County to that of West Virginia, Kentucky, and Tennessee. See, Elizabeth Chiang, "The Great Storm that Swept Through: The Effects of Globalization on Indiana County," *Pennsylvania History* 71, no. 2 (2004): 172.

4   Melvyn Dubofsky and Warren Van Tine, *John L. Lewis: A Biography*, abridged edition (Chicago: University of Illinois Press, 1986), 29, 63–67.

5   Beik, *Miners of Windber*, 276, 291–292, 303, 305–306.

6   Paul A. Carter, *Another Part of the Twenties* (New York: Columbia University Press, 1977), 145–165; Peter Fearon, *War, Prosperity, and Depression: The U. S. Economy, 1917–45* (Lawrence: University Press of Kansas, 1987), 15–19.

7   "Statement of the President to Anthracite and Bituminous Operators and Mine Workers' Representatives at the White House," 10 July 1922, United Mine Workers of America, District 2 Papers, MG 52, Box 31, Special Collections, University Archives, Indiana University of Pennsylvania (hereafter UMWA and IUP).

8   Carter, *Another Part of the Twenties*, 145–165; Fearon, *War, Prosperity, and Depression*, 15–19.

9   David Brody, *In Labor's Cause: Main Themes on the History of the American Worker* (New York: Oxford University Press, 1993), 147–148.

10   Historian John Higham clearly identified the nativist undercurrents in the postwar United States. He defines nativism as "intense opposition to an internal minority on the ground of its foreign (i.e., 'un-American') connections," through which "runs the connecting, energizing force of modern nationalism." See John Higham, *Strangers in the Land: Patterns of American Nativism, 1860–1925* (1955; repr., New Brunswick: Rutgers University Press, 1998), 4. Citations are to the reprint edition.

11   Stephen Meyer III, *The Five Dollar Day: Labor Management and Social Control in the Ford Motor Company, 1908–1921* (Albany: State University of New York Press, 1981), 193; Higham, *Strangers in the Land*, 261.

12   Higham, *Strangers in the Land*, 255, 356–357.

13   Richard M. Abrams, "The Failure of Progressivism," in *The Shaping of the Twentieth Century*, ed. Richard Abrams and Lawrence Levine, 2d ed. (Boston: Little, Brown, and Company, 1971), 69–70; Gary Gerstle, *Working-Class Americanism: The Politics of Labor in a Textile City, 1914–1960* (Cambridge: Cambridge University Press, 1989), 3.

14   John C. Hennen, *The Americanization of West Virginia: Creating a Modern Industrial State, 1916–1925* (Lexington: University of Kentucky Press, 1996), 4.

15   U.S. Bureau of the Census, *Census of Population, Composition and Characteristics of the Population, for Counties, 1920*, 864; U.S. Bureau of the Census, *Census of Agriculture, Agricultural Reports by States, 1920*, 264–265.

16   Sally McMurry, *From Sugar Camps to Star Barns: Rural Life and Landscape in a Western Pennsylvania Community* (University Park: Pennsylvania State University Press, 2001), 116–124.

17   Pennsylvania Department of Mines, *Report of the Department of Mines of Pennsylvania* (Harrisburg, PA, 1920–1927), annual reports, 1919–1926.

18   U.S. Bureau of the Census, Somerset County, PA., Manuscript Census, 1920.

19   Blankenhorn, *Strike for Union*, 5; Beik, *Miners of Windber*, 308.

20   Blankenhorn, *Strike for Union*, 58.

21   "Official Call Issued for Suspension of Work by the United Mine Workers at Midnight on March 31, 1922," *United Mine Workers Journal*, April 1, 1922 (hereafter *UMWJ*; article hereafter "Official Call Issued," *UMWJ*).

22   Brody, *In Labor's Cause*, 149–150; "Ohio and Western Pennsylvania Operators Repudiate Section of Agreement and Refuse to Meet Miners in Preliminary Conference; What Correspondence Shows," *UMWJ*, January 15, 1922 (hereafter "Ohio and Western Pennsylvania Operators Repudiate," *UMWJ*); "Operators Still Refuse to Meet the United Mine Workers of America in Joint Interstate Conference," *UMWJ*, February 1, 1922; and "Coal Operators Again Repudiate Pledged Word to Meet with Miners to Negotiate Wage Agreement," *UMWJ*, March 1, 1922 (hereafter "Coal Operators Again Repudiate," *UMWJ*).

23 "Pennsylvania Miners on Strike Determined to Win Union Recognition Despite Hardships," *UMWJ*, November 1, 1922; Beik, *Miners of Windber*, 59, 69, 273.

24 Fair Oak Coal Company to Lynn G. Adams, 18 March 1922, Records of the Pennsylvania State Police, General Correspondence, Coal Strike 1922, Pennsylvania Historical and Museum Commission, Harrisburg (hereafter PHMC).

25 W. E. Ambrose to Lynn G. Adams, 24 March 1922, Records of the Pennsylvania State Police, General Correspondence, Coal Strike 1922, PHMC, Harrisburg.

26 J. F. Huff to Lynn G. Adams, 18 March 1922, Records of the Pennsylvania State Police, General Correspondence, Coal Strike 1922, PHMC, Harrisburg.

27 Berwind White Coal Mining Company to Lynn G. Adams, 29 March 1922, Records of the Pennsylvania State Police, General Correspondence, Coal Strike 1922, PHMC, Harrisburg.

28 *The Consolidation Coal Company, a Corporation, Plaintiff v. John Brophy, President District No. 2, United Mine Workers of American, et. al.*, No. 3 Equity Docket, 1922, Somerset County Courthouse, Somerset, Pennsylvania.

29 *Lochrie Price Coal Company, a Pennsylvania Corporation v. John Brophy, President District No. 2, United Mine Workers of American, et. al.*, No. 15 Equity Docket, 1922, Somerset County Courthouse, Somerset, Pennsylvania.

30 The Somerset County Courthouse contains the bills of complaint for most coal companies in the county, filed between April 19 and May 2, 1922 against John Brophy and other officers in District 2.

31 Brothers Valley Coal Company to Lynn G. Adams, 24 March 1922; Romesberg Smokeless Coal Company to Lynn G. Adams, n.d.; and Tri-State Colliers Company to Lynn G. Adams, 24 March 1922, all found in Records of the Pennsylvania State Police, General Correspondence, Coal Strike 1922, PHMC, Harrisburg. These letters provided a list of "troublemakers" to the police in their initial response to the superintendent's request, and all three coal operations in question were in Macdonaldton or Garrett.

32 *The Consolidation Coal Company, a Corporation, Plaintiff v. John Brophy, President District No. 2, United Mine Workers of American, et. al.*, No. 3 Equity Docket, 1922, Somerset County Courthouse, Somerset, Pennsylvania.

33 *The Consolidation Coal Company, a Corporation, Plaintiff v. John Brophy, President District No. 2, United Mine Workers of American, et. al.*, No. 3 Equity Docket, 1922, Somerset County Courthouse, Somerset, Pennsylvania; *Lochrie Price Coal*

Company, a Pennsylvania Corporation v. John Brophy, President District No. 2, United Mine Workers of American, et. al., No. 15 Equity Docket, 1922, Somerset County Courthouse, Somerset, Pennsylvania. (Other coal companies in Somerset County filed similarly worded bills of complaint in April and May of 1922.)

34   "Miners' Families Living in Tents," *Somerset Herald*, May 24, 1922.

35   "Somerset County's Mines in 100 Per Cent Operation," *Somerset Standard*, April 6, 1922.

36   "The Strike," *Boswell News*, April 26, 1922.

37   "Coal Strike Enters 4th Week," *Somerset Herald*, April 26, 1922.

38   "Somerset County's Mines in 100 Per Cent Operation," *Somerset Standard*, April 6, 1922.

39   "Mining Industry Partially Paralyzed," *Somerset Herald*, May 10, 1922.

40   "Trouble Makers at Mines Restrained by the Court," *Somerset Standard*, April 20, 1922.

41   "The Strike That Failed," *Somerset Herald*, January 3, 1923.

42   "Miners Resent Activities of Organizers," *Somerset Herald*, September 20, 1922.

43   "Organizers Appeal for Aid for Strikers," *Somerset Herald*, October 25, 1922.

44   "Union Abandons Coke Region," *Somerset Herald*, January 24, 1923.

45   "What is That We Call 'Americanization?'" *Somerset County Star*, December 7, 1922.

46   "Dynamiting Felt for Miles," *Boswell News*, April 4, 1923; "Local Miner Injured," *Boswell News*, April 11, 1923; "Bridge Dynamited," *Boswell News*, July 18, 1923; "Alleged Dynamiters Held for Court," *Boswell News*, July 25, 1923; and "Bail Increased in Dynamiting Case," *Boswell News*, August 1, 1923.

47   "Windber Mine Tipples Dynamited," *Boswell News*, March 28, 1923.

48   "Operation of Mines Retarded by Union Interferences," *Somerset Standard*, May 4, 1922.

49   "Will Not Yield to Outside Influences," *Somerset Standard*, May 25, 1922; "All Classes are Affected," *Somerset Standard*, June 1, 1922.

50   "Social Fabric is Threatened," *Somerset Herald*, May 3, 1922.

51   "Union Abandons Coke Region," *Somerset Herald*, January 24, 1923.

52   "Miners' Strike Weakens," *Somerset Standard*, August 3, 1922.

53   "Coal Strike Enters 4th Week," *Somerset Herald*, April 26, 1922; Blankenhorn, *Strike for Union*, 23–24.

54 "Social Fabric is Threatened," *Somerset Herald*, May 3, 1922; "Mining Industry Partially Paralyzed," *Somerset Herald*, May 10, 1922.

55 "Resolution of Cambria County Pomona Grange," n.d., UMWA, District 2 Records, MG 52, Box 31, Folder 6, Special Collections and University Archives, IUP.

56 Minutes of the Somerset County Pomona Grange, October 21, 1922, Grange Records, Somerset Historical Center, an affiliate of the Pennsylvania Historical and Museum Commission, Somerset, Pennsylvania (hereafter cited as SHC, PHMC).

57 "Cambria County Pomona Grange Press Release," n.d., UMWA, District 2 Records, MG 52, Box 31, Folder 6, Special Collections and University Archives, IUP.

58 Minutes of the Somerset County Pomona Grange, June 26, 1920, January 22, 1921, and January 24, 1924, Grange Records, SHC, PHMC.

59 James Read to John Brophy, April 1922, UMWA, District 2 Records, MG 52, Box 29, Folder 9, Special Collections and University Archives, IUP.

60 Blankenhorn, *Strike for Union*, 16.

61 Blankenhorn, *Strike for Union*, 234.

62 Blankenhorn, *Strike for Union*, 235–241.

63 James Read to John Brophy, April 1922, UMWA, District 2 Records, MG 52, Box 29, Folder 9, Special Collections and University Archives, IUP.

64 Blankenhorn, *Strike for Union*, 20, 32, 40–44.

65 "Propaganda Spread Among Foreign-Born Miners Answered by Union," *UMWJ*, May 1, 1922 (hereafter "Propaganda Spread Among Foreign-Born," *UMWJ*).

66 "Propaganda Spread Among Foreign-Born," *UMWJ*.

67 "Propaganda Spread Among Foreign-Born," *UMWJ*.

68 "Letter to Somerset County Operator Repudiates Charges Against Union," *UMWJ*, April 1, 1923 (hereafter "Letter to Somerset County Operator," *UMWJ*).

69 "Propaganda Spread Among Foreign-Born," *UMWJ*.

70 Irving Bernstein, *The Lean Years: A History of the American Worker 1920–1933* (1966; repr., Baltimore: Penguin Books, 1970), 126. Citations are to the reprint edition.

71 "Letter to Somerset County Operator," *UMWJ*.

72 "Ohio and Western Pennsylvania Operators Repudiate," *UMWJ*.

73 "Coal Operators Again Repudiate," *UMWJ*; "Government May Force Operators to Meet Miners in Conference According to Terms of 1920 Agreement," *UMWJ*, March 15, 1922; "Official Call Issued," *UMWJ*; and Warren G. Harding, "Inaugural Address," 4 March 1921, Yale Law School Library, http://avalon.law.yale.edu/20th_century/harding.asp.

74 Warren G. Harding, "Inaugural Address."

75 Consolidation Coal miners' petition, 9 October 1922, UMWA, District 2 Records, MG 52, Boxes 29 and 31, Special Collections and University Archives, IUP; Robert Bussel, *From Harvard to the Ranks of Labor: Powers Hapgood and the American Working Class* (University Park: The Pennsylvania State University Press, 1999), 46.

76 "Propaganda Spread Among Foreign-Born," *UMWJ*.

77 Blankenhorn, *Strike for Union*, 46.

78 Press release from John Brophy, n.d., MG 52, Box 31, Folder 4, UMWA, District 2 Records, Special Collections and University Archives, IUP.

79 George Gregory to Andrew Mellon, U.S. Secretary of the Treasury, 11 November 1922, UMWA, District 2 Records, MG 52, Box 31, Special Collections and University Archives, IUP.

80 Wilbur K. Thomas, American Friends Service Committee, to John Brophy, 5 July 1922, UMWA, District 2 Records, MG 52, Box 31, Special Collections and University Archives, IUP.

81 Warren G. Harding, "Inaugural Address."

82 John Brophy to Hon. John Hays Hammond, US Coal Commission, 28 May 1923, UMWA, District 2 Records, MG 52, Box 31, Special Collections and University Archives, IUP.

83 Bernstein, *Lean Years*, 127–136; Brody, *In Labor's Cause*, 152; and Dubofsky and Van Tine, *John L. Lewis*, 93–112.

Separate But Never Equal:

Dewey W. Fox and the Struggle for Black Equality in the Age of Jim Crow

Connie Park Rice

ON JUNE 29, 1898, in the small community of Kearneysville, West Virginia, Dewey William Fox was born. Fox was the youngest of four children born to John H. and Mary E. Fox, former slaves of the Dandridge family from Jefferson and Berkeley counties. Although the recorded history of the Fox family begins in slavery, a history shared by many African Americans, it is a story of survival, struggle, and success. The birth of Dewey Fox symbolized the dawn of a new generation. Born two years after *Plessey v. Ferguson*, he lived most of his life with the reality of socially constructed segregation and a dream for equality. In the age of Jim Crow, Dewey Fox's dream for equality would have a major impact on the lives of African Americans in the coalfields of Monongalia County, West Virginia.

Fox's story epitomizes the struggles of numerous African Americans who fought to make economic, political, and educational gains within the confines of the segregated society south of the Mason and Dixon line. West Virginia had two written segregation laws, both imbedded in the constitution of 1872. One prohibited the teaching of black and white students in the same school; the other banned interracial marriages. African Americans in West Virginia were never denied the right to vote, nor were they excluded from railroad cars. Yet segregation existed. De facto segregation led to limited economic opportunities; the segregation of restaurants, hotels, and movie theaters; segregated housing; and constant diligence to keep that separation from becoming law.

In West Virginia, economic, political, and educational gains were possible, but they never came without strife.

The Fox Family in Slavery and Freedom

Following the Revolutionary War, General Adam Stephen, who designed the town of Martinsburg, built a plantation home on property located on the border line between Berkeley County and Jefferson County. Stephen called the plantation "The Bower." Stephen's daughter, Ann, married Alexander Spotswood Dandridge. When General Stephen died, he left the property to his grandson, Adam Stephen Dandridge.[1] The Dandridge family was one of several slave owners in the region. In 1821, Virginia required the slave masters in the state to submit a list of their slaves for taxation purposes. The name of Dewey Fox's nineteen-year-old grandmother appears on that list as a slave owned by the Dandridge family of Jefferson County. By 1850, the total population of Jefferson County was 15,357, and included 4,341 slaves.[2] That year, John Fox's owner, Lem Dandridge II, sold John's father and mother, leaving John and his siblings to fend for themselves. The sale of their mother upset Mrs. Dandridge who instructed the slave women to bring the Fox children up to the main house for one meal a day. Their meal consisted of corn bread and skim milk, eaten under the kitchen table in a large washtub that they shared with the cats and dogs. According to the 1860 slave census, the Dandridge family owned sixty-six slaves; thirty-six males and thirty females.[3] Only one male was listed as being thirteen, John Fox's age. As they grew up, the siblings became active in the Underground Railroad route that ran between Winchester and Martinsburg, Virginia.[4]

During the Civil War, the majority of Jefferson County residents sympathized with the South. While the county organized ten companies of Confederate soldiers, there are no records of Union companies in the county—although a few individuals joined the Union army. Strategically located between the North and the South, Jefferson County saw a considerable amount of fighting. Taking advantage of the confusion created by these skirmishes, John Fox ran away from the plantation to join the Union army.

He served as a laborer, cooking and cleaning for the army, until Confederate forces captured him at nearby Antietam on September 17, 1862. Fox remained a prisoner until the spring of 1865 when the Confederate army released him at Appomattox, Virginia, with "nothing but his clothes."[5]

Immediately following the war, Fox worked his way toward the East Coast, hiring himself out to farmers for food along the way. When he arrived at Hampton, Virginia, he obtained work on a boat that was plying goods between Hampton and Baltimore, Maryland. On the trip, the boat struck a Confederate mine in the Chesapeake Bay, and limped into Baltimore. From there, Fox worked his way back to the Jefferson County-Berkeley County line and the home of his former owner. Throughout Reconstruction, the need for lumber to rebuild the region soared, so Dandridge gave Fox a job, paying him two dollars and one meal a day for cutting timber at Pine Hills, a small community four miles outside of Martinsburg. While he worked on the timber, Fox used a double barrel musket to hunt food.[6]

Millard Kessler Bushong claimed in his history of Jefferson County that African Americans were one of the county's biggest problems during Reconstruction. He stated that because blacks did not understand their new "condition," they did not work and expected the government to support them. This created a labor shortage when the county needed able-bodied men to rebuild. He claimed that "former masters were required to take their ex-slaves and pay prescribed wages" and that this created "a hardship on the farmer, who could scarcely support his own family."[7] There are no records that indicate why Dandridge consented to hire his former slave, but in addition to hiring Fox, Dandridge eventually agreed to sell him one acre of land for five dollars. Fox quit working for Dandridge and cut the timber on his own land; he used the money to buy eight more acres.[8]

By 1870, the Fox farm in Averill Township, Jefferson County, was worth $500.00 and Fox had $100.00 in personal money. Fox had a wife, Hannah (age 23), and a daughter, Lucinda (age 1). He also had a farm laborer, Humphrey Fox (age 16).[9] Before long, John Fox sold his eight acres and used the money to buy three good farms between Kearneysville and Bardane in Jefferson County: the White farm, the Moore farm, and the Johnson farm. The Fox family lived on the White farm and rented the Johnson farm to a tenant.[10] The family continued to grow and prosper. In 1880, the family consisted of John, his second wife—Lucy E., Lucinda, Robert Edward, and Berthy [Bertha]. Fox had two farm laborers, cousins Oliver and Howard Dennis, and a fourteen-year-old servant named Emily Colsman.[11] John Fox continued to accumulate money and increase his holdings. By 1887, he had enough money to purchase additional livestock and a significant amount of farm equipment. The farm now encompassed one hundred and sixty-eight acres.[12]

When Dewey Fox was born in 1898, the Fox family included Dewey's father John, his mother Mary, and his siblings Lucinda, Robert Edward, and Charles Washington. Six months after his birth, tragedy struck. A train rammed the family carriage, and the impact broke Dewey Fox's back. Doctors placed him in a plaster jacket to support his back, leaving him a semi-invalid. Fox did not start school until he was eight years old. At that time, he had to walk one and a half miles to a school in Johnsontown, the first all-black community in West Virginia.[13]

The schools in Jefferson County were in poor condition throughout Reconstruction and into the first part of the twentieth century. In 1910, when Dewey Fox was twelve years old, attendance at the Jefferson County schools was low. Less than 58 percent of the white children were enrolled in school, and only one out of every three African-American children attended school, despite the large number of African Americans in the county. According to the 1910 census, Jefferson County had 12,390 whites and 3,499 blacks.[14] In comparison, the African-American population in the north central West Virginia county of Monongalia, where Fox would eventually live and work, was only 294.[15]

Fox continued his primary school education at the African-American grade school in Kearneysville. He passed his eighth-grade courses while he was still in the seventh grade, but the county refused to give him a diploma until the next year. About the same time, a steel jacket from John Hopkins Hospital in Baltimore, Maryland, replaced the plaster jacket that supported his back. Throughout the early years of Dewey Fox's education, John Fox worried about his son's health and how Fox would earn a living if he survived. In fact, in a will written in 1911, John Fox provided an inheritance for Dewey Fox on the condition that he reached the age of twenty-one. Obviously, he felt that his son's health was in jeopardy, because there were no similar conditions on his other two sons' inheritances. In addition, because Fox could not perform physical labor, his father made plans to send his son to a private black school in Luray, Virginia. Funded by six wealthy African Americans from Virginia and West Virginia, the Valley School hired black educators with degrees from nearby Storer College at Harpers Ferry, Howard University in Washington, D.C., and Tuskegee Institute in Alabama.[16]

John Fox died in 1911, and Dewey Fox continued to attend the Valley School. Fox's mother died in December of 1916, and the following February, Fox began attending Delaware State College at Dover, Delaware. His strong education from Valley School contributed to his success at the college. Within weeks of starting school, officials moved him out of the freshman class and into the sophomore class. Before the end of the semester, they advanced him to the junior class. Fox graduated with honors from Delaware State College in 1919 with a Bachelor of Science degree in political economics.[17]

Life in the Coal Camps

While Fox was attending college in Delaware, the community of Scotts Run, located in Cass District of Monongalia County, West Virginia, was becoming one of the centers of the coal industry in West Virginia. Coal companies established mines throughout the Scotts Run area and along the line of the Baltimore and Ohio Railroad on the west side of the Monongahela River through Grant District. Due to the increase in jobs, Monongalia County experienced a large population influx. Between 1910 and 1920, the population in Monongalia County increased from 24,334 residents to 33,618. The mining districts of Cass and Grant experienced the greatest increases. The population in Cass rose from 1,173 in 1910 to 3,160 in 1920, while the population of Grant rose from 2,495 to 4,807.[18]

Among the miners entering the coalfields of Scotts Run were African Americans from the southern states of Alabama, Georgia, and North Carolina. With a large increase in the African-American population, local boards of education struggled to provide the "separate but equal" school facilities required by the state constitution. In 1921, African-American schools opened in Chaplin (near Scotts Run) and Everettville in Grant District, and in Stumptown on Scotts Run in Cass District.[19]

Turmoil in the coalfields of Monongalia County made it a tense environment. Constant strikes and the importation of African-American strikebreakers into the coalfields were creating problems in the new schools. The previous spring, a fight between the teacher and the students closed the school at Chaplin three weeks before the end of the term. County supervisors, eager to maintain the discipline that the coal companies required, began looking for teachers who could maintain control. W. W. Sanders, the first African-American assistant superintendent of schools in West Virginia, visited the Fox farm after William Pickett of the NAACP told Sanders that Fox "was a midget," but would "get the job done." After meeting with Fox, Sanders made arrangements with Lynn Hastings, superintendent of Monongalia County Schools, for Dewey Fox to teach at Chaplin, a mining community in Scotts Run. Dewey Fox accepted Sander's offer and boarded a train for Morgantown. He was twenty-three years old.[20]

Dewey Fox was something of an anomaly in West Virginia in the 1920s. He certainly was not the typical black elementary school teacher in West Virginia; despite his credentials, it was most likely his gender and authoritative manner that earned Fox a position in the coal mining community at Scotts Run. According to Joseph S. Price, the typical black elementary teacher in West Virginia was a female graduate of at least a four-year high school between the ages of twenty-eight and thirty-one with no physical defects. Although 58 percent of the 119 teachers interviewed graduated from an approved normal school, only 3 percent held a bachelor's degree, and only 18 percent of the black teachers hired in the state secured their positions through a school superintendent. However, Fox, along with nearly 50 percent of the black teachers working in West Virginia, was born in the state. The majority of those teaching were born to farmers and laborers rather than professionals, as was Fox. Although small, with a high pitched voice from years of wearing his back brace, the rarity of a male teacher combined with his fierce determination to overcome any obstacles increased his authority in the classroom.[21]

When Fox's train arrived in Morgantown at nine o'clock in the evening, he knew no one in town and had nowhere to spend the night. A taxi driver took him to John Hunt's restaurant on Walnut Street. Mr. Hunt, a prominent African-American businessman in Monongalia County, found Fox a room for the night at Marian McNeill's house in the Greenmont section of Morgantown, a predominately African-American residential area. Fox boarded with the McNeills and traveled back and forth to Scotts Run every day of the week to teach.[22]

In 1921, Scotts Run was unlike any place Dewey Fox had ever seen. People crowded the streets in the mining communities along Scotts Run. Hills surrounding the mines were covered with hastily built houses that provided homes for the rapidly rising population. Throughout the winter, snow and rain turned the unpaved streets into deep, muddy ruts. Coal operators and miners struggled over control of the mines, and almost all of the men in the mining communities wore guns. Constant strikes led the coal companies to bring more strikebreakers into the mines with promises of good wages and lodging. Company trucks met them at the train stations and took them to the company store where they gave their names to the operators. They were then told to go and pick out a shanty. While the strikebreakers were at the company store picking up supplies, company guards threw the furniture and belongings of the striking miners out of the company owned homes and immediately gave the homes to the strikebreakers. All along both sides of the road from Osage to Cassville, the wives and children of striking miners sat next to their possessions, while the striking miners sat in the hills and watched the tipples with high powered rifles.[23]

The companies owned all of the school buildings in Scotts Run. Christopher Coal owned the school at Chaplin and established a "gentleman's agreement" with the local board of education. If anything went wrong in the schools, the teachers and the boards were to confer with the coal operators. The company had no influence over the curriculum, but it required order and discipline in the school. Occasionally, the company housed strikebreakers in the school. Children of the strikebreakers often arrived from areas that provided little or no education for African Americans and, therefore, the students tended to be older. The situation in all of the schools in the county became so bad between 1921 and 1924 that the county supervisor advised the teachers to use rubber hoses when disciplining the children. Fox stayed at Chaplin for two years. When Mrs. Furman, the teacher at the Jerome Park School in Morgantown, asked for a change, Fox switched schools with her. After teaching at Chaplin for one year, she quit.[24]

Meanwhile, new mines were opening up at National and Brady in Grant District, and the county superintendent asked Dewey Fox to go to National to teach. At that time, National was no place for families or a teacher, according to Fox. It was difficult to get into the area and there were very few accommodations. Fox had to travel from Morgantown to Opekiska, approximately fifteen miles up the east side of the Monongahela River. At Opekiska, Fox was met by a student named Willie, who rowed Fox across the river. Then, he had to walk over a hill to get to the school, which consisted of a one-room shanty with a Burnside stove in the center and a few desks on the sides. Every Friday, Fox had to shovel two and a half to three tons of coal into the school to keep the fuel from being stolen over the weekend.[25]

During this period, the New England Fuel and Transportation Company imported large numbers of black strikebreakers from the South to work in its mines and by 1925, 50 percent of the 550 men employed at its mines in Everettville and Lowesville were black.[26] Of course, the increase in African-American miners meant that there was also an increase in the number of black students attending the African-American school at Everettville.

In 1924, a black teacher named Jessie Holland arrived at the Everettville School. Miss Holland wanted to stay at Everettville during the week and take the train home to Morgantown on the weekends, but no one would give her a place to stay. Finally, the coal company gave her a coal shed. She spent the summer cleaning the coal out of the shed and lining the wall with newspaper held by copper tacks. The parents of one of her students gave her a Burnside stove, and she stayed there on weekdays throughout the winter. Three weeks before the school term ended, a male student attacked and beat her as she walked along the railroad tracks to catch a train home to Morgantown. Lynn Hastings, the county superintendent, called Dewey Fox and asked him to transfer to Everettville to teach. Conditions at Everettville were similar to the conditions he had encountered at Scotts Run. The conflict in the school was a product of the trouble found in the coalfields. People were constantly moving in and out as miners went on strike and strikebreakers moved in to take their jobs. Everyday, the school had three or four new faces. An African American named Charlie Clark worked as a "yellow dog," or guard, for the coal company. Clark's job was to stop any discussion on unions, but he frequently beat up anyone in the community who displeased him. Any violent act that he committed was supported by the coal company and the local sheriff.[27]

The children carried the fighting out of the communities and into the school. After three months of switchings and fist-fighting, Fox established order in the classroom. Reverend Johnny Linear from the Friendship Baptist Church helped Fox to restore order in the school. The company also backed his efforts, and often, even the white school in Everettville called on Fox to help maintain discipline. For the most part, parents, black and white, supported the teachers in their attempt to provide quality education for their children.[28]

The Quest for Equal Conditions

By 1930, the mining districts of Cass and Grant had more than doubled their population from 1920. Cass District now had 6,854 residents, and Grant District had 10,324 residents. At the same time, the rise and fall of the coal markets kept the mining economy of the county in an unstable condition. When the Great Depression hit the coalfields, it hit quickly and severely. In July of 1930, over 500 miners from northern West Virginia met in Morgantown in an attempt to solve the economic problems in the coalfields. Throughout the 1920s, work had been irregular in the coalfields, and following the Depression, companies resorted to reducing wages. Many of the miners faced starvation, and most of them depended upon charities to survive. District officials for the United Mine Workers called numerous strikes to improve the condition of the miners. However, the strikes usually led to the miners losing their jobs and being evicted from their homes. By 1931, most miners were working only one or two days a week, and were only paid between eighteen and twenty-seven cents per ton of coal.[29]

Such conditions prompted Fox to action. He worked with Alice Davis from the County Welfare Board and the American Friends Service Committee to help mining families. Davis and the American Friends Service Committee arranged for clothing, shoes, books, and food to be sent into the schools. The groups helped Fox set up one of the first hot lunch programs in the county at his Everettville School. The Friends also organized the first African-American 4-H Club at Everettville under Fox's leadership.[30] Fox created an organization called the Social and Economic Planning Council and served as its president. According to Fox, the organization's primary concern was the "rehabilitation of those that are needy and the development of our people, generally, into a better citizenship." Fox believed that blacks in Everettville were "socially and economically prostrate" with an absence of "respect and ambition." The council extended 4-H programs throughout the county under the full-time leadership of Marguerite Franklin, another African-American teacher in Monongalia County.[31]

Fox also worked with the Federal Emergency Relief Administration (FERA) to establish social and cultural programs for the mining communities. FERA established daycare programs, mothers' clubs, sewing clubs, garden clubs, and school playgrounds during the summer. Mothers came everyday to the school to help with the children. After the county welfare organizations started the feeding programs in the schools, they made the children take cod liver oil daily. To get the children to take it, Fox had cod liver oil for breakfast every morning as well.[32]

When FERA took over the work projects in 1934, Fox was hired as an administrator. He was responsible for placing African Americans in FERA projects. Identified in Monongalia County as the Negro Rehabilitation Project, Fox not only supervised the work relief project, but also worked with the federal government under the auspices of the Social and Economic Planning Council to plan and develop a black homestead for African-American miners in north central West Virginia. Fox used the position and his office on Beechurst Avenue in downtown Morgantown to assist local school authorities in obtaining funds from the Works Progress Administration (WPA) and private organizations to build a centrally located black high school in Monongalia County. In 1934, the Social and Economic Council of Northern West Virginia had a membership of approximately 2,000. Fox and the council began calling for a new black high school in 1933. After the eighth grade, African-American students who wanted to continue their education had to travel to the only black high school in the county located on White Avenue in the Greenmont section of Morgantown. The students had no bus transportation and were forced to pay a fifteen dollar tuition fee each term. As a result, few of the students from the mining communities attended or finished high school.[33]

Moreover, conditions at the black high school in Morgantown were rapidly deteriorating. Located in a house owned by James A. G. Edwards, a prominent African-American businessman in Morgantown, and leased to the school board, the school on White Avenue was insufficient. In a letter to Charles Houston, special counsel for the NAACP, Fox reported, "It is overcrowded. It is condemned by local authorities as a fire trap. It is unsanitary, unsafe for children or adults to move about in, it cannot be put in proper shape and the Negro family still lives in this house, actually occupying some classrooms in turns with the school children—bed pallets being made up on the teacher's desks after school to accommodate members of the family." In his letter, Fox focused on the need for a new high school accessible to all of the students, but he also reported to Houston that a movement was well under way to segregate all of the blacks in Morgantown into the White Avenue section. Fox hoped that "beginning with the Negro school location we intend to check this movement, and I believe we will succeed."[34]

Those who desired to maintain or increase racial segregation were not the only ones opposed to the new school. Many of Morgantown's black residents, whose families arrived in Morgantown before and after the Civil War and had created the first black high school in the county, did not want their children to travel outside of Morgantown to attend school, particularly one dominated by children from the outlying coal mining communities. In particular, Fox faced opposition from Edwards, who blocked several attempts the school board made to find a temporary site with better conditions. Referring to Edwards, Fox claimed, "I am determined that the Negroes of this County shall not be discriminated against longer in things educational and I am equally determined that we shall not be segregated. I am determined that those Negroes of the old school, the typical 'stool pigeons' well known in this country, shall be driven to the wall as far as Monongalia County is concerned."[35]

Fox wanted a new school, and he wanted it to be accessible to all the students in the county. Desperate for funding, the county superintendent of schools, Floyd B. Cox, suggested that Fox begin contacting government authorities to get them interested in the project. Fox approached F. W. McCullough, Works Progress Administrator for West Virginia; Prof. Arthur D. Wright, president of the John F. Slater Fund; West Virginia Senator Matthew Neely (1931–1941); West Virginia Congressman Jennings Randolph (1933–1947); and the Rockefeller and Ford foundations. Eventually, Fox came into contact with two African-American officials from the Resettlement Administration in Boston, R. D. Reid and J. R. Otis, who knew Mrs. Roosevelt personally. Once the plans for the building were completed, Superintendent Cox sent them to the state WPA office in Charleston, West Virginia. Cox also suggested to the school board that they buy a seven-and-one-half acre tract of land in Grant District in the Westover Addition, a suburb on the west side of the Monongahela River that was the geographic center of the black population in Monongalia County.[36]

When the board of education purchased the land for $5,000.00, citizens of Westover immediately began protesting the construction of a black school in their community. Fox and several other black citizens who lived in Westover met with the protestors in a joint meeting with the Morgantown and Westover mayors and their respective city councils. At the meeting, Fox explained the plan and felt that everyone was "extremely cordial." However, the protesters continued to meet with members of the board of education, signed petitions, and even offered to buy the land back from the board. Fearing that the protest would bring a halt to the project, Fox began writing to the NAACP headquarters in New York City in August of 1935 asking if the organization would help them win the fight for a black school in Monongalia County. Specifically, would the NAACP provide legal counsel if the protesters turned to the courts?

In October of 1935 the WPA allotted $57,619.00 for the construction of a black six-year central high school in Westover. Immediately, 375 residents of Westover signed a petition against the construction of Monongalia High School, claiming that the school would lower property values and stating that, if necessary, they would pass a bond issue to purchase the land from the Board of Education. In spite of their protest, construction began on the school in 1936. Opposition diminished when the Board of Education obtained an additional WPA grant to pave the roads surrounding the school site.[38]

Correspondence between Fox and the NAACP also led to the creation of an NAACP branch in Morgantown. As a result, Fox frequently consulted with NAACP officials over the discrimination and conditions facing blacks in the county. In doing so, Fox presents a vivid picture of black life in the West Virginia coalfields during the Great Depression. In a letter dated May 20, 1938, Fox wrote:

Coal is the dominant industry in Monongalia County. Other industries are glass manufacture, tin mills, shirt factories, pottery, farming and several others. Negro labor cannot obtain employment outside the mining industry except as domestics where wages are below a subsistence scale and the field is narrowing. Now the coal operators are gradually closing the field to Negro labor, despite the fact the local field is completely unionized and Negroes are loyal union men. Even the United Mine Workers officials refuse to intercede on behalf of this discrimination against their own members.

Most of the local coal mines are operated by out-of-state interests. Next to Western Pennsylvania, they engage in cut-throat competition against those operators whose proximity to the Great Lakes manufacturing centers gives them advantages in freight rates. 71 percent of the Negro miner's dollar goes back to the company store, mine camps are mere "patches" of rotten prostitution, disease and crime. Scotts Run, a mining center in Monongalia County, is the blackest spot on the industrial map of this Country. We invite investigation and comparison.

Another "bad" center is known as John Y. Eviction proceedings are now pending in a local justice's court against fifteen families embracing about 60 men, women and children. I have demanded from this justice full legal protection for these families and have taken advantage of every legal technicality. We have taken a stay of execution until July 1st. I have also conferred with the sheriff and have his pledge of cooperation. It has been suggested that we compromise by offering to pay rent (this is an "abandoned" mine camp) but the miserable shacks—isolated and unsanitary—are not fit for human habitation and I feel that some repairs and improvements should be made before rent is agreed upon.

I refuse to discuss the social or moral status of some of these families. They are the product of this contemptible system of exploitation practiced in these fields, they have nowhere to go and certainly there exists somewhere a moral responsibility to consider their well being.[39]

Poor conditions in the coalfields led to Fox's participation in another project with Otis and Reid from the Resettlement Administration in Boston and First Lady Eleanor Roosevelt. In October of 1933, the Secretary of the Interior publicly announced the purchase of land in Arthurdale for a subsistence homestead for unemployed coal miners from Monongalia County. A committee then began selecting homesteaders for the community. The committee required eligible homesteaders to have a good knowledge of farming, be physically fit, have a certain level of education, and show proper attitudes and ambitions. Although several African-American miners qualified for selection, the committee chose no African Americans for the project, partly due to the meetings in nearby Reedsville where citizens protested the inclusion of blacks or foreigners. In addition, since this was the first homestead project, officials were eager to ensure the success of the community.[40] In 1934, with the cooperation of FERA, Fox made an application to the Subsistence Homestead Division of the Department of the Interior for an African-American Subsistence Homestead in Monongalia County.[41]

In November of 1934, the State Relief Administration optioned 350 acres in Cass District for the project. On May 10, 1935, two representatives from Washington inspected the preliminary plans and completed the final plans for the development of the proposed homestead. Officials took surveys of the African-American population in Monongalia County and made plans for school facilities. Each homestead family was to receive four acres of land. Politics, however, scuttled the project. Local Democrats, in particular an African-American minister from Scotts Run, went to Washington to complain about the project and Dewey Fox's involvement in it. Dewey Fox was a Republican. In 1935, officials allowed the option on the land to expire.[42]

According to Fox, politics and discrimination influenced more than just the homestead project in Monongalia County. In the same letter that Fox wrote to NAACP officials concerning conditions in the coal camps, he added, "WPA authorities are yielding to the demands of prejudiced whites, and discriminating Negro WPA workers. Negro labor is being systematically excluded from private industry and whole-sale evictions of poverty stricken families are being attempted by local coal operators." Claiming that "Local Negroes are experiencing unjustified foreclosures against their homes by the federal H.O.L.C.," Fox reported that more blacks lost their homes in the past twelve months because of the Home Owners' Loan Corporation (HOLC) foreclosures than in the fifteen years preceding them. Stating that, in every case, an additional thirty to sixty days would have enabled the home owner to save his home, Fox insisted that "special consideration should be extended to our local groups simply because special attention is being given to the more and more rigid discrimination against our labor." As an example, he pointed to a black foreman named Frank Dale who did excellent work on the new high school but was then transferred to a white project where white WPA workers signed and circulated a petition against working for a black foreman. As a result, officials demoted Dale to a position of labor, breaking stones, at reduced wages. At the same time, white foremen on the same project were paid higher wages. Dale, an older man with a large family, who was also treasurer of the Monongalia branch of the NAACP, lost his home, which represented his life earnings, to foreclosure by the HOLC.[43]

Fox's own experiences with discrimination and political pressure made him well aware of disadvantages blacks faced in Monongalia County. Local Ku Klux Klan members began harassing Fox and his wife, Alma, at their home on Jackson Street in Westover. In the summer of 1936, they burned a cross on his lawn because of his efforts to obtain the high school and a subsistence homestead for African Americans. The cross-burning incident led former Illinois Congressman Oscar De Priest to visit Morgantown. De Priest, the first African American elected to Congress in the twentieth century and the first ever elected outside the South, delivered a speech on the courthouse square. African Americans from the surrounding communities traveled in trucks to hear his speech. The cross burning on Fox's lawn was the last overt action by the KKK in Morgantown for several decades, but the Klan had a major impact on Fox's life. Because of his long hours at work, the controversy involved, and her fear of the KKK, Fox's wife divorced him.[44]

Nevertheless, Fox achieved significant results in Monongalia County. Throughout the 1930s, Fox continued to press for improvements in the Everettville community and school. In the 1934–1935 school year, Fox became principal of the school. The county hired one teacher to assist him, followed by several more as the black population at Everettville grew. Fox also played the organ at the Friendship Baptist Church, taught music lessons to children in Morgantown, and became president of the first NAACP group in the county. He and Alma remained friends of John Hunt—Fox's first friend in Morgantown, and often traveled to Preston County on weekends to help Hunt at his resort, Indian Rocks.[45]

Fox's greatest achievement, however, was the new high school. After receiving Fox's letter to the NAACP, Walter White contacted Eleanor Roosevelt who replied that she would "be very glad to see Mr. Fox," and that she would see what could be done by the WPA and the HOLC in Washington concerning the racial discrimination and housing foreclosures Fox described. "As to the mine owners," the First Lady replied, "I am afraid there is nothing that we can do about them." Fox then received a telegram from White saying that Mrs. Roosevelt had agreed to meet with Fox on May 27, 1938, when she dedicated the new black high school named Monongalia High. But because Fox was unable to get away from his school in Everettville, Mrs. Roosevelt decided to visit him there. Hearing a commotion, he looked outside and saw Mrs. Roosevelt walking up from the playground with a little girl on each side holding her hand. All of the black schools in the county closed early on the day of the dedication so the students and teachers could attend the dedication. Fox described his meeting with Mrs. Roosevelt and the dedication of the new high school as "the highlight of my life."[46]

Despite all of his difficulties, Fox continued to work to improve black education. The school at Everettville was in deplorable condition. Fox, Parent Teacher Association members, and the local NAACP began requesting that the board of education build new primary schools for African Americans. In June of 1938, the board of education minutes show that the board received a letter from the NAACP regarding "requests which the Negroes of Monongalia County were asking concerning the school program as it affects their race." Early in 1938, the county started applying to the WPA for the funds to build new schools throughout the county, and in the summer of 1938, Monongalia County passed a bond issue to match the funds provided by the WPA for twenty-one building projects. As a result, the board made plans for three new black schools in Second Ward, Osage, and Everettville.[47]

The problem for the new school at Everettville was finding a satisfactory location. In order to obtain the land for a new white school at Everettville, the board of education deeded the property for the old black school back to the coal company. When the board of education contacted the company about obtaining a piece of property for the new black school, the company offered them a site behind a hill, away from the road, called "New Hill." Dewey Fox objected strenuously.[48]

African Americans in the community became upset with Fox's refusal to take the land. They feared that his stubbornness would prevent them from getting a new school. In a few weeks, local members of the Catholic Church showed up at Fox's house. They owned a tract of land that they were going to use to build a church. Ultimately, they had decided not to build a church on the property, so they offered Fox three acres for his school. Fox and the board of education were both pleased with the property. It was right along the road, directly across from the white school, and it had plenty of room for a playground.[49]

The new school at Everettville was built for a cost of $18,730.75. Constructed of red brick, it was a two-room school with indoor toilet facilities, a kitchen, a central hall, and folding doors between the rooms that allowed the rooms to be combined into an auditorium. Not long after the school was built, the white principal across the road asked Fox if his students could use the African American playground. Fox hesitated, and then said, "What about that segregation law?" The two principals eventually made an arrangement to share the playground—albeit at separate times.[50]

Conclusion

Dewey Fox had a major impact on the educational system for African Americans in Monongalia County. John Fox's strong desire to provide an education for his son led to Dewey Fox's dedication to provide quality education and educational facilities for African-American students in the county. Yet, despite Fox's efforts to ensure equal education and provide opportunities for black students, Fox insisted that there was "never any such thing as equality in education . . . not then and not now."[51]

Fox quit teaching school and left Monongalia County for Marion County, West Virginia, in 1948. He was fifty years old. Fox worked at several different jobs, but he never taught school again. In the early 1970s, he became the first black justice of the peace in Marion County and served for two terms. He continued to be active in the state and Marion County chapters of the NAACP. The Marion County KKK continued to cross his path occasionally. Independent in spirit, Fox remained active in political and community affairs, devoting his time to his church, local youths, and the schools. Winning numerous awards for community service, his friends and neighbors often fondly referred to Professor Fox as "Little Man." A pioneer black teacher, known for reciting classical poetry and his love of Mozart, Handel, and Beethoven, Fox died on February 23, 2003, at the age of 104; it was a life far longer than his father thought possible.[52] Born at the dawn of the Jim Crow era, Fox lived to see its demise. His determination, his dreams, his work, and his struggles, like thousands of black teachers in the Jim Crow south, offered hope and opportunity in a world segregated solely on the basis of color.[53]

NOTES

1   Jefferson County Historical Society, *Between the Shenandoah and the Potomac: Historic Homes of Jefferson County, West Virginia* (Winchester, VA: Winchester Printers, Inc., 1990), 103.

2   Millard Kessler Bushong, *A History of Jefferson County West Virginia* (Charles Town, WV: Jefferson Publishing Co., 1941), 97; and Dewey Fox, interview with author, Fairmont, West Virginia, September 20, 1995.

3   Mary R. Furbee, "A Living Legacy: 98-year-old Dewey Fox," *Fairmont Times West Virginian*, February 25, 1996; U.S. Bureau of the Census, Jefferson County, WV, Manuscript Census of Slaves, 1860.

4   U.S. Bureau of the Census, Jefferson County, WV, Manuscript Census, 1870; Furbee, "A Living Legacy."

5   Bushong, *History of Jefferson County*, 157, 198; Fox interview.

6   Fox interview.

7   Bushong, *History of Jefferson County*, 202.

8   Fox interview.

9   Fox interview; U.S. Bureau of the Census, Jefferson County, WV, Manuscript Census, 1870.

10   Fox interview.

11   U.S. Bureau of the Census, Jefferson County, WV, Manuscript Census.

12   "Deed Book Q," County Clerk's Office (Jefferson County Courthouse, Charles Town, WV), 253. (hereafter JCCCO).

13   Fox interview.

14   Bushong, *History of Jefferson County*, 251, 258.

15   Connie Park Rice, *Our Monongalia: A History of African Americans in Monongalia County, West Virginia* (Terra Alta: Headline Books, Inc., 1999), 194.

16   Fox interview; "Will Book C," JCCCO, 479.

17   Fox interview.

18   Earl L. Core, *The Monongalia Story: A Bicentennial History Vol. IV: Industrialization* (Parsons, WV: McClain Printing Co., 1982), 368, 456, 486, 490; "Dewey Fox dead at 104," *Fairmont Times West Virginian*, February 26, 2003.

19   Lynn Hastings, "Monongalia County Schools and Local History," Book 1, A&M 1375, West Virginia and Regional History Collection, West Virginia University, Morgantown, West Virginia, 71. (hereafter cited as WVRHC).

20   Fox interview.

21   Joseph S. Price, *The Negro Elementary School Teacher in West Virginia* (Institute, WV: West Virginia Collegiate Institute, 1924), 32–39.

22   Fox interview.

23   Fox interview; Core, *Monongalia Story*, vol. 4, 458.

24   Fox interview.

25   Fox interview.

26   State of West Virginia, Bureau of Negro Welfare and Statistics, *Report 1924–1925*, 36.

27   Fox interview.

28   Fox interview.

29   Earl L. Core, *Monongalia Story: A Bicentennial History Vol. V: Sophistication* (Parsons, WV: McClain Printing Co., 1984), 38, 41–42, 52.

30   Fox interview; Monongalia County Federal Relief Records, Progress Report, A&M 890, Box 3, F11, WVRHC.

31   Fox interview; Dewey Fox to Charles H. Houston, August 23, 1935, Papers of the NAACP, Part 3: The Campaign for Educational Equality, United States Library of Congress; Papers of the NAACP, Part 3, Series A: Legal Department and Central Office Records, 1913–1940, United States Library of Congress.

32   Fox interview; Core, *Monongalia Story*, vol. 5, 74; and Monongalia County Schools–Board of Education Minutes, Book 9, WVRHC, 71.

33   Fox interview; Dewey Fox to Walter White, August 16, 1935, Papers of the NAACP.

34   Dewey Fox to Charles H. Houston, August 23, 1935, Papers of the NAACP.

35   Dewey Fox to Walter White, August 16, 1935, Papers of the NAACP.

36   Dewey Fox to Walter White, August 16, 1935, Papers of the NAACP; Dewey Fox interview.

37   Fox to White, August 16, 1935, Papers of the NAACP; Fox interview.

38   *Dominion News*, October 3, 1935; Fox interview; *Dominion Post*, October 13, 1938.

39   Dewey Fox to Charles H. Houston, August 23, 1935; Dewey Fox to the NAACP headquarters, May 20, 1938, both in Papers of the NAACP, Branch Files (Morgantown), Part I Series G, Box 218, Library of Congress.

40   Stephen Edward Haid, "Arthurdale: An Experiment in Community Planning, 1933–1947" (PhD diss., West Virginia University, 1975), 74, 76, 79–80; Monongalia County Federal Relief Records, Progress Report, A&M 890, Box 3, F11, WVRHC.

41   Fox interview.

42   Fox interview; and Haid, "Arthurdale," 81. The officials were most likely Otis and Reid.

43   Dewey Fox to the NAACP headquarters, May 20, 1938.

44   Dewey Fox to the NAACP headquarters, 20 May 1938; Charles D. Lowery and John F. Marszalek, *Encyclopedia of African-American Civil Rights from Emancipation to the Present* (Westport, CT: Greenwood Press, 1992), 144.

45   Fox interview.

46   Eleanor Roosevelt to Walter White, May 25, 1938 and telegraph from Walter White to Dewey Fox, May 27, 1938, both in Papers of the NAACP, Morgantown Branch Files, Part I Series G, Box 218, Library of Congress; Fox interview.

47   Monongalia County School Board Minutes, 1 June 1938, 86, in WVRHC; Hastings, "Monongalia County Schools and Local History," Book 10, 136–137.

48   Fox interview.

49   Fox interview; and Hastings, "Monongalia County Schools and Local History," Book 10, 136–137.

50   Fox interview; Hastings, "Monongalia County Schools and Local History," Book 10, 136-137.

51   Fox interview.

52   "Dewey Fox dead at 104," *Fairmont Times West Virginian*.

53   Fox interview.

Section II

Class

"Sadly in Need of Organization":

Labor Relations in the Fairmont Field, 1890 to 1918

Michael E. Workman

I must still further pause, however, to note the fact that in a few short months we shall feel the throbbings of one of those great arteries of internal communication which have done so much to develop the resources of all countries where they exist. The long wished for railroad is at our door. We welcome its advent. And yet, to us, there comes with it some regrets. The venerable homogeneity of our society will be broken. Our old time hospitality and our earliest family-like social relations will be marred. —*Waitman Willey, 1885*

THIS WARNING WAS issued by one of the founders of West Virginia in an address at the celebration of the Municipal Centennial of Morgantown. With the Fairmont, Morgantown & Pittsburgh extension of the Baltimore & Ohio Railroad fast approaching, Willey was optimistic that Morgantown would prosper if capital could be attracted and "invested in profitable industries."[1] But, he was wise enough to realize that industrialization would entail a cost.

At the onset of its industrial takeoff, the society of the Upper Monongahela Valley region was indeed homogenous. Both the rural and urban segments of the population were composed largely of white, Anglo-Saxon Protestants. There had been no influx of immigrants since the 1850s, when the Irish arrived to build the first railroad, the Baltimore & Ohio Main Stem. The Irish either moved on or were assimilated, and by 1870 the region had a population that was 95.1 percent native white. During the next two decades, as the region remained in a state of arrested development, there was very little immigration, and the native white population increased slightly to 96.4 percent in 1890.[2] The people of the region spoke a common dialect, had a common heritage, and had similar ideas on law and the purpose of government. There were distinctions in society, but they were based on wealth, property, prestige, and intelligence rather than race, religion, language, or class. Social relations were indeed "family-like" because the major economic unit remained the family farm or business. Despite the growth of towns and some industry, social relations were largely personal and cordial. The "surplus extraction relationship," which divided producers and non-producers into distinct classes in industrialized societies, had not yet developed.[3]

Industrialization would change the "venerable homogeneity" of society. Industry required large numbers of workers who would adopt the regimen of the mines and factories and readily give up some of the value of their labor to the region's captains of industry. For the most part, coal miners rather than factory workers would make that sacrifice, at least until the much-sought industrial diversification occurred during the first two decades of the twentieth century. If the export-based economy were to succeed, miners would have to sell their labor cheap, because low wages were the key to penetrating markets. The region had no "home" market for its coal, and because of its late start in the industrial game, could find a place for its product only by undercutting competition and shipping to distant markets. Thus, labor's sacrifice would be great in order to compensate for the region's initial disadvantages vis a vis other regions—a late start, higher transportation costs, and no home market. Who would bear the cost? What groups would constitute the mining labor force from 1890 to 1920?

Native whites from the region would supply about half of the labor requirements. Although many served as common miners, natives had a near-monopoly on positions of trust inside the mine—machine runners and safety inspectors—and a complete monopoly on supervisory and managerial positions. For many native mining families, occupational and social mobility and the American dream of home ownership were distinct possibilities.

Immigrants from Italy and Eastern Europe supplied the second half. They differed from the natives in language, religion, and cultural values. Along with African Americans, whose numbers were very small until after World War I, these newcomers loaded coal and performed other common labor in the mines. Because of their stark differences, they had far fewer opportunities to move up the occupational ladder. Until the 1920s, they remained largely an alien population. Only a small minority was able to buy property or homes. The vast majority became tenants of the coal companies. Huddled in ethnic conclaves in the mining towns, the immigrants remained unassimilated, a proletariat in, but not of, society.

This essay focuses on labor relations in the coal industry in the Fairmont Field region, a six-county (Barbour, Harrison, Marion, Monongalia, Preston, and Taylor) area that was the industrial core of the larger ten-county Upper Monongahela Valley (add Randolph, Upshur, Lewis, and Tucker), during industrialization. A solid body of scholarship exists on this topic for the Central Appalachian region, especially southern West Virginia, but other than the pioneering work of Glenn Massay and Charles McCormick's provocative article, little has been written about the Fairmont Field.[4] Conditions were different in this part of Appalachia. In a 1993 interview for the *Appalachian Journal*, writer Meredith Sue Willis explained,

> On second thought, after looking at [Jack] Weller's book [*Yesterday's People*], maybe I'm not really a Southern Appalachian. In Shinnston [Harrison County], West Virginia, a third of the people were first-generation Italians who came to work in the mines, and that's a whole different world than Weller is talking about. From the very beginning, I was well aware of this group of people had a different background than my people. And there were people up in the hollers who were different still. There was the middle-class, small-town Americans, and the people who lived up there where you could hardly get there, and then there were the Italians and some other various groups. I was aware of a cultural clash . . . my part of West Virginia is not the *Night Comes to the Cumberlands* part of West Virginia. It was an industrial area. It was mixed culturally.[5]

As Willis suggests, cultural and ethnic divisions, far more than class or race distinctions, were at the root of a "cultural clash" during this period of rapid industrialization. The region's elite, its great captains of industry, and "middle-class, small-town Americans" were native whites, to be sure, but the coal-mining labor force was segmented in a unique manner. Although other Appalachian coalfields had a segmented labor force that was divided into native white, African-American, and immigrant elements, the ethnic mix of the Fairmont Field differed. Compared to West Virginia as a whole, a larger number of eastern and southern European immigrants lived and worked in the Fairmont Field.[6] While native miners, with opportunities for occupational and social mobility, tended to accept regional boosterism and were conservative to the point of resisting unionization, immigrant miners, once aroused from their quiescence, became the main source of labor militancy in the 1910s and during the great struggles of the 1920s. This is in marked contrast to southern West Virginia, where, for the most part, native miners became a radical element and provided leadership in the labor movement.

Yet another important distinction of the Fairmont Field was the ownership of the region's mines by indigenous figures, principally the Watson-Fleming interest or the so-called "Fairmont Ring" that controlled Consolidation Coal Company. The Fairmont Ring was nothing more than an extension of the Watson family coal dynasty. Patriarch James Otis Watson (1815–1902), in partnership with Francis Pierpont, had shipped the first coal from Fairmont on the just-completed Baltimore & Ohio Main Stem in 1852. After parting with Pierpont in 1874, he brought his three sons, Sylvanus Lamb (b. 1848), James Edwin (1858–1926), and Clarence Weyland (1864–1940), as well as his son-in-law, A. B. Fleming, into the business. Under the leadership of Clarence, the Watsons consolidated the region's most productive mines into the Fairmont Coal Company in 1901, and then merged this company with the Consolidation Coal Company of Maryland in 1903. With Clarence Watson (C. W.) at the helm, "Consol" developed a vast Appalachian coal empire that stretched from Eastern Kentucky to Central Pennsylvania. By 1928, it was the largest independent coal company in the United States.[7]

The Fairmont Ring forged a non-union regime as formidable as those that dominated southern West Virginia during this same period. While the extent of outside control in southern West Virginia has been exaggerated, it is nonetheless true that absentee coal operators played leadership roles in developing the industry and thwarting unionization. By their residence within the region and their hold on key political offices, the Fairmont Ring was as fearsome as Don Chafin's anti-union regime in Logan County.[8] They controlled not just the mines and mining towns, but also the courts, politics, and even the grounds of public discourse. They unleashed the "stick" of raw power to discipline rebellious miners and thwart periodic invasions of union organizers, but the "carrot" of high wages, steady work, and well-kept mining towns often made such heavy-handed actions unnecessary.

Fairmont Field: The "Capitalist Property" of C. W. Watson

Labor relations in the Fairmont Field have been characterized by long periods of relative calm followed by bouts of fierce struggle. The Fairmont Field's productivity and its geographic position—at the boundary between the largely unionized North and the nonunion South—made it an important theater in the conflict between capital and labor in the 1890 to 1933 period. In the early years, from 1852—when the first railroad shipments were made on the Baltimore & Ohio Railroad—to 1890, labor relations were largely free of conflict, judging by existing documentation. Strikes were rare, and they were usually short-lived and confined to a single mine or company. Cooperation among the mine workers of the Fairmont region was retarded by the fact that most miners lived on farms and viewed their jobs as seasonal.[9] In addition, the absence of a well-organized national miners' union limited the possibilities for miners to unite.

With the organization of the first large mining companies in the late 1880s and early 1890s, the demand for labor rose. The traditional source of labor no longer supplied enough workers, so they were imported, chiefly from Italy and the nations of the Austro-Hungarian Empire. These new workers did not have farms to help support themselves and their families or to return to when an economic downturn came. They were dependent upon the coal companies for their income and the company town for housing, food, and nearly every other aspect of life.

The arrival of the first immigrants coincided with the organization of the United Mine Workers of America in 1890. The UMWA was a merger of two earlier miners' unions, the National Progressive Union and the National Trade Assembly #135 of the Knights of Labor. Organized at Columbus, Ohio, the new union aimed to "unite mine employees and ameliorate their condition by means of conciliation, arbitration of strikes." The main goals of the miners' union were to secure higher wages, eliminate payment with scrip, make the mines safer, and institute the eight-hour day.[10]

After the UMWA was organized in 1890, union organizers made regular visits to the Fairmont field to gain recruits and recognition for the union. Significant strike efforts were mounted in 1892, 1894, 1897, and 1902. Only the 1897 and the 1902 efforts made much of an impact, and all four were failures. The coal operators used detectives and injunctions to defeat the UMWA. As Glenn Massay explained, regionalism was a much stronger force than "labor solidarity" in the Fairmont Field:

> The operators of the Fairmont field prospered when other fields were strike-bound, and they could afford to entice their men back to work by simply raising wages. Would-be organizers found themselves without followers and then in jail. Perhaps it is significant that the unionizing drive originated in the national labor movement, not, apparently, among the local men . . . There was too much obvious truth in the argument that union organizers were "emissaries of rival regions," and it was too clear that the best gains for Fairmont men came when other fields were on strike. Local miners probably sympathized with the national labor movement, but they behaved in a spirit of regionalism and economic individualism; they took work and wages when they could get them.[11]

There would be no strikes or labor difficulties at the mines of the Fairmont Coal Company (part of the Consolidation Coal Company after 1903) until the 1920s. Even the great Paint Creek-Cabin Creek strike of 1912–1913 failed to kindle a spark among the company's miners. In fact, the miners of the Fairmont Field continued to act as *de facto* strikebreakers during the strikes of the 1900s and 1910s; they not only resisted the call for organization, but also produced extra tonnage during the national strikes to supply markets that were otherwise served by union mines. The miners seemed content with their lot and willing to work without a union. The miners' quiescence can be explained, in part, by the welfare capitalism of the company and the acceptance of the spirit of regionalism. Yet, the political power of the company was just as important. Not only did the company employ a secret service, which, like Sheriff Don Chafin's "thugs" in Logan County, kept the UMWA out by intimidation, but it also had the support of the business community of the region, as well as key politicians and judges. An anecdote reported by *Fairmont Times* columnist C. E. "Ned" Smith in 1934 reveals the power that the company possessed even over speech. Smith wrote,

> The Elks were giving a burlesque on Uncle Tom's Cabin. Sam Leeper, we think, was Simon Legree and Ed Mayers Uncle Tom. In the original script Simon Legree says: "An't I your master? Didn't I pay twelve hundred dollars, cash, for all there is inside your cussed old black shell? An't you mine, body and soul?"
>
> Tom replies: "No, No! My soul an't yours, mas'r; you haven't bought it—ye can't buy it; it's been bought and paid for by one that is able to keep it, and you can't harm it."
>
> In the Elk's version, which caused the trouble, Uncle Tom's reply was in these words: "Yes, Massa, you might own dis ole body of mine, but my soul belong to de Fairmont Coal Company!
>
> After the first performance and before the boys had removed their blackface and make-up, a coal company bull brought in the orders to change that speech or else. The speech was changed. The next night it was God and not the Fairmont Coal Company who owned Uncle Tom's soul.[12]

The UMWA mounted yet another failed effort in the Fairmont Field in 1907. After some limited success in Tunnelton, Preston County, the effort was squelched by an injunction. The *United Mine Workers Journal* blamed the "great combinations of capital which control the affairs of the state" for holding back the union.[13] The Fairmont Field remained the "capitalist property" of C. W. Watson, who was elected to the U.S. Senate in January 1911.[14]

Labor Turns to the Left

The UMWA's impotence, along with the failures of its political allies in state electoral politics, led those who were sympathetic to labor and reform in the state to turn to the left. The *Labor Argus*, a weekly first published in 1906 from Charleston, was taken over by Socialists in 1909. With the rise of Frank

Keeney and Fred Mooney, District 17 came under Socialist influence. In the 1910 elections, Socialists in Kanawha County elected three of their members to county offices.[15] In Marion County, a Socialist candidate for the legislature nearly beat a Republican in the same election.[16] By 1911 Clarksburg had a strong Socialist party, with a socialist newspaper, the *Clarksburg Socialist*. The town's party included a Finnish Branch composed of forty-five workers from the tinplate mill, and it hosted a "great mass convention" of state Socialists in December.[17] In Fairmont, Socialists fielded a whole slate of candidates for election to county offices in 1911. On the eve of the election, Dr. J. C. Broomfield, minister of the Methodist-Protestant Church delivered an address: "Is Socialism as a Protest Justified?" A large crowd heard his answer, "protest spells progress," but none of the Socialist candidates prevailed.[18]

The Socialist Party of the state achieved its greatest success between 1912 and 1915. The Paint Creek-Cabin Creek strike of 1912–1913 radicalized miners in the Kanawha Field. In the 1912 presidential election, the Socialist candidate Eugene Debs garnered 5.7 percent of the vote in the state. The center of Socialist strength in the election in the Fairmont Field was Preston County, which gave 30 percent of its vote to Debs; Marion and Harrison counties recorded 9 percent for Debs.[19] Socialist strength in Harrison and Marion counties can be attributed to the large number of skilled workers—particularly glassworkers—living there. Preston County's strong support for Debs is more difficult to explain. The large number of railroad workers in the county may account for some of the vote, but not all. Perhaps the UMWA organizational drive and the long strike at Tunnelton radicalized some of the miners living there. Socialists also gained victories in several of the region's towns. In Morgantown, Socialists—with support from glassworkers in the Second, Fifth, and Fourth wards—elected a councilman in April 1912.[20] In the same year, Socialists won the entire municipal ticket at Adamston, an industrial town near Clarksburg with a concrete block factory and large glass industry.[21] In the municipal elections in Fairmont in 1913, a Socialist, along with several Progressives, landed seats in the City Council.[22] Eugene Debs spoke in Clarksburg in 1914,[23] drawing a large crowd of "miners, mill-workers, laborers, businessmen, lawyers, doctors, and preachers who received the message with great enthusiasm."[24] In 1915, the Socialists of Star City, a small town near Morgantown with a large glassworker population, won control of the town.[25]

Conservatives in the region were quick to react. Clarksburg and Fairmont passed laws that forbade public addresses without the special permission of the mayor or police chief. A Socialist, Fred Strickland, was arrested in Clarksburg when he spoke without gaining the mayor's permission.[26] The Consolidation Coal Company continued to use its "secret service" to bully both UMWA organizers and Socialists. In 1913, special agents Bob and Frank Shuttlesworth shot UMWA organizer John Brown in the stomach. Later, when J. Verve Johnson, a Harrison County Socialist, and J. H. Snider, a Marion County Socialist leader, were beaten, the Socialist press blamed the coal company's "thugs."[27]

Growing Tensions: The "lawless element of the foreign population"

The repressive tactics of the coal operators and their political allies thwarted the organizational efforts of the UMWA and the Socialist party alike. In the next few years, the main threat to their regime came not from outsiders, but from the largely unassimilated ethnic proletariat living within their own mining towns. As ethnic tensions rose, the great capitalists and their middle-class, WASP allies came to associate socialism and radicalism of any stripe with ethnic rebellion.

Although most of the immigrants lived in the mining towns, the trolley system permitted them to shop and attend social functions in Clarksburg and Fairmont. On payday and on Saturday, many of the miners and their families would take the interurban line to Fairmont, Clarksburg, or one of the smaller towns such as Farmington, and spend the day shopping. At Fairmont, they might also go to the races or the circus, spend time at "Traction Park," or gather in front of the courthouse to socialize.[28] Paid in cash by most coal companies after 1900, miners contributed to the prosperity of the incorporated towns.[29]

The immigrants' regular visits to the towns brought them in contact with the native middle class, which did not always approve of their behavior. The attitudes of natives toward foreigners were often less than hospitable. Newspapers frequently demonstrated a nationalist or local bias. Whenever an accident occurred in a local mine, the name of the victim was generally given—except if it was a foreigner. Thus, "a Pole" was injured or "a Slav" killed. The *Fairmont Index* reported on June 9, 1893, "A Polander, whose name we did not learn, got his foot badly mashed last Wednesday in No. 1 by a fall of coal." On June 5, 1914 the *Clarksburg Daily Telegram* related that a foreign miner with an "unpronounceable name" was admitted to the Miners' Hospital with a broken leg.

Some efforts were undertaken toward Americanization, either from the bottom-up by the immigrants themselves or by the coal companies or other civic organizations. Italians were conspicuous in Americanization efforts. Columbus Day was an opportunity for Italian immigrants to express their love for their native land as well as their allegiance to their "American homes." In October 1911, the city of Fairmont gave the "compatriots of Columbus" the "freedom of the city." Under the auspices of the *Societa Cristoforo Colombo* of Fairmont, which promoted citizenship and "higher ideals" among Italians, a parade of 1,000 "sturdy sons of Italy" was held. The *Societa Vittorico Emmanuele II* of Monongah, as well as the *Societa Unione e Fratettanza* of Morgantown, sent delegations that marched in the parade.[30]

However, the newspapers related far more episodes of immigrant disorder and crime than celebrations of solidarity. Incidents of immigrant-on-immigrant crime were reported with some ire, such as the case of an Italian inflicting "fourteen hatchet wounds" on a compatriot at a fight in Farmington in 1911,[31] or with scorn, as in 1914 when six foreigners at the Grant Town mine with "names that sound like a sausage mill," not liking the "efforts of their culinary artist, roasted the cook on the stove."[32] However, the greatest indignation was reserved for cases of immigrant-on-native crime. In July 1911, an Italian, who was "almost a stranger," stabbed a mine foreman at Monongah. Bloodhounds were brought in, and the culprit was captured.[33] According to the *Fairmont Times*, the severe stabbing "created more than the usual condemnation among the Americans as well as the law-abiding foreigners of the region." The newspaper also took the opportunity to apprise its readers of the need for stricter law enforcement:

> For some time in the region as well as in Fairmont and Clarksburg, it has been generally remarked that a certain lawless element of the foreign population was being allowed too much latitude, and that trouble would result unless they were checked. It is now hoped that . . . those foreigners who believe they

can do as they please in this country will be taught their proper place in no uncertain manner by the authorities . . .

While conditions are said to be worse in the small mining towns, it often occurs in Fairmont that foreigners become unruly, and rude upon the streets, much to the discomfort and disgust of the pedestrians. It has been noticed of late that gangs would take up the entire sidewalks, compelling ladies and others to pass around them.[34]

In less than two weeks, this incident was followed by a more gruesome case of immigrant-on-native crime. On July 21, a Monongah police official "sent a bullet into his own brain" after receiving two mortal wounds at the hands of an Italian. The "murder" was believed to have been planned and executed by the Black Hand.[35] Other incidents involving the "lawless element of the foreign population" were related by the newspapers with alarming frequency between 1911 and 1915.[36]

The whole matter of native-foreigner relations came to a head in late-1914 and 1915 as a result of three concurrent developments. In July 1914, the Yost Law, which provided for the enforcement of prohibition, went into effect. Very few immigrants backed the dry law, which denied them one of the essential elements of their Old World culture. Many continued to imbibe, getting their spirits from native moonshiners or from adjacent Pennsylvania.[37] In August, the European war started. Some immigrants—probably less than 500—returned to their homelands to fight.[38] The sympathies of the region's natives were with the Allies, and this gave them yet another reason to distrust the immigrants from Italy and the Austro-Hungarian Empire. The final development was a recession that struck in the last months of 1914, which made work scarce and added to the growing tensions.

During the winter of 1914–1915 conflict between natives and the "lawless element of the foreign population" reached a peak in Marion County with a miners' strike at Farmington and the Kilarm Black Hand trials. The Farmington strike has been ably described by Charles H. McCormick, so there is no need to provide more than an outline here. It began on February 15, 1915, at the three mines of Jamison Coal and Coke Company near Farmington, along the B&O main stem west of Fairmont. A dispute over explosives triggered the strike. The coal company, which had been supplying free powder to the miners, was unhappy about the excessive breaking of the coal into small sizes, so it ordered the miners to pay for their own in order to limit use. The company granted a pay raise of four cents per car to offset the cost. As a result,

> ... three to six hundred Farmington strikers took to the county roads. Side-by-side from mine to mine to the accompaniment of drum and bugle marched Italians, Poles, Russians and a large contingent of Croatians or Serbo-Croations. Brandishing stout hickory clubs or pick handles studded with lethal spikes or iron bolts, and perhaps carrying concealed pistols and daggers, they followed an American flag mounted above a red flag and a banner inscribed "United We Stand; Divided We Fall, Give Us Justice; or Nothing at All."[39]

Four days after the walkout began, reports of violence were brought to the attention of Marion County Sheriff, C. D. Conaway, who led a force of a dozen armed officers to the scene with "John Doe" warrants to arrest the culprits. After arrests were made, the posse found itself surrounded by a force of angry foreign miners, who swarmed over them, clubbing them as they fired their guns wildly. All of the officers were badly beaten, and Constable William Riggs later died. According to McCormick, Riggs was targeted for the most severe beating because he was well known to the immigrants as an enforcer of the Yost Law. After Riggs died on February 24, a veritable army of 500 "special officers" rounded up the supposed culprits. A total of 134 were indicted for murder or assault. McCormick noted that "every name on the list appeared to be Slavic or Italian." In mid-April the rioters were tried at the Marion County Courthouse for murder and lesser offenses. By May 18, convictions, confessions, and plea bargains had sent almost fifty immigrants to the state penitentiary for terms of ten months to life on a range of offenses.[40]

The Farmington riot was a spontaneous outburst rather than a well-planned rebellion or part of a union campaign. The UMWA claimed no role whatsoever in the strike. As the *Fairmont Times* pointed out, Riggs, as well as another injured deputy, were union members.[41] The episode was clearly a case of ethnic conflict, as McCormick demonstrated. It provided some hard lessons for both sides. The natives were frightened, and, as a result, became even more determined to enforce the laws. And, as McCormick wrote,

The legal proceedings stemming from the Farmington riot taught the Italian and Croatian communities hard lessons about American law and who had life and death power over them. At the same time, rigorous enforcement of prohibition laws against them demonstrated another way by which the native, middle-class community intended to control them and lay down the rules for becoming American.[42]

A second manifestation of the ethnic conflict of 1915 was the arrest and trial of the Kilarm Black Hand gang for murder and the violation of the Red Man act, a state law that made it a misdemeanor to belong to any organization that conspired to commit personal injury or robbery. Although the Black Hand preyed mostly on Italians, it terrified natives nearly as much as the "red flag" of the Farmington strikers. Both represented the threat of anarchy to the law and order-minded native middle class. The Black Hand gained a presence in Marion and Harrison counties as the result of a large influx of Calabrian and Sicilian immigrants, a few of whom brought with them the old world crime syndicate, the Mafia. The natives called the organization the Black Hand because that was its characteristic signature, but members called it the *Famalie Vagabonda*. There were Black Hand organizations in Fairmont and Clarksburg and in some of the mining towns, such as Kilarm, Monongah, and Enterprise. The group engaged in extortion, operated brothels, and sold alcohol and narcotics. John McKinney, who was a police detective for the city of Fairmont and active in the Black Hand investigations of the 1920s, explained that the Fairmont organization operated a "resort" in that town. Its headquarters was in the rear of a barbershop on Water Street along the Monongahela River. Both the Fairmont and Clarksburg organizations had ties to a Baltimore group.[43] But as the occasional turf wars that broke out between the local groups indicate, the Black Hand was hardly the monolithic organization that many natives imagined.

The Black Hand made its first appearance in Marion County in 1908. Its first headquarters was in Kilarm Hollow, an "out-of-the-way" mining town on a spur of the old Monongahela River Railroad on the Harrison-Marion county border. Kilarm, according to the *Fairmont Times*, held the "title for badness" in Marion County. Bootlegging was done in the open; there was gambling and "Sunday violations."[44] The Black Hand organization there engaged in extortion, dynamiting homes, and beating and torturing those who dared to resist. It was temporarily broken up in 1908 by a sting operation undertaken by the Marion County prosecuting attorney, but it was subsequently revived.[45] The episode in 1915 involved a murder committed by a trio of Black Handers from Kilarm and nearby Enterprise, another mining town. A total of eleven were arrested on the murder charge or for violation of the Red Man law. The acknowledged head of the Kilarm organization was *domo* Sam Palma, described by the *Fairmont Times* as a "handsome, dark-haired Italian."[46]

The trial began in March 1915, shortly after that of the Farmington strikers. The *Times* noted that the sessions were well attended. The audience included "a bevy of [Fairmont] Normal school girls," who studied the case. Palma was tried first on the charge of second-degree murder. His testimony indicated that the Black Hand was headquartered in a boarding house in Kilarm, and that the murder came about as a result of an ongoing battle between it and another Black Hand boarding house in Enterprise.[47]

The most damaging testimony was given by John Torcha, one of the defendants who turned state's evidence. Torcha was a true vagabond, having changed his address twelve times since coming to the region. He belonged to two orders of the *Famalie Vagabonda*, one in Preston County and a second in Harrison County. Torcha described the manner in which he was initiated into the organization. He came to a meeting at the boarding house in Kilarm, where the oath of secrecy was administered. The members of the society formed a circle around a "colored hanky on the floor covered with stilettos." Torcha took the oath as he leaned over and placed the palm of his right hand on the blade of the dagger. The trial ended with five convictions for second-degree murder, including that of Palma, who was sentenced to life imprisonment, and a conviction for violation of the Red Man act.[48]

Neither the convictions of the Fairmont strikers or the Kilarm Black Hand gang ended the menace of the "lawless element of the foreign population," however. A minority of foreigners continued to toy with various shades of radicalism until 1919, and the Black Hand maintained its grip on the region until the 1920s. The impact of this ethnic rebellion was to place a great fear in the hearts of the old stock middle class, a sentiment that led to the great blossoming of the Ku Klux Klan in the early 1920s. Even the coal barons, the most powerful element in society, were not exempt from the terror. In 1916 a plot was discovered to blow up the homes of prominent Fairmonters, including the Watsons. The plot was frustrated by a boy who overheard the conspirators. The sheriff confiscated dynamite and blasting caps, and armed men were posted to guard the homes of the great capitalists. The trio of conspirators escaped, however, by spreading cayenne pepper to throw the hounds off their scent, finding refuge in the old O'Donnell mine in Fairmont, the traditional "haunt for murderers."[49]

The UMWA Returns:
Miners Enter the "Kindergarten of the Labor Movement"

The ethnic uprising also put a new light on the organizational efforts of a more conservative UMWA. According to Sam Montgomery, who continued to play a leadership role in the affairs of District 17, the Farmington strike led to a change in sentiment in the Fairmont region in favor of the union. At a November 22, 1915, meeting of District 17 field workers at Charleston, Montgomery told the organizers that the time was ripe to "invade" the Fairmont Field. He declared that "lawyers, doctors, preachers, professional and business men in general" were in sympathy with the union movement. The change in sentiment had come both because of the "hoggishness" of the Consolidation Coal Company and because, after the Farmington strike, the "people did not want a repetition of the Colorado situation." International Executive Board member Thomas Haggerity, who chaired the meeting, told the group that he and President Cairns had recently returned from Flemington, in Taylor County, where they had held an enthusiastic meeting with a large crowd "headed by a brass band." They conferred with the superintendent of the mine, who agreed to sign a contract with the UMWA for eight hundred men. When they returned later to finalize the agreement, they found that the detectives of the Consolidation Coal Company had intimidated the superintendent from doing any further business with the union. Haggerity agreed with Montgomery that the time was right to organize the Fairmont Field. It was decided to pursue the organization effort and appeal to the International Executive Board for funds.[50]

The 1916 UMWA campaign bore fruit because the time for organization was opportune. Prosperity had returned to the Fairmont Field, and there was a labor shortage. In May, the union sent organizers to Taylor and Preston counties, an area where they had attained some success in 1907–1909. After holding a meeting with three hundred men at Grafton, the organizers succeeded in getting an agreement with the Maryland Coal Company, a medium-size company that had operations at Wendell in Taylor County.[51] One possible reason for the success was the liberal views of the president of the company, J. W. Galloway, who later became the director of the Bituminous Coal Distribution Division of the U.S. Fuel Administration. Consolidation Coal Company and many of the independents announced a wage increase to allay miners and defuse the union campaign.

The entry of the United States into the war in April 1917 changed everything, however. As George T. Watson, who headed the West Virginia operations of Consolidation Coal Company said, "the get together spirit" was in the air. Labor and capital were working together in the "common cause of winning the war."[52] The ethnic conflict of previous years seemed to fade away as miners demonstrated their patriotism by purchasing Liberty bonds, and joining in patriotic rallies.[53] Middle-class values of work and dedication to God and country became paramount. The Americanization efforts of the coal companies and civic organization were stepped-up. Work, the quintessential middle-class virtue, was made mandatory for all. In June 1917, the state of West Virginia passed a vagrancy law that required all able-bodied men above 16 to work 36 hours a week. Sam Montgomery, state commissioner of labor and friend of the working man, announced that the anti-vagrant law would be enforced.[54]

The high demand for coal, the scarcity of labor, and the sympathetic attitude of the Wilson administration toward unions, along with the "get together spirit," permitted the UMWA to gain a hold in the Fairmont Field. On May 1, 1917, about eight hundred miners in the so-called "Thin Vein Field" (with operations in the Upper Freeport and Lower Kittanning seams) in Preston and Taylor counties, along with about 1,200 in the Flemington Field between Grafton and Clarksburg (both "fields" were actually sub-fields of the greater Fairmont Field), went on strike. Seventeen coal companies were affected. The miners demanded union recognition, increased pay, the eight-hour day, and payment by the short ton.[55] The Central West Virginia Coal Operators Association, which represented the coal companies in the Fairmont

Field, responded to the strike by granting all of the miners in the field a wage increase and the eight-hour day.[56] The miners demanded more and continued their strike. Governor John J. Cornwell telegrammed Secretary of Labor W. B. Wilson, to ask for a representative to mediate. The two Department of Labor representatives found that the miners were firm in their insistence upon union recognition. District 17 of the UMWA, which claimed no responsibility for the strike, sent organizers into the field. An agreement was reached on May 5 between the miners and the operators in the Flemington Field, but the operators in the Thin Vein Field in Preston County, claiming economic hardship, refused to grant the miners' demands. Finally, after the intercession of Sam Montgomery and District 17 President Frank Keeney, the coal operators signed a contract providing for recognition of the UMWA, the eight-hour day, payment by the short ton, and higher wages. A total of eight companies in the Flemington Field and seven in the Thin Vein Field signed union contracts.[57]

The coal companies of the more central part of the Fairmont Field, Monongalia, Marion, and Harrison counties, dominated by Consolidation Coal Company, remained nonunion. This situation changed in 1918, however, after C. W. Watson made his historic change of heart. It all started in March, when C. W. applied for a commission in the U.S. Army Ordnance Reserve Officers Corps. Watson, widely known as the "leading businessman and capitalist of West Virginia," was accepted for service as a colonel.[58] As the *Fairmont West Virginian* editorialized, Watson "could not fight for the government with one hand and oppose government plans for efficiency [which included unionization] with the other."[59] On May 10, C. W.'s nephew, George T. Watson, announced that Consolidation Coal Company's West Virginia division had signed a new compact with its men designed to "ensure continuous operation and promote amicable relations." Watson said the company's West Virginia mines would be run on the open-shop: the compact included a provision that permitted employees the right to "membership or non-membership in labor or other organization." He noted that "this is a step between the present conditions and straight unionism, which is likely," and said that the coal company, which had opposed the UMWA for nearly three decades, "removed the barriers which had been placed in the way of organizers in this region."[60] *Coal Age*, a leading coal journal, noted the significance of the about-face:

Consol has always been a model employer, but it has resolutely combated the unionization of its workmen, feeling that a union meant an organization of discontent. They feared discontent would become chronic wherever an institution was formed with the purpose of keeping such discontent alive.... Much water has passed over the wheel since the U.S. entered the world conflict.[61]

C. W. Watson's motive for recognizing the UMWA came under scrutiny later in the month of May when he announced that he would enter the race for the U.S. Senate. It was later charged that the agreement with the UMWA was the result of a deal, trading Watson's acceptance of the union for the union's political support.[62] The UMWA did show its appreciation by supporting Watson in the November election. Frank J. Hayes wrote a letter to Frank Keeney endorsing Watson because he had "adopted a broad, liberal attitude toward labor."[63] However, the support of the UMWA was not enough. In the eyes of his fellow operators Watson was guilty of a gross betrayal of their common cause, and they joined together to help defeat him.[64]

Following the announcement of Consol's decision, the Central West Virginia Coal Operators Association met on May 14 to consider adopting the open-shop plan. Some of the operators were "not exactly enthusiastic" about the plan, but it was approved.[65] On May 26, the officials of Consol met with UMWA officials at the Watson Building in Fairmont to hear the union's plans for organization. In this historic meeting, Clarence Weyland, James Edwin, and George T. Watson gave International President Frank Hayes and District 17 President Frank Keeney approval to launch a campaign. According to the *United Mine Workers Journal*, what followed was the "greatest organization drive in the history of West Virginia."[66]

The organization drive was directed not only to coal miners, but to practically every other craft in the region. Fairmont, as well as Clarksburg, had a union craft tradition that dated to the 1890s. Masons, letter carriers, carpenters and joiners, painters, railroad workers, and flint glass workers had previously formed locals, but many craft workers remained unorganized. On the same day as the historic meeting between the Watsons and the UMWA officials, the Monongahela Valley Trades and Labor Council, American Federation of Labor, was formed to coordinate the various organization efforts. Soon, barbers, street car employees, butchers, carpenters, clerks, machinists, and workers of other crafts were organized into the appropriate AFL unions.[67]

The UMWA campaign got underway in late May. The union sent six or seven international organizers, including "Mother" Jones, to help Frank Keeney and Fred Mooney of District 17 with the effort. The miners were generally receptive to the union pitch. However, as Keeney related, at some mines the miners "did not want to organize, but the superintendent got busy and told the miners they could organize if they wanted." Some miners also balked at the UMWA initiation fee, which was regularly ten dollars, but was reduced to one dollar. Another problem was that the AFL also sent organizers, and there was some conflict with the UMWA men.[68]

Mother Jones, who was eighty-eight years old, played a predominant role in the campaign. The meetings that she led were always attended by a bevy of reporters. She spoke out on a variety of issues at a June 9 miners' meeting at Haywood Junction, near Lumberport in Harrison County. She declared that she would not leave without success, as she had done after the 1902 strike. She poked fun at middle-class women reformers, the Women's Christian Temperance Union, and suffragettes: the "labor movement has done more to humanize and Christianize this country than all the churches, universities, and capitalistic institutions."[69] At a scale convention held in Fairmont on July 25, Mother Jones informed the miners that they had a long way to go:

> You have only entered the kindergarten of the labor movement as yet. You are going to have a full course of school. You will graduate from one class to another. I want you to study hard, to read, to think. The labor movement is the most instructive school in the country.[70]

At scores of meetings in mining towns across the region—Monongah, Watson, Simpson, Pinnickinnick, Haywood Junction, Owings, Enterprise, Bingamon, Middleton, Riverdale, Viropa, Glenn Falls, Serepta, Opekiska, Annabelle, Gypsy, Wilsonburg, Wolf Summit, Worthington, Rivesville, Reynoldsville, Everson, Carolina, Laura Lee, Dakota—the UMWA organizers stood before the polyglot masses of miners and proclaimed the union gospel. The union was performing its historic role of uniting the miners, traditionally divided into ethnic enclaves, into a single working-class organization. Rather than radical ideology, the union leaders proclaimed doctrines that Fairmont's old-stock Protestants would heartily approve. Keeney stressed patriotism: "The war has got to be won at all hazards." His platform was "organize, educate, negotiate, conciliate, and arbitrate." Keeney spoke against strikes, and both he and Mother Jones were lauded by the newspapers for stopping wildcat strikes.[71] The miners were called "industrial soldiers." They were instructed not to be a "slacker, the worse that a man can be called." They were urged to keep away from whiskey and to practice thrift, industry, integrity, and patriotism—the virtues of American Protestantism.[72]

By September, a large part of the 17,000 miners in the Fairmont Field had been brought into the UMWA. The union celebrated the "greatest year in the history of the labor movement in West Virginia" with a grand Labor Day. A "monster parade, miles in length," was formed at Monongah and the miners, nearly 10,000 strong, marched to Traction Park in Fairmont. Everyone was on foot except Mother Jones, who rode in an auto. At the park, the miners were addressed by William Rogers, president of the West Virginia Federation of Labor, and Mother Jones. Following the morning session, lunch was spread under the shade trees. After hearing another round of addresses, the miners spent the afternoon watching a baseball game, listening to concerts by the Greater Fairmont band and the Polish band of Monongah, or dancing.[73] Much organizing work remained to be done. As it turned out, the entire field was not organized until 1922.[74] The region's miners, as Mother Jones pointed out, were still in the "kindergarten of the labor movement." They would receive a "full course of school" in the tumultuous 1920s.

NOTES

1 *Celebration of the Municipal Centennial of Morgantown; Historical Oration by Waitman T. Willey* (Morgantown, WV: New Dominion Steam Print, 1886).

2 Oscar Chapman, *Report of the Committee on the Upper Monongahela Valley*, Table 22: Population, Number, and Percentage Distribution, by Color and Nativity for the United States, West Virginia, and Upper Monongahela Valley, 1930–1870, U.S. Census. The U.S. figure was 73.0 percent; the state's was 93.2 percent.

3   Robert Brenner, "Agrarian Class Structure and Economic Development in Preindustrial Europe," *Past and Present* 70 (1976): 30–75.

4   Glenn Frank Massay, "Coal Consolidation: Profile of the Fairmont field of Northern West Virginia, 1852–1903," (PhD diss., West Virginia University, 1948); Charles H. McCormick, "The Death of Constable Riggs: Ethnic Conflict in Marion County in the World War I Era," *West Virginia History*, 53 (1993): 33–58.

5   Thomas E. Douglass, "Interview of Meredith Sue Willis," *Appalachian Journal* 20 (Spring 1993): 287–293.

6   In 1910 the coal mining labor force of the six-county Fairmont Field was composed of 53.7 percent foreign nationals; the state percentage was 36.2. While Italians represented 23.2 percent of the mining labor force of the Fairmont Field, they were only 8.9 percent in the state as a whole. Eastern European (Austrians, Bohemians, Hungarians, Lithuanians, Litvitch, Polish, Romanian, Russian, Slavish, Servian, Croatian, Magyar, Horwatt, and Bulgarian) miners made up 28 percent of the Fairmont Field's labor force, and 16.3 percent of the state's. The Fairmont Field held a much smaller proportion of black miners, 2.4 percent, than the state, 17.7 percent. Calculated from West Virginia Department of Mines, *Annual Report*, 1910–1911, "Nationalities of Persons Employed at the Mines and Coke Ovens, By Counties," 105–106.

7   Massay, "Coal Consolidation," 235–236.

8   See, for example, Ronald D. Eller, *Miners, Millhands and Mountaineers: Industrialization of the Appalachian South, 1880–1930* (Knoxville: University of Tennessee Press, 1982). Chafin's anti-union machine is described in Lon Savage, *Thunder in the Mountains* (Elliston, VA: Northcross House, 1986).

9   Massay, "Coal Consolidation," 235–236.

10 Maier B. Fox, *United We Stand: The United Mine Workers of America, 1890–1990* (Washington, D.C.: United Mine Workers of America, 1990), 22–23.

11 Massay, "Coal Consolidation," 281.

12 C. E. Smith, "Good Morning!" *Fairmont Times*, February 8, 1934.

13 Quoted in *Labor Argus,* Sept. 19, 1907.

14 *Labor Argus*, June 13, 1911.

15 Frederick Allan Barkey, "The Socialist Party in West Virginia From 1898 to 1920: A Study in Working Class Radicalism," (PhD diss., University of Pittsburgh, 1971), 46.

16 "Marion County makes plutes uneasy," *Labor Argus*, November 10, 1910.

CULTURE, CLASS, AND POLITICS

17   "Clarksburg Finlanders dedicate Socialist Hall," *Labor Argus*, July 20, 1911; "Great Mass convention of Harrison County Socialists," *Labor Argus*, December 7, 1911.

18   "Socialists," *Fairmont Times*, July 15, 1911; "Dr. Broomfield's strong sermon to Socialists heard by large crowd," *Fairmont Times*, November 4, 1911.

19   Barkey, "The Socialist Party in West Virginia," 120.

20   *Morgantown Post-Chronicle*, April 1, 1913; quoted in Barkey, "The Socialist Party in West Virginia," 124.

21   Stephen Cresswell, advisor, *Socialists in a Small Town: The Socialist Victory in Adamston, West Virginia* (Buckhannon, West Virginia: Ralston Press, 1992), 3.

22   "Fairmont Municipal Election," *Fairmont Times*, December 10, 1913.

23   See *West Virginia History* 52 (1993) for a discussion of the visit of Eugene Debs to West Virginia in 1913 and 1914.

24   "Socialist leader draws full house," *Clarksburg Daily Telegraph*, April 3, 1914.

25   "Socialists of Star City celebrate hilariously," *Morgantown (WV) New Dominion*, January 11, 1915. See Stephen Cresswell, "When the Socialists Ran Star City," *West Virginia History* 52 (1993) for a full discussion of the Socialist victory in Star City.

26   Barkey, "The Socialist Party in West Virginia," 170.

27   *Wheeling Majority*, July 3, 1913; quoted in Barkey, "The Socialist Party in West Virginia," 170.

28   "Payday brings out crowds," *Fairmont Times*, August 23, 1909; "Circus will attract miners," *Fairmont Times*, April 24, 1912; "Monster Merry throng of Folks here Saturday," *Fairmont Times*, May 5, 1912.

29   "Payday at Monongah," *Fairmont West Virginian*, August 2, 1916.

30   "Italian allegiance to their American homes shown by celebration," *Fairmont Times*, October 14, 1911; "Sturdy sons of Italy pay tribute," *Fairmont Times*, October 13, 1913; McCormick, "The Death of Constable Riggs," 36.

31   *Fairmont Times*, January 14, 1911.

32   "Foreigners placed their cook on red-hot stove," *Fairmont Times*, November 11, 1914.

33   "Bloodhounds trail Italian," *Fairmont Times*, July 12, 1911.

34   "Feeling is High," *Fairmont Times*, July 14, 1911.

35   "Officer sent bullet into own brain," *Fairmont Times*, July 22, 1911.

36   "Two men butchered on streets of Clarksburg," *Fairmont Times*, August 17, 1911; "Foreigner killed, others wounded in fusillade following card game at Monongah," *Fairmont Times*, March 17, 1912; "Dragnet thrown out for other Black Handers," *Fairmont Times*, June 27, 1913; "Clarksburg has Black Hand Mystery," *Fairmont Times*, July 28, 1913; "Foreigner nearly lynched near Farmington," *Fairmont Times*, November 14, 1913; "Italian merchant dying from attack," *Fairmont Times*, December 18, 1913; "Keg party victim dies of gunshot wound," *Clarksburg Daily Telegram*, March 12, 1914; "Race riot on Washington St as Negroes and Italians mix it up with Irish Confetti," *Clarksburg Daily Telegram*, December 3, 1914.

37   McCormick, "The Death of Constable Riggs," 36; "Demise of John Barleycorn," *Clarksburg Daily Telegram*, July 1, 1914.

38   "Local men going back to fight," *Clarksburg Daily Telegram*, August 4, 1914; "Many Italians go home for Christmas," *Clarksburg Daily Telegram*, December 5, 1914.

39   McCormick, "The Death of Constable Riggs," 38.

40   McCormick, "The Death of Constable Riggs," 40–48.

41   *Fairmont Times*, February 26, 1915. Riggs was a member of the Brotherhood of Railroad Conductors "for twenty years" and Deputy Buckley was a union stogie maker in Fairmont.

42   McCormick, "The Death of Constable Riggs," 48.

43   John C. McKinney, "Black Hand!" *Marion County Centennial Yearbook* 42.

44   *Fairmont Times*, March 27, 1917.

45   *Fairmont Times*, March 27, 1917.

46   "11 accused in Mafia Murder," *Fairmont West Virginian*, March 8, 1915.

47   "Witnesses for Palma," *Fairmont West Virginian*, March 19, 1915.

48   *Fairmont West Virginian*, March 23, 1915. The Red Man law was passed by the state legislature in 1882 in response to the activities of the "Red Men" of Wetzel County. This secret vigilante group engaged in various acts of intimidation in Wetzel and Marion counties during the 1870s and 1880s. See Waitman T. Willey, "Law and Lawyers—Their Relation to a Republican Form of Government," *Constitution and By-Laws of the West Virginia Bar Association* (Morgantown, West Virginia: New Dominion Steam Printing House, 1886), 118, for a brief reference to the "Red Men."

49   *Fairmont Times*, July 17, 1916.

50   "A Meeting of the Field Workers," Charleston, West Virginia, November 22, 1915, UMWA, International Archive, District 17, 1915. (When the author used

these archives they were still in the possession of the UMWA; since that time they have been transferred to the Historical Collections and Labor Archives, Pennsylvania State University, State College, PA.)

51   "Grafton," *Fairmont Times*, May 9, 1916; "Strike of Maryland Coal Company miners ends," *Clarksburg Daily Telegram*, May 10, 1916.

52   *Fairmont West Virginian*, May 10, 1918.

53   "Fine spirit at Monongah," *Fairmont Times*, June 6, 1917; "Citizens of Grant Town raise nation's flag," *Fairmont Times*, June 25, 1917; "Consol Honor Roll," *Fairmont Times*, October 27, 1917.

54   *Fairmont Times*, July 21, 1917.

55   Benjamin F. Squires to W. B. Wilson, 18 June 1917, Department of Labor, Division of Conciliation, Records of the Federal Mediation and Conciliation Service (FMCS), Record Group 280, Dispute Case Files, File Folder 33/390, National Archives, Suitland, Maryland.

56   *Fairmont Times*, May 4, 1917.

57   Sam Ballantyne, J. M. Zimmerman, and P. F. Gatens to John P. White, 19 June 1917; Benjamin M. Squires to W.B. Wilson, 18 June 1917, both in UMWA, International Archive, District 17, 1917. The eight companies in the Flemington Field were the Maryland Coal Company, Pittsvain Coal Company, Simpson Creek Coal Company, Rosemont Coal Company, Stafford Gas Coal Company, White Horse Coal Company, Robinson and Phillips Coal Company, and the Harrison Coal Company. The seven in the Thin Vein Field were the Merchants Coal Company, Albright Smokeless Coal Company, Austin Coal and Coke Company, Gorman Coal and Coke Company, Hiorra Coke Company, Virginia-Maryland Coal Company, and Horchler Coal Mining Company.

58   *Fairmont Times*, March 12, 1918; March 15, 1918.

59   *Fairmont West Virginian*, May 10, 1918.

60   *Fairmont Times*, May 10, 1918; *Fairmont West Virginian*, May 10, 1918.

61   R. Dawson Hall, "The Labor Situation," *Coal Age*; quoted in *Fairmont Times*, May 21, 1918.

62   Howard B. Lee, *Bloodletting in Appalachia* (Morgantown: West Virginia University, 1969), 143. The suspicions of those who supposed it to be a political deal were raised by the fact that Consolidation Coal Company did not recognize the UMWA at its mines in eastern Kentucky, central Pennsylvania, or western Maryland.

63 "Labor pledges its support to Watson," *Fairmont Times*, October 28, 1918.

64 Lee, *Bloodletting in Appalachia*, 143.

65 *Fairmont West Virginian*, May 14, 1918; May 15, 1918.

66 "West Virginia Mines now being organized," *United Mine Workers Journal* 39, no. 6 (June 13, 1918): 5.

67 "Archives of Fairmont Labor Temple No. 1885," Huett Nestor Papers, West Virginia and Regional History Collection, West Virginia University, Morgantown, WV; "West Virginia Mines now being organized," 5; *Fairmont West Virginian*, May 27, 1918.

68 *Fairmont West Virginian*, June 10, 1918; "Mine Worker Notes," *Fairmont West Virginian*, June 24, 1918.

69 "Meeting addressed by Mother Jones," *Fairmont West Virginian*, June 10, 1918.

70 "Miners holding scale convention," *Fairmont West Virginian*, July 25, 1918.

71 "Watson mines hear UMW officers," *Fairmont West Virginian*, June 5, 1918.

72 "Pinnickinnick miners organized," *Fairmont West Virginian*, June 7, 1918; "Addresses miners at Gypsy," *Fairmont West Virginian*, July 17, 1918.

73 "Labor Day observed," *Fairmont West Virginian*, September 2, 1918.

74 Organization Reports, District 17, 1916–1935, UMWA, International Archives.

The Matewan Massacre:

Before and After

Rebecca Bailey

"The Worst Has Come"[1]

MAY 19, 1920, dawned dreary and overcast. Though rain drizzled from the clouds intermittently throughout the day, the small town of Matewan in Mingo County, West Virginia, teemed with miners, as union relief funds were being distributed. In the midst of the activity, at 11:47 a.m., a party of Baldwin-Felts agents disembarked from train #29, having come to Matewan to enforce eviction notices for the Stone Mountain Coal Corporation. According to Walter Anderson, one of the surviving agents, Albert Felts contacted Mingo County Sheriff G. T. Blankenship seeking his help in processing the evictions. He was denied, but managed to secure authorization from a local justice of the peace. Chief of Police Sid Hatfield and Mayor Cabell Testerman confronted the Baldwin-Felts agents as they made their way through town. Both Hatfield and Testerman contested the agents' authority to process the Stone Mountain evictions. Their primary arguing point, Hatfield later asserted, was that because the houses in question lay within Matewan's municipal limits, he and Testerman possessed the jurisdictional sovereignty to halt the evictions.[2]

At this point, the confrontation ended and the Baldwin-Felts agents crossed the railroad tracks and proceeded up Warm Hollow to process the evictions.

Hatfield and Testerman left to telephone county officials. Testerman allegedly called Mingo's prosecuting attorney Wade Bronson to enquire about the legality of the evictions. After Bronson read him "the Red Man's Act—the riot act," Testerman authorized Hugh Combs, a Methodist "exhorter" and local miner, "to obtain reliable men to protect the town." One of Matewan's two telephone operators later testified that Hatfield told either Blankenship or Deputy Sheriff "Toney" Webb that "those sonsabitches will never leave here alive." Throughout the afternoon, armed men arrived in Matewan, and the town became "a powder keg." The situation grew so tense that Matewan Grade School let out early, and the children were sent home to get them off the streets.[3]

Hatfield, Testerman, and a crowd of miners, men, women, and children spent the afternoon watching the Baldwin-Felts agents carry out the Stone Mountain evictions. At one point, Hatfield allegedly approached Albert Felts, who raised his gun and told Hatfield he was trespassing on private property. Smiling, Hatfield replied, "That's alright, I'm a private man," and continued his advance on Felts. When Hatfield drew near, Felts told Hatfield that he had been shot at during an ambush at Paint Creek but had refused to back down or be "bluffed out." Hatfield assured Felts that if there were any trouble, "here, no one will go to the hills on you . . . They will come face to face." Testerman again asked Felts to desist. Felts refused but offered to stop and return to town if Testerman could prove that he was acting illegally.[4]

At approximately 3:30 p.m., the Baldwin-Felts agents completed their work, came back across the tracks into Matewan, and checked into the Urias Hotel. Although miners and Police Chief Hatfield had observed the six evictions, there had been no more confrontations. A surviving Baldwin-Felts agent later recalled that the proceedings had gone smoothly, citing as proof the agents' transfer of one family's belongings to another location at the evictee's request. In contrast, local recollections of the evictions present the agents' actions as the primary stimulus for the gun battle. The agents allegedly arrived heavily armed and proceeded to bully everyone they encountered, callously and haphazardly piling the belongings of a miner's sick wife in the rain, for example. However the agents comported themselves, they made their way to the Urias unmolested.[5]

Once at the hotel, the agents disassembled and repacked their large firearms, a legal necessity since only three of them (the Felts brothers and C. B. Cunningham) possessed the required licenses to carry pistols in Matewan. Several individuals reported to Albert Felts that trouble was brewing and that armed miners were milling about the town. Felts gathered his men and told them that if a conflict erupted, they were not to fight or resist arrest but to go quietly because bail would be posted and the situation resolved peacefully. After repacking their weapons, all but one of the agents sat down to a meal. As time approached for the 5:15 p.m. train to Welch, the agents thanked the hotel manager, Anderson "Ance" Hatfield, for his hospitality, again brushed aside concern for their safety, and made their way across Mate Street (main street) to the railroad depot.[6]

While the agents waited for the train, Sid Hatfield approached Albert Felts and requested that he accompany him to a meeting with Mayor Testerman. Standing just inside the doorway of the Chambers's Hardware store, Hatfield, Testerman, and Felts again began to argue. Hatfield threatened Felts with arrest, to which Felts responded that he too possessed a warrant—for Hatfield. Mayor Testerman offered to post bond because, as police chief, Hatfield was needed in Matewan. Felts demurred, and Testerman asked to see the warrant papers. After reading them, he declared the papers "bogus," whereupon (depending on the witness) either Hatfield or Felts pulled a gun and fired.[7]

Within seconds, a blaze of gunfire erupted. Mayor Testerman staggered away clutching his stomach while Albert Felts fell where he stood, mortally wounded. Pandemonium ensued; African-American laundry owner John Brown, who was standing at the depot waiting on a shipment, led state senator M. Z. White and his wife, who were awaiting the train, to safety in a nearby basement. According to one child witness, a thousand shots rang out in less than ten minutes. The only other armed Baldwin-Felts agents, C. B. Cunningham and Lee Felts, who were standing nearby, drew their guns, but neither made it to Albert Felts's side. The other agents scattered, seeking cover. One agent, as he ran past Mary Brown who was looking for her husband John, asked, "What's the best way to get out of this town?" Pointing to the river, she shouted, "Split the creek!" Two agents (the wounded Anderson brothers) managed to climb aboard the waiting train before it quickly pulled out of the station. Another agent, who later claimed to have hidden in a coal shed, slipped out of town undiscovered. The other five agents received no mercy. J. W. Ferguson fled wounded to the back porch of Mary Duty's home and begged her to hide him. On the verge of hysteria herself, Mrs. Duty fled back into her house as armed men approached. She later claimed to have heard Ferguson say, "Gentlemen, I have not fired a shot in your town." Despite his pleas, Ferguson was shot again, allegedly by Fred Burgraff, who told him, "You S.O.B., you've gone too far. You're going to die."[8]

As quickly as it had started, the fighting ended. Several men loaded Mayor Testerman on a train bound for Welch, where he died later that night, his only words being: "Why did they shoot me? I can't see why they shot me." The bodies of the dead agents lay where they had fallen until Sheriff Blankenship, accompanied by Williamson Mayor W. O. Porter, arrived from Williamson at 7:15 p.m. Mayor Porter supervised placing the corpses onto a Williamson-bound train, which left the men who undertook the grisly task "literally covered with blood." For the rest of the night, the armed men of Matewan patrolled the town, tensely watching the trains as they sped by on the way to Williamson. Unbeknownst to them, one train carried a contingent of West Virginia state police, who thought it best to arrive in Matewan the following morning.[9]

The Matewan "Massacre," as it came to be called, was the flashpoint for West Virginia's most infamous mine war. Before an oppressive peace returned to southern West Virginia in 1922, nearly two dozen people died in and around Mingo in a twenty-eight month long struggle to organize the region. Meant to be a reckoning, what happened in the streets of Matewan on May 19 launched a war. What follows is a reevaluation of the Massacre's causes and impact.

"The Gathering Storm"

During the early months of 1920, a series of local, state, and national events began a slow implosion that resulted in tragedy on the streets of Matewan on May 19 of that year. The external causes of the terrible tragedy included the uncertain economic state of the coal and rail industries as federal wartime regulation finally ended; the escalation of tension between the UMWA and southern West Virginia's coal operators; and the political contests that played to and focused the discontent of various voting groups. However, the

Matewan Massacre also simultaneously vented long-brewing local political, economic, and social resentments.

Although 1920 later proved to be another banner year for national coal production, in the early months of the year sluggish economic conditions complicated by inadequate railcar supplies fostered tension in Mingo County. Preparations for the general election in November had also begun early. The Republicans hoped to recapture control of the county government, while Democrats aspired to retain their primacy. In scrambling for ways to improve their respective positions, the local coal and political leaders of Mingo made decisions that directly influenced their subsequent response to the violence in Matewan on May 19, 1920.

Generally poor employment, production, and transportation conditions, combined with the escalating conflict between the union and the operators did not present any new danger that would have alerted observers to the potential for the raw violence of the Massacre. Familiar with the vicissitudes of the economic situation and labor relations rhetoric, no one noticed the danger until the fateful blast. However, an examination of the status of the coal industry nationally, and in southern West Virginia in early 1920, reveals how it provided the backdrop to the events that occurred on the streets of Matewan on May 19.

"Before I Ever Work A Union Man"

In the late winter and early spring of 1920, resolving the leftover tensions from the national coal strike of 1919 preoccupied the federal government. Despite the general effectiveness of wartime federal regulation, the fitful and uneven dismantling of government controls created a climate of resentment and distrust. While the government had released the coal industry from price controls and distribution directives, it denied the UMWA request for a wage increase with a specious semantic argument over whether the war had actually ended. On January 22, 1920, when the chairman of the Senate Committee on Education and Labor introduced a bill to create a National Labor Board, (which would have extended the federal role in labor relations) the measure received little support or attention. Both the UMWA and the leaders of the coal industry were too busy shoring up their own positions and presenting their case to the Bituminous Coal Commission, the federal investigative body created as part of the settlement of the 1919 strike.[10]

While UMWA representative Van Bittner testified before the commission that miners' earnings had lagged far behind both the inflation of the cost of living and the wages paid to other industrial workers during the war, Acting President John L. Lewis traveled to West Virginia to announce that the UMWA was launching an organizational drive that would, once and for all, unionize southern West Virginia. Lewis had foreseen the potential of a unique situation. First, it was highly probable that the Coal Commission would find that union miners deserved a wage increase. Second, Lewis would also have seen that given the prevailing economic conditions in southern West Virginia, the miners employed in the non-union fields were likely to clamor for the same concession. By renewing the union's pledge to organize southern West Virginia before the Bituminous Coal Commission issued its report, John L. Lewis positioned the union to materially benefit from both the federally sanctioned wage increase and the discontent rampant in southern West Virginia.[11]

During the late winter of 1919–1920, national coal production had been curtailed by at least 25 percent. In the Williamson-Thacker coalfield, production fluctuated between 30 and 60 percent, well below the national average. A major contributing factor to the dismal coal situation was a transportation crisis. The national rail systems were still under federal regulation and there were not enough cars to move southern West Virginia's coal. Those hardest hit by the prevailing conditions were the miners. Working fewer and shorter days, the miners faced increasing tangential pressures that heightened resentment against their employers. Shortly after a neighboring coalfield reported an outbreak of influenza, some companies raised the miners' compulsory payment for medical services.[12]

While struggling to survive the closing days of winter with less income and high living expenses, miners in the Williamson-Thacker Field watched helplessly as their employers belligerently defied local governments. The Hunt-Forbes Coal Company refused to compensate the city of Williamson for the damage its coal hauling had inflicted on the city streets. In Kermit, the Gray Eagle Coal Company concluded a three-year fight with the town council and the county court over a public road by hiring guards to protect company property that occupied the contested land. As the actions of these companies illustrated, the coal companies of the Williamson-Thacker Field either blindly or arrogantly disregarded the potential impact of their actions.[13]

Between February 6 and April 16, 1920, a series of seemingly disconnected events underscored why the miners of Mingo County responded, after nearly twenty years' indifference, to the union's call. Taken together, the events illuminate why the miners came to believe that their best hope for economic security and political freedom lay with joining the UMWA.

First, just one week after John L. Lewis "invaded" Bluefield, Governor Cornwell addressed the West Virginia Lumber and Builders' Supply Association in the same city. The governor, while claiming to be a trade unionist, praised the role of southern West Virginia's non-union operators in breaking the 1919 strike. Cornwell went on to explain that if the union succeeded in organizing these same non-union fields, nothing would stop the nationalization of the coal industry. When Cornwell pledged to resign if he failed to keep the peace during the pending effort, his audience leaped to its feet in a standing ovation and shouted its support. Deliberately or not, the governor exposed whose ally he would be.[14]

If the miners had any doubts about Cornwell's position, just one month later, on March 13, the hearings of the governor's commission investigating the 1919 Armed March on Logan ended. Headed by Governor Cornwell's hand-picked proponents of law and order, the commission had lost the cooperation of District 17 officials and concluded not with substantive recommendations for addressing the endemic problems of labor relations in southern West Virginia, but with a blanket condemnation of the miners' actions and those of District 17's leadership in particular. The governor of West Virginia not only allied with their employers, but the miners of southern West Virginia were also forced to acknowledge that he had no interest in giving their grievances an impartial public hearing.[15]

The public announcement on March 29, 1920, that the Bituminous Coal Commission had recommended that coal miners receive a 27 percent wage increase only underscored the benefits of union membership. To add insult to injury, one of Mingo County's two newspapers, the *Williamson Daily News*, printed an editorial proclaiming coal a mismanaged industry, that a wage increase for miners was "only right," and that the miners should not have to suffer because of "the mismanagement of others." Less than a week after the *Daily News*'s apparent statement of empathy for the miners' cause, strikes broke out among the railroad workers of the Norfolk & Western.[16]

Despite the building pressure these incidents represented, there was one more culminating event that drove the miners of Mingo to action. In late March or early April 1920, the miners at the Howard Colliery at Chattaroy asked for a ten cents per car raise, to offset an increase in the cost of living. After calling in his superintendent from Bluefield, the Howard manager asked the men to give them a week to consider the request, and also asked the men to continue working, which they did. Three days later, the miners of the Howard Colliery were offered a nine-cent raise to which they agreed and were "well satisfied." At the end of their shift that same day, however, the men exited the mines only to find a notice that the prices of all their necessary work articles had been raised five to twenty-five cents; moreover, it would now be compulsory that the miners buy exclusively in the Howard Colliery company store. The miners of the Howard Colliery had negotiated in good faith and felt betrayed.[17]

On the morning of April 16, the miners of Burnwell Coal & Coke arrived at the mines and found the following sign posted on the drift mouth: "To the miners of Burnwell Coal Company: We shall have this 27 per cent raise; we want this 27 per cent raise which the Government has granted us." The Burnwell miners refused to enter the mines. Eighty of the ninety-two Burnwell employees signed a letter and sent two of their own to Charleston to request a charter from District 17. When asked why they undertook this action, one of the men replied that it was because the miners wanted "to belong to an organization of [their] own craft." District 17 Secretary-Treasurer Fred Mooney sent the Burnwell miners back to Mingo with a promise to send organizational assistance if the men returned to work. The Mingo organization drive had begun.[18]

Within days, large gatherings of miners were orchestrated. Miners flocked by the hundreds to take the union obligation. On April 22, approximately 500 men attended a meeting at the Baptist Church in Matewan and between 275 and 300 joined the union. The next day, also in Matewan, between 700 and 800 miners congregated. The response seemed to portend that the miners of Burnwell would get their wish. For four heady days, W. E Hutchinson, and the other employees of the Burnwell Coal & Coke Company went into the mines as union men. However, dismissal notices were delivered on April 27 and the union miners and their families were given just three days to vacate company housing. Their employer, Mr. Pritchard, declared "he would let his mine go until moss grows over it, until it falls into the huckleberry ridge, before he would ever work a union man." The line of confrontation had been drawn.[19]

Shortly after the organizational flurry began, the coal companies of the Williamson-Thacker Field launched their effort to stem the tide. The day the Burnwell miners received their eviction notices, the *Williamson Daily News* reported that it had "talked to both sides, but got no hope as to a reasonable settlement." The *News* pleaded with both sides to "remember that people will suffer," and to "keep within the bounds of the law." Just two days later, on April 29, reports of operator activity revealed that the *News*'s appeals had fallen on deaf ears. The companies began evicting the men who were assisting in the organization of the miners. The operators also imported detectives, machine guns, and high-powered rifles. Although the operators would later deny any culpability in the escalation to violence, the same story illuminated the first clue to the immediate cause of the Matewan Massacre, now less than three weeks away. On April 27, 1920, Mingo County Sheriff G. T. Blankenship arrested Albert C. Felts, brother of the head of the Baldwin-Felts Detective Agency, for illegally processing evictions. Felts posted a two thousand dollar bond and appeared in magistrate's court two days later, where he and twenty-seven other agents were placed under a peace bond.[20]

Hoping to de-escalate the situation, Sheriff Blankenship called a meeting with miners before the courthouse in Williamson. Blankenship asked the miners if they would peacefully comply with their evictions if he and his deputies processed the writs. According to Blankenship, the miners agreed. Although Blankenship later claimed to have processed hundreds of evictions for the coal companies, less than two dozen were undertaken between mid-April and May 19.[21]

There were at least two reasons for the low number of evictions. First, despite a promise to process any evictions lawfully ordered by the circuit court, Blankenship in fact engaged in stalling tactics, asking companies to obtain both ten-day notices from the court and then three-day notices from justices of the peace. Second, in early May, after holding a strategy session with Baldwin-Felts Detective Agency Chief Thomas L. Felts, the operators began compelling their employees to sign "yellow dog" contracts. The contracts not only precluded union membership, but also contained a clause that must have been particularly galling to many miners. In addition to agreeing not to join the union, the men who signed the contracts also endorsed a denunciation of the union. At least two of the companies within the vicinity of Matewan, the Stone Mountain Coal Corporation and the Red Jacket Consolidated Coal & Coke Company, instituted this policy in early May.[22]

Despite the evictions and the imposition of the "yellow dog" contracts, the organization of Mingo's miners continued unabated during the first two weeks of May. Emboldened by Sheriff Blankenship's vow to protect them from unlawful acts by the companies, the miners held mass public meetings. Three thousand attended a union meeting at Matewan on May 6, 1920. Eight days later, two hundred miners took the union pledge at Williamson. On that same day, May 14, another meeting was held at Matewan, although no organizers or representatives of the UMWA were present. The speakers included Hugh Combs, a miner and Southern Methodist "exhorter," an African-American minister, and George Allen, a merchant from Thacker, all of whom endorsed the union as the miners' advocate. Combs informed the miners that joining the union was a just act of self-protection against the operators who were themselves organized. The African-American minister, identified only as "Johnson," informed his fellow black miners that Abraham Lincoln had given them their freedom, but now the union would "give them their liberty." Allen told those assembled that he had once been a miner, and because he was in sympathy with their cause, pledged the use of his store if the miners needed a place to meet. That the miners spoke so freely because of Blankenship's protection, and had an open forum in his hometown of Matewan, soon received public acknowledgment. During the annual meeting of the West Virginia State Federation of Labor, District 17 Secretary Fred Mooney informed his fellow conventioneers that "the UMW had the support and backing of the Mingo County officials, a condition which had never existed before in the history of the organization." Reported in the *Bluefield Daily Telegraph*, Mooney's comments further galvanized the anti-union sentiments of Mingo's coal operators.[23]

While Fred Mooney celebrated the establishment of the union beachhead in Mingo, the *Williamson Daily News* made increasingly ominous observations about conditions in Mingo County. The May 8 issue noted, "There is a general restlessness among our laboring people . . . what will be the outcome of this discontent no one can predict." Just six days later, the *Daily News*, after reporting on yet another miners' meeting, noted that "the operators have made nothing public . . . as to what position they will take . . . secrecy seems to prevail on both sides."[24]

Just what the *Daily News* expected the two opponents to reveal is not clear. The operators' intractability concerning the UMWA had been openly admitted to their employees. In addition to Pritchard of Burnwell, other companies publicly advertised their position. On May 5, Borderland Coal Company posted a notice that paralleled Pritchard's comment in clarity. The notice read, "This is a free country . . . but . . . no union men shall be employed by this company." Stone Mountain posted a similar notice on the front window of its company store in Matewan.[25]

For their part, District 17 and the national leadership of the UMWA also moved forward with the formalities of the organization effort. After the state Federation of Labor convention ended, the first international organizers arrived in Mingo. By Monday, May 17, 1920, the union had initiated efforts to set up a tent colony near Matewan for evicted union miners. District 17 official C. H. Workman arrived that day with tents and instructions to lease all land available for a tent colony. Just two days later, on Wednesday, May 19, Albert and Lee Felts, leading a contingent of Baldwin-Felts agents, stepped off the train at 11:47 a.m. Their assignment: to ensure the eviction of several families from Stone Mountain company housing in the town of Matewan. Less than six hours later, the Felts brothers, five of their men, Matewan's mayor, and two bystanders would lay dead or dying on the streets of the town. However, as other studies of violence in Appalachia have shown, economic oppression is not sufficient unto itself to spark an encounter like that which occurred in Matewan on May 19.[26]

"The Threats of Evil Doers"

As 1919 drew to a close and the political season of 1920 opened, at the national, state, and local levels, Republicans dominated the stage. After orchestrating the defeat of President Wilson's internationalist agenda (personified by the League of Nations), and driving him to utter physical collapse, the nation's leading Republicans turned on each other. At issue was the direction an internally focused nation would follow. The leading liberal contender for the presidential nomination was Hiram Johnson, a Progressive railroad trust-busting senator from California, whose chief allies were Senators William E. Borah and William S. Kenyon of Iowa. Although Kenyon, leader of the United States Senate's "agricultural bloc," was sympathetic to the complaints of organized labor, he had openly denounced civil disorder as a means of addressing these issues. The party's conservative wing backed United States Army General Leonard Wood, a law and order advocate. In part because both Johnson and Wood claimed to be the heir of Theodore Roosevelt, the 1920 Republican national convention deadlocked, which resulted in the nomination of Warren G. Harding. However, the election-year politics of Wood, Johnson, and Kenyon still affected the labor-dominated politics of West Virginia. Wood, who drew substantial support from southern West Virginia's Republican coal elite, concluded his primary stumping in Williamson, Mingo County, on May 22, 1920, just three days after the Massacre. Both Johnson and Kenyon later served on the Senate's Investigating Committee into conditions in the West Virginia coalfields in 1921.[27]

Apparently united in their determination to regain the governorship after Cornwell's anomalous tenure, West Virginia's Republican party suffered from a schism that mirrored the national conflict. Samuel B. Montgomery, a former leader of the West Virginia State Federation of Labor, UMWA attorney, and commissioner of labor under Governor Hatfield, had the support of the liberal wing of the party and the laboring classes of the state. Judge Ephraim F. Morgan of north-central West Virginia had the support of the conservative wing of the party, except for the ultraconservative coal elite of southern West Virginia, who supported Colonel Paul Grosscup, an oil and gas executive from Charleston. The three-way contest between Montgomery, Morgan, and Grosscup revealed that the factional legacy of Henry D. Hatfield's gubernatorial tenure not only had persisted, but also had expanded. In 1916, Hatfield had forced his party to nominate Ira Robinson for governor; Robinson, who had undermined the Paint Creek-Cabin Creek operators' efforts to kill the 1912–1913 strike, splintered Republican hegemony. Democrat John J. Cornwell won the governor's office in 1916—the only time this happened between 1896 and 1932. Montgomery, Morgan, and Grosscup each represented a statewide Republican faction that refused, yet again, to line up behind a single leader. More important, two of the counties soon to be swept up in the crisis between the UMWA and West Virginia's non-union operators, Mingo and McDowell, were the focal point of this state-level political controversy.[28]

The intrafactional Republican fight did not play out in smoke-filled backrooms, but on the front pages of southern West Virginia's newspapers. On January 18, 1920, the *Bluefield Daily Telegraph* revealed the first indication that disharmony ruled in the southern coal counties. According to the *Telegraph*, Mingo Republican party chairman, M. Z. White, issued an endorsement letter for Morgan, while his counterpart in McDowell, Colonel Edward O'Toole, had done the same for Colonel Grosscup.[29]

In the weeks before the primary election, White utilized traditional machine tactics to ensure Mingo's support for Morgan. The March 9 issue of the *Williamson Daily News* asserted that White convened the county's nominating convention by reading a list of candidates he hoped would be considered. When a smattering of objections was raised, White "angrily" adjourned the meeting without considering alternative candidates. The *Daily News* concluded its report by noting that what remained in question was whether "the better citizens in the Republican Party . . . resented their continued sale and delivery" enough to act.[30]

Led by Edward O'Toole and T. E. Houston, two of the most powerful coal executives in the smokeless fields, McDowell's political factions turned to the West Virginia Supreme Court of Appeals. Less than three weeks before the primary, the court issued a *mandamus* ruling that ordered both the unseating of McGinnis Hatfield (a Grosscup partisan) as county chairman, and the selection of the Morgan list of election officers. While these actions appeared to indicate politics as usual for southern West Virginia, the fight over the Republican gubernatorial bid released two new forces that profoundly influenced events in southern West Virginia for the next two years.[31]

For the first time, West Virginia's laboring classes, especially its coal miners, had a legitimate advocate-candidate, and their votes could swing the gubernatorial election. The regular party machinery's infighting over Morgan and Grosscup meant that the deciding bloc of votes belonged to

Montgomery. The anticipated loyalty of the "organized faithful" inspired what the *Williamson Daily News* referred to as an "entente cordiale" between Grosscup and Montgomery. But in southern West Virginia, the single largest group of potential Montgomery supporters, the miners, *were not organized*.[32]

It was at this point that the simmering political and labor unrest in southern West Virginia coalesced in Mingo County. According to S. D. Stokes, a Williamson Democrat, it was the three-way fight between Morgan, Montgomery, and Grosscup that actually spawned District 17's effort to unionize southern West Virginia, starting with Mingo. Stokes wrote to a friend that Mingo's Republican chairman, M. Z. White, with the collusion of former governor Hatfield's brother, Greenway Hatfield, had sold the county's primary returns to Morgan-supporter T. E. Houston for fifty thousand dollars. However, because White could not keep quiet, and publicly proclaimed Mingo for Morgan, Grosscup's campaign manager visited Montgomery's campaign manager in Charleston and reminded him of two important things. First, that Montgomery could carry southern West Virginia if the miners were free to vote their consciences, and second, that the way to "liberate" these voters was to organize them. Thus, Stokes claimed, "in about ten days, the coal operators in this field observed unrest among their laboring men and woke up to the fact that the ordinary labor agitator was . . . abroad in the land." What prevents dismissal of Stokes's version of events as partisan gossip is a review of the timing of White's public actions and the launch of District 17's organization drive. White's letter of support for Morgan had appeared in the January 18 issue of the *Bluefield Daily Telegraph*. Thirteen days later, John L. Lewis announced that the UMWA would initiate a unionization offensive in southern West Virginia.[33]

The turmoil in Republican ranks in McDowell and Mingo counties extended into their most loyal constituency—the African-American community. On the same day that the *Bluefield Daily Telegraph* revealed the schism between the leaders of the party, it also noted that a "New Emancipation Movement" had been started by the "Colored Republican Laboring Men's Organization of McDowell County." The avowed intent of the organization was not hostile to the operators or any other business interests, but had as its only goal "to make an effort to end negro political slavery." By May, the movement had spread to Mingo County. At a meeting held at the county courthouse in Williamson on May 14, the African-American voters of Mingo County were exhorted to vote independently because, the "Republican bosses hang on their necks at election time," but promptly forget them after. As the *Williamson Daily News* observed, the speakers and their audience also spoke enthusiastically in favor of the organization efforts of the UMWA.[34]

Besieged by the union organizers, who also stumped for Montgomery, and threatened with the defection of the black vote, Mingo's Republican machine intensified its efforts to regain control over the pending primary election. County Chairman White sued the Democrat-controlled county court because it required the list of registered voters in Mingo to exclude voters whose registrations were disputed by precinct officers. The West Virginia Supreme Court of Appeals granted White a *writ of mandamus* on May 18, 1920, one day before the Matewan Massacre.[35]

Although the report of the supreme court of appeals does not reveal that the disputed registrations included any from those in Matewan's precinct, there is evidence that indicates how the town came to be the focal point for the political and labor agitation in Mingo. First, in the spring of 1920, the county's two most powerful officials, Sheriff Blankenship and the president of the county court, Blankenship's brother-in-law E. B. Chambers, were democrats from Matewan. Second, the town's mayor and Chambers's ally, C. C. Testerman, was targeted for criminal prosecution shortly after winning a second term. The statement he issued after being found not guilty of prohibition violations reveals not only the town's political tensions, but also his determination to oppose the forces arrayed against him. Testerman asserted:

> I am mayor of Matewan, elected by the people for a second term, which speaks for itself. They can say what they please, law when they please, fight when they please, but I am going to run the town according to the laws of this state and the best of my ability. Instead of our town having the name it has possessed these many years, you are going to hear of Matewan representing the highest order of good government. Our elections are held quietly now. We have no drunks, and our people at last can go to church unmolested and worship as they please. I intend to do what is right the best I know how and the threats of evil doers won't alter me in my course one inch.

Under Testerman's tenure, during the early spring of 1920, Matewan became a free assemblage haven for the organization of Mingo's miners. Testerman's allies, Blankenship and the county court, also provided protection for the union cause. Half of the county's miners lived in Magnolia District, of which Matewan was the center. To effectively disenfranchise these voters, the protective power of the Matewan Democrats—led by Sheriff Blankenship on behalf of his Chambers' in-laws—had to be neutralized. The Matewan Massacre occurred six days before the primary election was held.[36]

When the Republican primary was held, the newspapers reported on the unnatural state of quietness and the low voter turnout. District 17 officers fumed to no avail about the number of ballots floating in the river. Although he lost the state by more than 160,000 votes, pro-labor candidate Montgomery lost Mingo County to Morgan by fifteen votes in the Republican gubernatorial primary race. When the Matewan Massacre trial ended in acquittal in March 1921, four of the remaining sixteen defendants were members of the Chambers family of Matewan; a fact that had led at least one Hatfield family descendant to refer to the Massacre as "the Chambers' War."[37]

On May 19, Matewan's Chief of Police Sid Hatfield and a group of deputized citizens and miners engaged in a gun battle with representatives of the Baldwin-Felts Detective Agency. Ten men, including Mayor Testerman, died as a result. Within weeks, the Operators' Association of the Williamson-Thacker Field initiated a lockout in an attempt to defeat the rampant success of the UMWA's organization drive. In retaliation, the union issued a strike call, which it rescinded only after twenty-eight months of bloody conflict and violent oppression, including the retaliatory assassination of Hatfield, and the Battle of Blair Mountain, the largest armed civilian insurrection since the United States' Civil War. History has recorded the impact of this second "mine war" on the future of the United Mine Workers of America in West Virginia. What has been overlooked is the local effect of the Massacre and subsequent events, and also how the story of those events was transformed to serve the agendas of people and interests far from Mingo County.

"Remember Your Enemies . . . and Your Friends"
In the space of a few minutes the event that became known as the Matewan Massacre transformed the social, economic, and political relations of Mingo

County. Prior to May 19, 1920, class or occupation had not been the determining factor in an individual's access to political power or social status. However, how the local citizens viewed the Massacre reconfigured the course of their public associations. To the striking miners and their allies, the Massacre was a glorious instance of retributive justice; Sid Hatfield and his compatriots had given the Baldwin-Felts agents what they had coming. For other members of the community, the incident did not reflect the "identity" of Matewan and was a source of outrage and shame. As one contemporary reporter observed, "the difficulty has now become a bigger proposition than any ordinary strike . . . it is the ranging of a community into opposing factions."[38]

Social affiliations directly influenced, and, in turn, were affected by the Massacre and strike. The Red Cross and the local chapter of the YMCA refused humanitarian aid to the inhabitants of Mingo's tent colonies. The Red Cross denied assistance because "no Act of God" was responsible for the miners' plight, while the YMCA denounced the colonies as centers of immorality. After another "battle" occurred in Mingo shortly before the first anniversary of the Massacre, the local chapter president of the YMCA backed a vigilante organization that helped restore "law and order" to Mingo County.[39]

In a similar vein, the Massacre and strike divided people of faith in Mingo County. Before May 19, 1920, religious affiliation had not automatically characterized an individual's class status. For example, area natives—including miners' families—and operators together had founded and attended the Matewan Methodist Church. However, when Church of Christ minister Mose Alley "went bond" for Massacre defendant Reece Chambers, at least one member of his congregation denounced him. Mrs. Mary Duty, on whose porch agent J. W. Ferguson had been killed, made her condemnation of Alley succinctly by stating, "If Preacher Alley will swallow and shield this I am through with him."[40]

The social-religious cleavage over the Massacre rapidly took on class overtones. After the Massacre, one woman remembered, "the churches gave no haven to the poor miners." This comment reflects the social differentiation based on religious affiliation that had been brought into sharp focus by the labor conflict in Mingo. "Mainstream" Protestants, whose churches were led by educated clergy, remained aloof from the situation, with the prominent exception of Williamson's Presbyterian minister, J. W. Carpenter, who condemned the strike from his pulpit and also helped organize the Mingo Militia. By contrast, the miners drew strength and inspiration from their evangelical Christian beliefs. The inhabitants of the striking miners' tent colonies, in addition to singing hymns, used religious imagery to express their commitment to the struggle in letters to the *United Mine Workers Journal*. The centrality of the miners' faith was underscored by the attention religious affiliation received during the Matewan Massacre trial in 1921. When Hugh Combs, the man entrusted by Mayor Testerman to select men to protect Matewan on May 19, testified during the Massacre trial, he was forced to explain his vocation as a Methodist exhorter because prosecuting and defense lawyers argued about whether he was a "Holy Roller." In the wake of the 1920–1922 strike in Mingo, membership in mainstream Protestant churches or Pentecostal and Holiness churches became an automatic indicator of an individual's place in the local social order.[41]

Because it elicited such a divisive reaction from the inhabitants of Mingo County, the Matewan Massacre also transformed the atmosphere of everyday public intercourse. For the first year of the conflict, Sheriff Blankenship's and Sid Hatfield's support for the strike made Matewan the miners' town. Non-cooperative merchants were boycotted. Without fear of official retribution, many miners harassed and intimidated those who failed to support them. Two weeks after murdering an abusive former deputy sheriff and railroad guard, miners erected a mock effigy of his gravesite on a sandbar in the river at Matewan. The slain man's widow also claimed that one of the men acquitted of the murder confronted her on the street, raised her veil and laughed in her face. After the murder of Ance Hatfield, the proprietor of the Urias Hotel and star witness against Sid Hatfield and the other Massacre defendants, the *New York Times* observed that an "exodus" from Matewan began that lasted for more than a year. Although the *Times* judgmentally declared that it was the "good people" who left, the reality was far more complex. Those who remained in the town either supported the miners for their own political or material reasons, or sought to remain neutral. The merchants who had aided the miners during the strike were rewarded with the miners' patronage for as long as they remained in business. Although some businessmen, such as John Brown, eventually returned to Matewan, others, such as brothers Joseph and Samuel Schaeffer, left forever.[42]

The most profound and lasting impact of the Massacre lay in its effect on local politics. Mingo County had never been governed by a single party funded by absentee industrial interests. There is also much to suggest that District 17 chose Mingo County to launch the 1920 southern West Virginia organization drive specifically because the union hoped to turn the factional divisiveness of the county's political elite to its own advantage. Moreover, testimony from the Massacre trial suggests that old political animosities might have played a role in the escalation to violence on May 19, 1920. In the Massacre's aftermath, at least initially, the actions of certain Mingo County leaders indicate that the union strategy had succeeded. E. B. Chambers and A. C. Pinson, the leaders of the two Democratic factions in the county, both contributed to the bonds of the Matewan men indicted for the Massacre. Judge Damron, a maverick Republican who won over the miners by overseeing the indictments of local coal officials, guards, and Baldwin-Felts detectives, received the miners' endorsement for his 1920 reelection bid. Even the former chairman of the County Republican organization, Greenway Hatfield, signed a union contract for his mine.[43]

However, by the early fall of 1920, most of Mingo's leaders had abandoned their pro-union stance. James Damron resigned from the bench and withdrew from the judge's race. By the time the Massacre defendants went to trial in January 1921, Damron had joined the prosecutorial team. A. C. Pinson, after defeating Greenway Hatfield in the 1920 sheriff's race, cooperated wholeheartedly with the efforts of Governor Cornwell and the operators to secure the miners' defeat. One miner later alleged that Pinson had accepted the bribes Blankenship refused, and in return for fifteen thousand dollars, allowed the operators to mount machine guns on the rooftops in Matewan. Even E. B. Chambers evaded public identification with the miners' cause when members of the senate committee traveled to West Virginia on a fact-finding mission in 1921.[44]

After the UMWA abandoned the Williamson-Thacker strike in October 1922, it appeared that the pre-strike political equilibrium of machine rule had been restored in Mingo County. Fellow Williamson "City Ring" man, Alex Bishop, succeeded Pinson as sheriff in 1924, and was in turn followed by Greenway Hatfield. However, with the long-delayed liberation of the miners by Franklin Roosevelt's New Deal legislation, the pro-union stand taken by G. T. Blankenship and his Chambers in-laws was finally rewarded. Although Blankenship never reentered Mingo County politics, Thurman "Broggs" Chambers and several other Chambers family members or their allies served terms as Mingo County Sheriff. For the next three decades, the politicians who ruled Mingo held power as a result of tactics pioneered by the Chambers family, or had connections to families that had also sided with the miners during the 1920–1922 strike. However, the politicians the miners rewarded with control of Mingo County proved as exploitive as the old Williamson and Hatfield machines. Remarking on the corruption uncovered during a Justice Department investigation of Mingo County politics in the late 1960s, a United States Attorney observed, "Freedom has been lost in Mingo County. There is a government of the organization, by the organization, and for the organization." How little times had changed along the banks of the Tug Fork River.[45]

The abandonment of the Williamson-Thacker strike nearly two and a half years after the Matewan Massacre underscores the fatal tragedy of May 19, 1920. By the mid-1920s, Mingo's miners, who had been earning nearly seven dollars for an eight-hour shift in 1920, were now laboring for a mere two dollars for twelve or more hours of work—if they worked at all. In 1924, District 17 was stripped of its autonomy and union membership among the state's miners fell to a negligible number. Despite John L. Lewis's vigorous and forceful leadership, or maybe because of it, the UMWA nearly collapsed in the decade between the Williamson-Thacker strike and the dawn of the New Deal. By 1925, only six of the mining companies operating in Mingo County in 1910 were still running, and they were all subsidiaries of industry giants such as U.S. Steel. In spite of the success of the anti-union campaign of the early 1920s, the national coal industry, as one historian noted, "went from riches to rags" even before the Great Depression hit in 1929. How the union, the operators, and the government reacted to the Matewan Massacre and Williamson-Thacker strike provides insight into how this state of affairs transpired.[46]

The Matewan Massacre was more than just a deadly encounter between outraged miners and the agents of their oppression. Previous scholars, by reducing the active players in the story to two institutions, the company and the union, diminished the Massacre's historical significance by relegating it to the long struggle for labor rights in West Virginia. An examination of the reactions of the miners' union, the coal industry, and the state and national governments to the chain of events set in motion by the Massacre fosters a new understanding of the relationship between these groups.

In the aftermath of the First World War, the United States' largest labor union squared off against the country's most powerful corporation. Beginning in 1919, the United Mine Workers of America and the United States Steel Corporation fought repeatedly in a monumental struggle over the direction of the nation's industrial development. In the decades that followed, the struggle between these two giants transformed American labor relations, the role of government in the economy, and the meaning of individual rights in a modern society.[47]

One of the first battlegrounds was West Virginia, where the armies that took the field were motivated by philosophies forged in an era that had already passed. The majority of District 17's officers were West Virginians who strongly believed in the customary rights that came from an elemental connection to their native soil. Their desire to promote the collectivization of their fellow miners reflected a devotion to protecting the inherent dignity of men who toiled in the bowels of the earth. Their opponents in the Williamson-Thacker strike of 1920–1922 believed just as strongly in the absolute right of property. Although these equally intractable groups fought for two and a half years over the unionization of southern West Virginia, neither side gained from the struggle; the benefits accrued to their external allies. To understand how both John L. Lewis and the corporate giants of southern West Virginia's coal industry profited from the bloody and protracted strike, one must first understand how the local combatants wound up fighting a war of attrition that allowed their allies to move in, take over, and transform the conflict.[48]

The venality of local politics ensured that at least one faction would ally with the union in the hopes of gaining primacy over the county. Should the situation turn dangerous, the proximity of the Kentucky border offered an escape route for beleaguered organizers. The Williamson-Thacker Field was a peripheral concern; the operators were focused on keeping the union out of Logan and McDowell, the largest coal producing counties in the state. If, or when, violence broke out, the union could disclaim responsibility because after all, Mingo was Hatfield-McCoy country, and the county's reputation spoke for itself.[49]

After spending nearly eight million dollars on the West Virginia unionization drive of 1920–1922, the union admitted defeat in October 1922. Although District 17's officers escaped legal punishment for strike-related activities in Mingo and charges of treason following the Battle of Blair Mountain, their days as leaders of West Virginia's miners were numbered. John L. Lewis, who had barely won reelection as UMWA president in 1922, forced Keeney and Mooney to resign in 1924. Lewis simultaneously stripped District 17 of the right to autonomously elect its officers. Until his own death in 1960, Lewis maintained his grip on West Virginia's miners; from District 17 headquarters down to the local level, union officers held their positions based on their loyalty to Lewis. Frank Keeney, Fred Mooney, and Bill Blizzard all died broken men, outcasts from their union. In contrast, John L. Lewis emerged a hero from the chaos of the 1920–1922 strike and the union's near-disintegration in the 1920s and early 1930s.[50]

Like the leadership of District 17, the operators of the Williamson-Thacker Field were also unwitting foot soldiers of a larger army. Their failure to contain the 1920 unionization drive resulted in Mingo's capitulation to the interests of neighboring competitors and ultimately, the absorption of the Williamson-Thacker Field into the U.S. Steel coal empire. A brief overview of the strike from the operators' perspective reveals how the corporate giants of southern West Virginia's coal industry used the strike, not only to defeat the union, but also to drive out their smaller competitors.[51]

Dismissed by other southern West Virginia operators as "gentry" who had to be "forced" into adopting anti-union policies, the Williamson-Thacker operators made several critical mistakes that ultimately cost them autonomy. First, they failed to enforce their "no mediation" stance throughout the field. Although Red Jacket, Borderland, and Stone Mountain moved quickly to evict union miners, other companies waited until federal troops had come to Mingo for the second time in December 1920. Second, even after the might of the state and federal governments had been brought to bear on the strike, the coal and political leadership of the county allowed itself to be led "astray on side issues" that prolonged the effort to restore law and order. A dispute over whom Governor Morgan would recognize as the commander of the Mingo Militia illustrates that not even the strike could prevent the habitual squabbling of the county's elites. Third, the operators resorted to tactics that the union could expose and manipulate in the court of public opinion; for example, several operators faced charges of bribery and complicity in agent provocateur activity.[52]

The most egregious and ill-timed act by the operators was the Lick Creek tent colony raid on June 14, 1921, which took place on the same day that the West Virginia State Supreme Court of Appeals ordered the release of three men who had been jailed for infractions against an illegal enaction of martial law. Since June 14 was also Flag Day, the attack seemed to embody the struggle over which side was fighting in defense of "American" values, the union or the force of state police and local vigilantes who trampled the colony's flag and killed one miner. Not long after the Lick Creek incident, the United States Senate finally approved an investigation into conditions in West Virginia. Distracted by the chaos that had swept over the county, the Williamson operators were forced to seek alliances with their more powerful neighbors.[53]

The extent to which Mingo's elites bungled their handling of the strike is most evident when their actions are contrasted with the reaction of operators in the neighboring fields. During the early heady days of the union drive in Mingo, the operators of the Guyan, Pocahontas, and Winding Gulf Fields quickly moved to erect a *cordon sanitaire* around Mingo. Wage hikes were ordered and compulsory donations to the operators' "insurance" fund were raised. To better coordinate their activities, the Tug River and Winding Gulf operators' associations moved their headquarters to Welch, where the Pocahontas operators were based and where a regional office of the Baldwin-Felts agency had been established. When strike-related disturbances spread to the surrounding fields, the operators again acted vigorously. In Logan County, Sheriff Chafin received the necessary funds to enlarge his army of deputies. When coal company property on the Mingo-McDowell border was attacked, McDowell Sheriff S. A. Daniel requested state police officers from Governor Cornwell. The most decisive, if ill-advised and reprehensible, action supported—if not orchestrated—by the operators, was the retaliatory murder of Sid Hatfield and Ed Chambers on August 1, 1921.[54]

Even while the operators of the surrounding fields worked to contain the Williamson-Thacker strike, they also scrambled to turn the Mingo operators' trouble to their own benefit. Union miners from McDowell County, West

Virginia, and Pike County, Kentucky, were driven into Mingo's tent colonies, exponentially raising the union's relief costs. When the UMWA launched the Mingo strike, the N&W coal cars assigned to the Williamson-Thacker district were diverted to the Tug River and Pocahontas Fields. Because the strike also coincided with a resurgence in the demand for coal, while the mines in Mingo were idle, or producing at diminished capacity, Tug River and Pocahontas operators commanded nine dollars a ton. The coup de grace for many Mingo operators came when giant corporate interests gobbled up their mines. In 1920, Williamson entrepreneur E. L. Bailey sold out to the Solvay combine, a subsidiary of the American Rolling Mills Company, but stayed on as superintendent. Later that same year, the Thacker Fuel Company purchased the Grey Eagle Coal Company of Kermit. The Williamson-Thacker strike's failure had not resulted from Mingo County's membership in the anti-union southern West Virginia monolith. Instead, the strike gave the monolith the much-needed opportunity to "quietly" but "effectively" move in and defend its Achilles' heel.[55]

The reaction of the state and federal governments to the 1920–1922 West Virginia conflict also illuminates much about post-World War I America. In fact, the second West Virginia mine war could hardly have been more ill timed. Wartime regulations had demonstrated that the bureaucratization of American society, and more important, the American economy, could result in efficient production expansion. Politicians and business leaders emerged from the war aware that a new era of business and government relations was dawning. Convinced more than ever that elites should be entrusted with directing the ship of state, politicians and business leaders set about "engineering" the "consent" of the society they governed. As a result, the national debate on civil liberties was transformed. Voices of dissent, especially those that rose from the working class, were denounced and stifled. After decades of upheaval and reform, the American public fell in behind the country's business and political leaders and entered the 1920s intolerant of challenges to the status quo. In 1920, the average American was more interested in the maintenance of a steady supply of coal than in the individual human rights of West Virginia's miners.[56]

Against this backdrop, the political and business elite of West Virginia declared themselves the guardians of the nation's most important fuel supply and proceeded to wage war on the rights of the state's working class. When a local jury acquitted Sid Hatfield and the other Massacre defendants, the state legislature passed a bill that abrogated defendants' rights. Although the West Virginia State Supreme Court of Appeals later criticized abuse of the law, it did not overturn it. A former West Virginia circuit court judge even advocated the abolition of juries because in his opinion, they all too often acted like mobs. The West Virginia legislature also passed a law, proposed by Mingo State Senator M. Z. White that empowered coal companies to seek compensation from unions for property damage suffered during strikes. Inspired by the example of the injunction cases used to kill the Williamson-Thacker strike, West Virginia's coal operators spent the 1920s slowly strangling and bleeding dry the UMWA. By 1928, over 200 injunction cases had been won by the state's coal elite; miners were prevented from parading on public highways and meeting on private property or even in churches. Denied the basic constitutional right of freedom of speech, West Virginia's miners believed that they not only had been silenced, but also enslaved.[57]

The collusion of the federal government in West Virginia's mistreatment of the miners stemmed primarily from inertia. While generally supportive of the rights of industrial workers, national leaders on both ends of the political spectrum perceived themselves primarily as the protectors of the public interest. As a result, the federal government became an often-unwilling collaborator in the oppression of West Virginia's miners. Unable to compel the anti-union operators to negotiate a settlement of the Williamson-Thacker strike, Washington could only step in and restore peace when the struggle lapsed into armed confrontation.[58]

The United States Supreme Court provided little guidance in the resolution of labor conflicts because, for decades, it simultaneously recognized a small number of union rights while primarily protecting the interests of the open shop. Only violent public upheaval seemed to motivate federal legislators into action, whereupon they empanelled investigating committees or commissions. By the time of the Williamson-Thacker strike, the efficacy of the seemingly endless string of senate committees and federal commissions was openly questioned. Known as "that puttering and futile investigating committee," the committee entrusted with the examination of conditions in West Virginia in 1921 not only failed to arrive at a solution for the state's labor woes, but its chairman also managed to antagonize both sides in the controversy. The inability or unwillingness of the federal government to guide the course of labor relations in the 1920s contributed to the arrival of "desolate days in America."[59]

Conclusion: "All the Demons in Hell"

While the Matewan Massacre resulted from a unique convergence of local tensions, it also stands as a tragic example of the inherent conflicts that have periodically erupted throughout our nation's history. In the spring of 1920, Mingo County, like a pressure-cooker, steamed under the accumulated weight of decades of political corruption and chronic economic instability. When violence swept through the streets of Matewan, the local community, West Virginia, and America were forced to confront the issues that had precipitated the fatal confrontation. Unable to arrive at a consensus over the meaning of property versus individual rights, the residents of Mingo County divided and perpetuated a cycle of mutual exploitation and denigration. After the bloody rampage, the survival of any semblance of community necessitated retreating behind a wall of silence. As one Matewan native observed, "how do you explain to a child that the nice old man who runs the post office was once indicted for murder?" It is not uncommon for current residents to count as an ancestor someone from both sides of the conflict.[60]

Beyond the boundaries of Mingo County, the meaning of the Matewan Massacre was surrendered to the agendas of whoever invoked it. The United Mine Workers of America transformed the ambush of the Baldwin-Felts agents into a righteous attack on the forces of oppression, which they hoped would inspire all of West Virginia's miners to rise. To the proponents of the open shop and welfare capitalism, the Massacre demonstrated the disruptive and destructive influence of trade unionism, which must be defeated in order for liberty to survive. Liberals and conservatives alike manipulated details of the story in order to frighten Americans into supporting industrial reform. However, the Massacre, and the two years of strife it ushered in, evaded ultimate definition and thus proved to be of limited utility to any one faction in the body politic. The American people were already accustomed to dismissing Appalachian violence as the outbursts of a backward subculture. Because both the defenders and the critics of the events in Matewan on May 19, 1920, felt compelled to note that the little town on the Tug was also the home of the "feudin' Hatfields and McCoys," the Massacre passed into the realm of myth. Secure in their own modernity and prosperity, the rest of the United States ignored the ravages of the war being waged between the forces of capitalism and unionism. Until the dark days of the Great Depression when the wail of another Appalachian community torn asunder echoed through the hills of Eastern Kentucky, Americans turned a blind eye and a deaf ear.[61]

The contrast between what was born of the suffering in "Bloody Mingo" and "Bloody Harlan" illustrates the callousness of contemporary American society. In a nation weary of reform and intoxicated by the wealth of the post–World War I economy, the miners of Mingo and southern West Virginia were acceptable casualties; after all, had not these people always been unfortunate reminders of our violent past? At one point during the struggle, the *New York Times* editorialized that the union and miners should be forced to abandon their unhealthy tent colonies for centrally located camps that could be watched and protected. Just one decade later, to a nation brought to its knees by utter economic collapse, the miners of eastern Kentucky became a symbol that America had to change. The head of yet another federal commission, liberal crusader Robert La Follette, sent Heber Blankenhorn, who had also documented events in West Virginia, into Harlan. Armed with the information gathered by Blankenhorn, the federal government indicted the coal operators and the deputies for their crimes. In search of votes, Franklin Roosevelt and the Democratic Party courted organized labor. Once in office, when Roosevelt and his congressional allies set about repaying their debt to the unions, they enlisted the aid of Jett Lauck, who in his 1921 Senate testimony had compared West Virginia's anti-union operators to slaveholders. Lauck would help draft the National Recovery Act.[62]

But what of Matewan, Mingo County, and the miners of southern West Virginia? The news of the passage of the National Industrial Recovery Act sent the United Mine Workers of America sweeping back into the coalfields. Inspired by the union's revival, the former "mayor" of the Lick Creek tent colony joyously observed, "All the demons in Hell can't keep us from organizing . . . now!" However, in the decades that followed, members of the generation traumatized by the events of 1920–1922 often fell victim to divorce, alcoholism, madness, and "the ugliness of their memories." For others, to forget was to heal. Because some healing did take place, the son of a Massacre participant preached the funeral of a woman driven from her home because of what she had seen that day. The 1987 release of John Sayles's film *Matewan* renewed interest in the story and facilitated funding for an oral history initiative. The rediscovery of the Matewan Massacre could have meant the reopening of old wounds. Instead, it became a path for a community in search of its own past and a way for the people of Matewan to reclaim their story.[63]

NOTES

1 *Williamson Daily News*, May 20, 1920. With the exception of "The Gathering Storm," all of the section headers in this essay are quotations from participants or observers of the events under examination.

2 The time of the Baldwin-Felts agents' arrival was noted in the *Charleston Gazette*, May 21, 1920. For a sample of the variety of accounts of the Massacre, see *Williamson Daily News*, May 20, 1920; *West Virginia Federationist*, May 20, 1920; "Testimony of Sid Hatfield" U.S. Congress, Senate Committee on Education of Labor, *West Virginia Coal Fields: Hearings . . . to investigate the recent acts of violence in the coal fields of West Virginia and adjacent territory and the causes which led to the conditions which now exist in said territory* (Washington, DC: GPO, 1921), 205–221 (hereafter *West Virginia Coal Fields*); *State of West Virginia v. Sid Hatfield, et al.*, H. C. Lewis Collection, Eastern Regional Coal Archives (ERCA), Craft Memorial Library, Bluefield, West Virginia (hereafter Lewis Collection). For the trial see "Trial Testimony of Hugh Combs," Lewis Collection; "Testimony of Sid Hatfield," *West Virginia Coal Fields*, 219; Walter Anderson to John J. Cornwell, telegram, 20 May 1920, John J. Cornwell Papers, West Virginia Regional History Collection (WVRHC), West Virginia University, Morgantown, West Virginia (hereafter Cornwell Papers); *West Virginia Legislative Handbook and Manual*, 1920, 763; "Testimony of Sid Hatfield," *West Virginia Coal Fields*, 206.

3 "Statement of Mr. E.C. Price," excerpt from Report of #9, 29 May 1920, Lewis Collection; "Trial Testimony of Jesse P. 'Toney' Webb," Lewis Collection; "Trial Testimony of Hugh Combs," Lewis Collection; see sections 10 and 13 of Chapter 35 "An Act Concerning Deadly Weapons, Etc," in *Acts of the West Virginia Legislature for the year 1882* (Wheeling, WV: W. J. Johnston, Public Printer, 1882), 421–424; *McDowell Recorder*, March 4, 1921; "Testimony of Jesse P. 'Toney' Webb,"

Lewis Collection; Hawthorne Burgraff, interview with John Hennen, summer 1989, Matewan Oral History Project (MOHP); and Venchie Morrell, interviews with John Hennen, summer 1989, and with Rebecca J. Bailey, summer 1990, MOHP.

4 "Trial Testimony of Dan Chambers" and "Statement of Miss Jennie Mullens," 21 August 1920, Lewis Collection; Rhodri Jeffreys-Jones, *Violence and Reform in American History* (New York: Franklin Watts, 1978), 91; "Trial Testimony of Mrs. Elizabeth Barrett," Lewis Collection.

5 Walter Anderson to John J. Cornwell, telegram, 20 May 1920, Cornwell Papers; Hawthorne Burgraff interview; Venchie Morrell interview, summer 1990; "Trial Testimony of Charlie Kelly," Lewis Collection; Dudley Williams statement, Lewis Collection.

6 Walter Anderson to John J. Cornwell, telegram, 20 May 1920, Cornwell Papers; Statement of Miss Jennie Mullens, 21 August 1920, Lewis Collection; "Trial Testimony of Joe C. Jack," February 25, 1921, unknown newspaper, Matewan Omnibus Collection, ERCA; *Charleston Gazette*, May 21, 1920. The Anderson Hatfield who ran the Urias Hotel was called "Ance" to differentiate him from his more famous kinsman William Anderson "Devil Anse" Hatfield.

7 Ernest Hatfield interview with John Hennen, summer 1989, MOHP; "Trial Testimony of C.E. Lively," February 25, 1921, unknown newspaper, Matewan Omnibus Collection; "Trial Testimony of Joe C. Jack," 25 February 1921, unknown newspaper, Matewan Omnibus Collection; "Testimony of Sid Hatfield," *West Virginia Coal Fields*, 206; Ernest Hatfield interview; *Charleston Gazette*, May 21, 1920; Hawthorne Burgraff interview; "Closing Argument of Wade Bronson," March 19, 1921, unknown newspaper, Matewan Omnibus Collection; John McCoy interview with John Hennen, summer 1989, MOHP; "Smokey" Mose Adkins interview with John Hennen, summer 1989, MOHP.

8 "Trial Testimony of Joe C. Jack," February 25, 1921, unknown newspaper, Matewan Omnibus Collection; Johnny Fullen interview with Rebecca J. Bailey, summer 1990, MOHP; Dixie Accord interview with John Hennen, summer 1989, MOHP; Howard Lee, *Bloodletting in Appalachia: The Story of West Virginia's Four Major Mine Wars and Other Thrilling Incidents of Its Coal Fields* (Morgantown, WV: West Virginia University Library, 1969), 54; *Charleston Gazette*, May 21, 1920; Ruby Aliff interview with Rebecca J. Bailey, summer 1989, MOHP; "Statement of Mrs. Billy (Mary) Duty," 9 September 1920, Report of #19, Lewis Collection; "Trial

Testimony of C. E. Lively," February 25, 1921, unknown newspaper, Matewan Omnibus Collection.

9   Notes taken (probably by T. L. Felts) following meeting with owner of the *Mingo Republican*, Dr. R. M. Musick, May 22, 1920, Lewis Collection; "Slight Flareup," *Charleston Gazette*, May 21, 1920; "Trial Testimony of Jack Gallion," Lewis Collection; J. W. Weir to John J. Cornwell, 21 May 1920, Cornwell Papers.

10   Frederick A. Barkey, "The Socialist Party in West Virginia from 1898 to 1920: A Study in Working Class Radicalism" (PhD diss., University of Pittsburgh, 1979), 214; U.S. Congress, Senate, Committee on Education and Welfare. *West Virginia Coal Fields: Personal Views of Senator Kenyon and views of Senators Sterling, Phipps, and Warren . . .*, 67th Congress, 2d Session, Senate Report 457 (Washington, DC: GPO, 1922), 20 (hereafter *Personal Views of Senator Kenyon*); Mildred Allen Beik, *The Miners of Windber: The Struggles of New Immigrants for Unionization, 1890s–1930* (1996; repr., University Park, PA: The Pennsylvania University Press, 1997), 260.

11   *Coal Trade Journal*, February 4, 1920, 13–14; "John L. Lewis, Head of United Mine Workers of America, Here for Purpose of Organizing Local Coal Fields," *Bluefield Daily Telegraph*, January 31, 1920.

12   *Coal Trade Journal*, January 21, 1920, 25; January 7, 1920, 13; and February 4, 1920, 120; E. F. Striplin, *The Norfolk and Western: A History* (Roanoke, VA: Norfolk and Western Railway Company, 1981), 153; *Coal Trade Journal*, February 18, 1920, 174; "Testimony of W. E. Hutchinson," *West Virginia Coal Fields*, 79.

13   City Attorney S. D. Stokes to Hunt-Forbes Coal Company, 8 January 1920, Samuel Davis Stokes Papers, WVRHC (hereafter Stokes Papers); Court Documents relating to the case, in Box 5 of the Stokes Papers.

14   "Flurry Caused By Invasion of Head of Mine Workers," *Bluefield Daily Telegraph*, February 1, 1920; *Bluefield Daily Telegraph*, February 7, 1920.

15   Merle T. Cole, "Martial Law and Major Davis as 'Emperor of Tug River,'" *West Virginia History* 43 (Winter 1982): 118–144, 126.

16   *Williamson Daily News*, March 30, 1920; *Williamson Daily News*, April 1, 1920; *Coal Trade Journal*, April 7, 1920, 368.

17   "Testimony of Frank Ingham," *West Virginia Coal Fields*, 29.

18   "Testimony of W. E. Hutchinson," *West Virginia Coal Fields*, 80–81.

19   "Testimony of W. E. Hutchinson," *West Virginia Coal Fields*, 83–85; *Williamson Daily News*, April 23, 1920.

20 *Williamson Daily News*, April 27, 1920; *Williamson Daily News*, April 29, 1920.

21 *Williamson Daily News*, April 29, 1920; *Charleston Gazette*, May 20, 1920; *Charleston Gazette*, May 20, 1920; "Testimony of C.[sic] T. Blankenship," *West Virginia Coal Fields*, 490; During the testimony of Harry Olmsted, coal company attorney S. B. Avis noted that prior to the beginning of the strike on July 1, 1920, only twenty evictions had taken place: a total of sixteen from two separate occasions at Matewan, two at Red Jacket, and two at Burnwell; by the time the Senate hearings began in July 1921, a total of 369 evictions had been processed in Mingo County. Avowal of S. B. Avis, during "Testimony of Harry Olmsted," *West Virginia Coal Fields*, 223–271, 257.

22 Avowal of S. B. Avis, during "Testimony of Harry Olmsted," *West Virginia Coal Fields*, 223–271, 257; "Testimony of T. L. Felts," *West Virginia Coal Fields*, 881–905, 891; Arthur Warner, "West Virginia: Industrialism Gone Mad," *Nation*, October 1921, 373; *United Mine Workers Journal*, February 15, 1921, 17; "Affidavit of C. L. McKinnon," *Red Jacket v. John L. Lewis, et al.*, in *Transcript of the Record: District Court of the United States for the Southern District of West Virginia at Charleston: Red Jacket Consolidated Coal and Coke Company, et al, plaintiffs, versus John L. Lewis, president, et al.* 3 vols. (Charleston, WV: Jarrett Printing Company, 1923): 788a–791a, 789a.

23 *Charleston Gazette*, May 20, 1920; Daniel P. Jordan, "The Mingo War: Labor Violence in the Southern West Virginia Coal Fields, 1919–1922," in *Essays in Southern Labor History; Selected Papers, Southern Labor History Conference, 1976*, eds. Gary M. Fink and Merle E. Reed, Contributions in Economics and Economic History, no.16 (Westport, CT: Greenwood Press, 1977), 107; *Williamson Daily News*, May 15, 1920; Report of Operative #24, 14 May 1920, submitted to George Bausewine by T. L. Felts, 17 May 1920, reprinted as part of "Hatfield Exhibit No. 2," in "Testimony of Sid Hatfield," *West Virginia Coal Fields*: 205–221, 215; *Bluefield Daily Telegraph*, May 15, 1920.

24 *Williamson Daily News*, May 8, 1920; *Williamson Daily News*, May 14, 1920.

25 Crandall Shifflett, *Coal Towns: Life, Work and Culture in Company Towns in Southern Appalachia, 1880–1960* (Knoxville: University of Tennessee Press, 1994), 123; Arthur Gleason, "Public Ownership of Private Officials," *Nation* 29, May 1920, 724–725, 724.

26 "Testimony of W. E. Hutchinson," *West Virginia Coal Fields*, 80; *Charleston Gazette*, May 20, 1920; "Testimony of Sid Hatfield," *West Virginia Coal Fields*, 219;

Dwight B. Billings and Kathleen M. Blee, *The Road to Poverty: The Making of Wealth and Hardship in Appalachia* (Cambridge: Cambridge University Press, 2000).

27   Wesley M. Bagby, *The Road to Normalcy: the Presidential Campaign and Election of 1920* (1962; repr., Baltimore: Johns Hopkins Press, 1968), 31–32; *Bluefield Daily Telegraph*, February 1, 1922; Robert K. Murray, *Red Scare: A Study of National Hysteria, 1919–1920* (New York: McGraw Hill, 1964), 30–31, 206; *Williamson Daily News*, May 20, 1920; *Charleston Gazette*, May 21, 1920. See member list of the committee that follows the title page, *West Virginia Coal Fields*.

28   Jim F. Comstock, ed. *West Virginia Heritage Encyclopedia*, 25 vols. (Richwood, WV: Jim Comstock, 1976), 15: 3302; Evelyn L. K. Harris and Frank J. Krebs, *From Humble Beginnings: West Virginia State Federation of Labor, 1903–1957* (Charleston, WV: Jones Printing Co., 1960), 119; *Williamson Daily News*, May 15, 1920; Comstock, *West Virginia Heritage Encyclopedia*, 15: 3340–3341 and 10: 2064–2065.

29   *Williamson Daily News*, May 15, 1920; *Bluefield Daily Telegraph*, January 18, 1920; A. D. Sowers, *Some Facts about McDowell County, West Virginia* (Keystone, WV: A. D. Sowers, 1912), in Rare Book Collection, WVRHC.

30   *Williamson Daily News*, March 9, 1920.

31   *Williamson Daily News*, May 15, 1920.

32   *Williamson Daily News*, May 15, 1920.

33   S. D. Stokes to Carl E. Whitney, 19 August 1920, Stokes Papers, WVRHC.

34   *Bluefield Daily Telegraph*, January 18, 1920; *Williamson Daily News*, May 15, 1920.

35   "Testimony of C.E. Lively," 25–26 February 1921, unknown newspaper, Matewan Omnibus Collection; *State ex rel. M. Z. White, Relator v. County Court of Mingo County*, in *Reports of the West Virginia Supreme Court of Appeals* 86 (March 16–September 21 1920): 517–518.

36   *Williamson Daily News*, January 30, 1920; Billings and Blee, *The Road to Poverty*, 281; and *Williamson Daily News*, May 25, 1920.

37   *Williamson Daily News*, May 25, 1920; John L. Spivak, *A Man in His Time* (New York: Horizon Press, 1967), 87; Houston G. Young, West Virginia Secretary of State, *State of West Virginia: Official Returns of the General Election, held November 2, 1920* (Charleston, WV: Tribune Printing Co., 1920), J. Hop Woods Bound Pamphlet Collection, volume 15, index no. 11476, WVRHC; Margaret Hatfield, undated correspondence with author, ca. 1997–1998, letter no.16.

38   *Mingo Republican,* January 2, 1914; *Williamson Daily News,* November 26, 1920; *Williamson Daily News,* June 2, 1916; Bertha Damron, interview with Rebecca J. Bailey, summer 1989, MOHP; Catherine McNicol Stock, *Rural Radicals: Righteous Rage in the American Grain* (Ithaca: Cornell University Press, 1996); William Lynwood Montell, *Killings: Folk Justice in the Upper South* (Lexington: University Press of Kentucky, 1986), 138; "Testimony of Mrs. Stella Scales," Lewis Collection; Margaret Hatfield, undated correspondence with author, ca. 1997–1998, letter no. 16; Roy Hinds, "The Last Stand of the Open Shop," *Coal Age,* November 8, 1920, 1037–1040, 1038.

39   McAlister Coleman, *Men and Coal* (New York: Farrar and Rinehart, 1943), 100; "Mine Dynamited On Matewan Day," *New York Times,* May 20, 1921.

40   Margaret Hatfield, undated correspondence with author, ca. 1997–1998, letters no. 16 and 24; T. E. Bowman, "From West Riding to West Virginia," unpublished manuscript, Matewan Development Center, Matewan, West Virginia; Sara Lubitsch Tudiver, "Political Economy and Culture in Central Appalachia, 1790–1977," (PhD diss., University of Michigan, 1984), 122; "Statement of Mrs. Billy Duty," 9 September 1920, Report of #19, Lewis Collection.

41   Jeannette Simpkins, interview with Rebecca J. Bailey, summer 1990, MOHP; "Testimony of J. R. Brockus," *West Virginia Coal Fields,* 344; Edward L. Ayers, *The Promise of the New South: Life After Reconstruction* (New York: Oxford University Press, 1992), 408; Jacquelyn Dowd Hall, et al., *Like a Family: The Making of a Southern Cotton Mill World* (Chapel Hill and London: The University of North Carolina Press, 1987), 126, 178–179; Virginia Grimmett, interview with Rebecca J. Bailey, summer 1989, MOHP; "From Nolan, W.Va.," *United Mine Workers Journal,* April 15, 1921, 15; Margaret Hatfield, undated correspondence with author, ca. 1997–1998, letter no.16.

42   *Williamson Daily News,* November 4, 1920; *Cincinnati Enquirer,* May 28, 1921, Matewan Omnibus Collection; "Report of # 9," 29 July 1920, Lewis Collection; "Record of Mrs. Sid Hatfield," Lewis Collection; *Williamson Daily News,* December 13, 1920; "Statement of Mrs. Pearl Hatfield," Lewis Collection; "Berman Hatfield Case," Lewis Collection; Bertha Damron interview; *Williamson Daily News,* August 16, 1920; Margaret Hatfield, undated correspondence with author, ca. 1997–1998, letter no. 5; "Affidavit of George H. Gunnoe," Lewis Collection; Lee, *Bloodletting,* 61; Johnny Fullen interview; "Good People Flee Matewan," *New York Times,* May 21, 1921; Dixie Accord interview; Rufus Starr, interview with John Hennen, summer

1989, MOHP; "Testimony of A. E. Hester," *West Virginia Coal Fields*, 826; E. K. Beckner to John J. Cornwell, 13 July 1920, Cornwell Papers; Jim Backus, interview with John Hennen, summer 1989, MOHP; Dixie Accord interview; Johnny Fullen interview; Abraham J. Shinedling, *West Virginia Jewry: Origins and History* 3 vols. (Philadelphia: Maurice Jacobs, Inc, 1963), 2:1038; "Statement of Eli Sohn," *State of West Virginia v. C.E. Lively, George Pence, and William Salter*, Lewis Collection.

43   *Williamson Daily News*, February 20, 1919; "Sid Hatfield by Isaac Brewer," Lewis Collection; "Trial Testimony of Elizabeth Burgraff," Lewis Collection; "Trial Testimony of Hugh Combs," Lewis Collection; Broadside, "Remember Your Friends—Also Your Enemies," Lewis Collection; "Report of #9," 6 August 1920, Lewis Collection; "Report of #19," 5 August 1920, Lewis Collection; *Coal Trade Journal*, January 5, 1921, 7.

44   *Williamson Daily News*, October 12, 1920; *Bluefield Daily Telegraph*, January 19, 1921. Lee, *Bloodletting*, 51; Richard Burgett interview; Anne Lawrence, ed., *On Dark and Bloody Ground: An Oral History of the U.M.W.A. in Central Appalachia, 1920–1935* (Charleston, WV: Miner's Voice, 1973): 105–109,107; *West Virginia Coal Fields*, 486.

45   *West Virginia Bluebooks*, 1925–1935; Tom Blankenship, interview with C. Paul McAlister, Summer 1990, MOHP; Rufus Starr interview; "Testimony of C. E. Lively," *West Virginia Coal Fields*, 355; Margaret Hatfield, undated correspondence with author, letter no. 16; Huey Perry, *They'll Cut Off Your Project: A Mingo County Chronicle* (New York: Praeger Press, 1972), 155–156, 209–210, 252, 255; James Washington interview; Lawrence, *On Dark and Bloody Ground*, 119–121; John H. M. Laslett, *Across the Sea: A Comparative Study of Class Formation in Scotland and the American Midwest, 1830–1924* (Urbana: University of Illinois Press, 2001), 231; A. D. Lavinder, interview with Bill Taft and Lois McLean, Matewan, WV, June 22, 1973, WVRHC; Venchie Morrell interview, Summer 1990.

46   Jim Backus interview; Fred Mooney, *Struggle in the Coalfields: The Autobiography of Fred Mooney* (Morgantown, WV: West Virginia University Library, 1967), 127–128; Irving Bernstein, *The Lean Years: A History of the American Worker, 1920–1933* (1960; repr., New York: Da Capo Press, 1983), chapter 10; Edmund Wilson, "Frank Keeney's Coal Diggers," pt. 1, *New Republic*, July 8, 1931, 195–199; Wilson, "Frank Keeney's Coal Diggers," pt. 2, *New Republic* 15, July 19 1931, 229–231; and John Hennen, *The Americanization of West Virginia: Creating a Modern Industrial State, 1916–1925* (Lexington, KY: University Press of Kentucky, 1996), 105; Phil M.

Conley, *History of the West Virginia Coal Industry* (Charleston: Education Foundation Inc., 1960), 262; James P. Johnson, *The Politics of Soft Coal: The Bituminous Industry from World War I through the New Deal* (Urbana: University of Illinois Press, 1979), 95.

47   Bernstein, *The Lean Years*, 127; Ron Chernow, *The House of Morgan: An American Banking Dynasty and the Rise of Modern Finance* (New York: Simon & Schuster, 1990), 82.

48   David A. Corbin, "'Frank Keeney is Our Leader and We Shall Not Be Moved': Rank and File Leadership in the West Virginia Coal Fields," in *Essays in Southern Labor History: Selected Papers from the Southern Labor History Conference, 1976*, eds. Gary M. Fink and Merle E. Reed, Contributions in Economics and Economic History, no. 16, (Westport, CT: Greenwood Press, 1976), 147; Mooney, *Struggle in the Coal Fields*, 13, 16; "Testimony of W. Jett Lauck," *West Virginia Coal Fields*: 1045; McIntosh Memoir, George C. McIntosh Papers, WVRHC; "Opening Statement of Z. T. Vinson," *West Virginia Coal Fields*, 8–15, 8; Spivak, *A Man in His Time*, 102.

49   45 percent of the men who worked in Mingo's mines in 1920 came from families that had lived in the greater Big Sandy Valley since before 1850 (assertion based on author's data analysis of U.S. Census records); Corbin, "Frank Keeney Is Our Leader,"147; *United Mine Workers Journal*, July 4, 1901, 2; *Mingo Republican*, June 3, 1920; "Seven Prisoner After Mingo Battle," *New York Times*, May 27, 1921; Richard D. Lunt, *Law and Order vs. The Miners: West Virginia, 1906–1933* (Charleston, WV: Appalachian Editions, 1992): 147, 152–153; "Testimony of C. F. (Frank) Keeney," *West Virginia Coal Fields*, 184.

50   Hennen, *Americanization*, 112; Ronald L. Lewis, *Black Coal Miners in America: Race, Class, and Community Conflict, 1780–1980* (Lexington, KY: University of Kentucky Press, 1987), 163; *New York Times*, October 28, 1922; Mooney, *Struggle in the Coal Fields*, x, 124–128; Melvyn Dubofsky and Warren Van Tine, *John L. Lewis: A Biography* (New York: The New York Times Book Company, 1977), 81; Melvin Triolo, interview with John Hennen, Summer 1989, MOHP; Cabell Phillips, "The West Virginia Mine War," *American Heritage* 25 (August 1974): 58–61, 90–96, 94; "William (Bill) Blizzard," *West Virginia Heritage Encyclopedia*, 3: 470–471.

51   Conley, *History of the West Virginia Coal Industry*, 262; David E. Whisnant, *Modernizing the Mountaineer: People, Power, and Planning in Appalachia* (Knoxville: University of Tennessee Press, 1994), 110.

52   Justus Collins to George C. Wolfe, 16 September 1920, Justus Collins Papers, WVRHC; *Bluefield Daily Telegraph*, July 2, 1920; F. A. Lindsay, on behalf

of Allburn Coal & Coke Company, to S. D. Stokes, 21 December 1920, Stokes Papers; S. D. Stokes to James P. Woods, 2 July 1921, Stokes Papers; M. Z. White to Greenway W. Hatfield, 28 August 1921, Ephraim F. Morgan Papers, WVRHC; S. D. Stokes to G. W. Coffey, 19 August 1920, Stokes Papers; J. M. Tully, W. A. Wilson, and G. R. C. Wiles, *Mingo Republican,* July 8, 1920; C. F. Keeney to Newton Baker, 4 October 1920, Cornwell Papers; "Statement of R. H. Kirkpatrick," *West Virginia Coal Fields,* 101; "Testimony of J. R. Brockus," *West Virginia Coal Fields,* 344–345.

53   "Testimony of Frank Ingham," *West Virginia Coal Fields,* 34; "Ex Parte Lavinder, et al," *Reports of the West Virginia State Supreme Court of Appeals,* 88 (February 22, 1921–June 24, 1921), 713–721, 721; "Testimony of Albert E. McComas," *West Virginia Coal Fields,* 303; "Keeney Exhibits, #18 and 19" in "Testimony of C. F. (Frank) Keeney," *West Virginia Coal Fields,* 166–167; *Mingo Republican,* May 23, 1913; Major Thomas B. Davis to Ephraim F. Morgan, 5 June 1922, Morgan Papers; Heber Blankenhorn, "Marching Through West Virginia," *Nation,* September 14, 1921, 288; "Testimony of Albert E. McComas," *West Virginia Coal Fields,* 304–305; *New York Times,* May 27, 1921; Winding Gulf Coal Operators' Association, Bulletin #62, 31 May 1920, Collins Papers; *Coal Trade Journal,* June 2, 1920, 588.

54   Justus Collins to George C. Wolfe, 16 April 1920 and "Bulletin F," 21 May 1920, Collins Papers; "Testimony of R. C. Kirk," *West Virginia Coal Fields,* 471; "Testimony of W. R. Thurmond," *West Virginia Coal Fields,* 867; "Testimony of C. [sic] T. Blankenship," *West Virginia Coal Fields,* 487; McDowell County Sheriff S. A. Daniel to John J. Cornwell, 2 September 1920, Cornwell Papers; William N. Cummins to John J. Cornwell, 17 November 1920, Cornwell Papers; G. T. Blankenship to John J. Cornwell, 6 November 1920, Cornwell Papers; Lee, *Bloodletting,* 67; *Charleston Gazette* August 2, 1921, quoted in Corbin, *Life, Work and Rebellion: The Southern West Virginia Coal Miners, 1880–1922* (Urbana: University of Illinois Press, 1981), 210.

55   Neil Burkinshaw, "Labor's Valley Forge," *Nation,* December 8, 1920, 639; "Report of #5," 19 February 1921, Lewis Collection; *Bluefield Daily Telegraph,* July 2, 1920; *Coal Trade Journal,* June 2, 1920, 588; S. D. Stokes to Thomas B. Garner, 26 May 1920; "Testimony of E. L. Bailey," *West Virginia Coal Fields,* 277, 284; Theodore Kornweibel, Jr., ed., "Reports by Informant C-61 to A. E. Hayes, for the Southern District of West Virginia, 9 October–30 October 1920," in *Federal Surveillance of Afro-Americans (1917–1925): The First World War, the Red Scare, and the Garvey Movement* (microfilm project of University Publications of America), 9 October 1920; "West Virginia's War," *Literary Digest,* December 18, 1920, 16.

56   William K. Klingaman, *1919: The Year Our World Began* (1987; repr., New York: Harper & Row, 1989), 548–549; Robert H. Wiebe, *The Search for Order: 1877–1920* (New York: Hill and Wang, 1967), 293–302; Hennen, *Americanization*, 1–2.

57   "Resign As Governor Whenever Unable To Preserve Order," *Bluefield Daily Telegraph*, February 7, 1920; "Governor Explains Why Federal Troops Brought Into West Virginia—Six Thousand Deluded Men In Insurrection, Reason He Offers," (reprint, New York Commercial, 1921), Pamphlet 7640, WVRHC; "The Jury Bill," Senate Bill #14, *Acts of the West Virginia Legislature for the Year 1921*, 183–184; S. D. Stokes to S. B. Avis, 22 March 1921, Stokes Papers; *State of West Virginia v. J. S. McCoy*, in *Reports of the West Virginia State Supreme Court of Appeals* 91 (April 25, 1922–October 17, 1922): 262–268; *Bluefield Daily Telegraph*, October 1, 1921; J. C. McWhorter, "Abolish the Juries," *West Virginia Law Quarterly* 29 (January 1923): 97–108; Senate Bill# 359, *Acts of the West Virginia Legislature for the Year 1921*, 322–329; *United Mine Workers Journal*, October 1, 1922, 3; "Testimony of William McKell," *West Virginia Coal Fields*, 941; Winthrop D. Lane, *Civil War in West Virginia: A Story of the Industrial Conflict in the Coal Mines* (New York: B.W. Huebsch, 1921;1994), 22; Lunt, *Law and Order vs. The Miners*, 96, 152–153, 172, 166–167, 145–149, and 154.

58   James M. Cain, "The Battleground of Coal," *Atlantic Monthly*, October 1922, reprinted in *The West Virginia Mine Wars: An Anthology*, ed. David A. Corbin (Charleston, WV: Appalachian Editions, 1991): 151–159, 155; William R. Trail, "The History of the United Mine Workers in West Virginia, 1920–1945," (master's thesis, New York University, 1950), 6; *Personal Views of Senator Kenyon*, 6; Clayton Laurie, "The United States Army, The Return to Normalcy in Labor Dispute Interventions: The Case of the West Virginia Coal Mine Wars, 1920–1921," *West Virginia History* 50 (1991): 1–24, 8; Kornweibel, "Reports by Informant C-61," in *Federal Surveillance of Afro-Americans*, 23 October 1920; Edward Eyre Hunt, F. G. Tryon, and Joseph H. Willits, eds., *What The Coal Commission Found: An Authoritative Summary By The Staff* (Baltimore: The Williams & Wilkins Company, 1925), 34.

59   Bernstein, *The Lean Years*, 190–191; Arthur Gleason, "Company-Owned Americans," *Nation*, June 12, 1920, 794–795, 795; *Bluefield Daily Telegraph*, January 28, 1922; *Personal Views of Senator Kenyon*, 4; Johnson, *Politics of Soft Coal*, 111; Arthur Gleason, *The Book of Arthur Gleason: My People, and "A.G." An Appreciation By Helen Hayes Gleason* (New York: William Morrow & Company, 1929), 183.

60   Margaret Hatfield interview.

61   Jane Dailey, "Deference and Violence in the Postbellum South: Manners and Massacres in Danville, Virginia," *Journal of Southern History* 63 (August 1997): 581–582; David Fowler and David Robb, "Statement of Conditions in Mingo County, W.Va.," *United Mine Workers Journal*, January 1, 1921, 10; Editorial, *National Coal Mining News*, September 29, 1921; Jeffreys-Jones, *Violence and Reform in American History*, 37 and 155–157; Charles Frederick Carter, "Murder to Maintain Coal Monopoly" *Current History* 15 (1922): 597–603; Allen W. Batteau, *The Invention of Appalachia* (Tucson: University of Arizona Press, 1990), 124, 115.

62   George J. Titler, *Hell in Harlan* (Beckley, WV: BJW Printers, n.d.); Hunt, et al., *What the Coal Commission Found*, 235–236; *New York Times*, May 27, 1921, 3; Batteau, *The Invention of Appalachia*, 123; Beik, *The Miners of Windber*, 267; Spivak, *A Man in His Time*, 59; Lunt, *Law and Order vs. The Miners*, 181.

63   Richard Burgett interview, Lawrence, *On Dark and Bloody Ground*, 108; Margaret Hatfield, undated correspondence with author, ca. 1997–1998, letter no. 29; Hiram Phillips, interview with John Hennen, Summer 1989, MOHP; Virginia Grimmett interview; Venchie Morrell interview, Summer 1990; Hawthorne Burgraff interview; John A. Velke, *Baldwin-Felts Detectives, Inc.* (Richmond, VA: s.n., 1997), 200; Paul Lively interview; Johnny Fullen interview; *Bluefield Observer*, May 13, 1992.

Progress and Persistent Problems:

Sixty Years of Health Care in Appalachia

Richard P. Mulcahy

THE UNITED STATES has a two-tiered health care delivery system. This fact has been documented in countless studies and reports dating back at least sixty years.[1] However, the public has largely ignored this point due to the fact that average statistics taken over the years have indicated a general increase in the nation's overall healthfulness. From 1900 to the present day, the media and official government documents have reported vast improvements in the following categories: life expectancy, infant mortality, and maternal mortality. In addition, these statistics have also indicated that Americans have reasonable access to an abundant supply of well-trained and highly competent medical professionals, as well as a wide array of modern hospitals and clinics.

Unfortunately, this picture is not accurate. While not denying that the average statistics are correct (they are), it must be understood that averages can be misleading since they can hide wide disparities between various groups and populations. This has certainly been the case with health care delivery in the United States. Thus, while various commentators have pointed to the improvements mentioned above with justifiable pride and thereby proclaimed our health care delivery system the best in the world, an important reality has been ignored. The current health care system has distributed its benefits neither equitably nor evenly.

When our health care delivery system is inspected more closely, a different picture emerges. Instead of our having a single uniform system, this country has two different and distinct systems: one for cities and one for rural areas. By all available standards, the urban system is the far superior of the two, having the very best in terms of access to health resources. The rural system, on the other hand, has traditionally lagged behind its urban counterpart in this regard: it has not simply been second-rate, but out-and-out deficient in almost every measured category.[2] Because of this, the numerical bulk of the average health improvements mentioned above have been enjoyed primarily by urban populations, with rural areas coming in a distant second.

This was the context in which health care delivery operated in the bituminous coal industry due to its rural nature prior to 1946. However, the difficulties mentioned above were only the beginning. In addition to those deficiencies were the problems associated with the industry's company doctor system. While some company physicians were dedicated and capable, many were incompetent and should not have been allowed to practice medicine.[3] A variety of parties, including the federal and state governments as well as private agencies, have made efforts to address these issues over the many years that have passed. One major contributor to those efforts was the United Mine Workers of America (UMWA) Welfare and Retirement Fund. The UMWA's program, over the course of a generation, significantly narrowed the gap between rural and urban health care.[4] However, despite those improvements, the gap remains. This essay will survey how such improvements were made, concentrating primarily on the contributions made by the fund, as well as offer recommendations on future policy.

Rural Heath Care and Its Discontents

In 1945, rural health care was the quiet scandal of American social policy. A basic problem was the fact that most health professionals neither practiced nor were located in rural areas. The statistics tell the story. While average figures gathered in 1940 indicated that there was roughly one physician for every 831 people, the reality was that urban areas enjoyed a ratio of one physician for every 700 persons, while rural areas had one for every 1500 to 2000. The ratio in even more remote areas, such as rural New Mexico, was far higher, exceeding one for every 3500.[5] Added to this was the fact that many rural physicians were under-trained. Although by 1945, the American Medical Association (AMA) had enthroned the four/four model of American medical education (four years of undergraduate education followed by four years of medical school), there were many rural physicians in practice who were not products of this system. The four/four model had been pioneered by Johns Hopkins University when it opened its medical school in 1893.[6] Other medical schools, such as Harvard, started raising their academic standards in the 1870s.[7]

This notwithstanding, by 1900, a number of two-year medical schools were still in operation. Not only were the degrees these schools offered deficient, they usually did not require any previous undergraduate study for admission. Concerned about this, the American Medical Association commissioned a study of medical training in the United States by the Carnegie Endowment for the Advancement of Education. Officially titled *Bulletin Number Four*, Professor Abraham Flexner directed the study, which came to be known as the *Flexner Report*. Among its recommendations, the report urged that the remaining two-year medical schools be compelled either to upgrade their programs, or close their doors.[8]

The last of these schools, however, did not close their doors until 1923, and then only with the stipulation that their final graduates be grandfathered and allowed to practice medicine.[9] Owing to the fact that these people were undereducated, they simply could not compete with their more competent colleagues in the cities, and so migrated to the countryside and practiced there.[10] In addition to these, there were also many practitioners who did not attend medical school at all. In these cases, the person involved knew something about medicine, and was therefore awarded an MD by his state legislature via a private bill.[11]

Hand-in-hand with this controversy was another problem. One of the major issues associated with rural practice was the question of social and professional isolation. Although some people had various ways of effectively meeting this challenge, many others did not. The result was that a number of rural practitioners found solace with either drug or alcohol abuse. This situation was little better with regard to hospitals and other treatment facilities. Many rural hospitals, especially in south-central Appalachia, operated on a for-profit, or proprietary, basis. Because they were money-making operations, these institutions had an incentive to offer minimal services at very high costs.¹² With this, most were deficient in the most basic of services, including maintenance of records and housekeeping.¹³ Many dispensaries and doctors' offices in the region were little better.¹⁴

The for-profit health care delivery model employed and promoted in the United States was the problem. This model tended to view physicians not as people engaged in a public service, but as individual entrepreneurs selling a service. The AMA endorsed this approach, along with fee-for-service payment, as the means for assuring quality care. Since the physician was an independent free standing professional, he could practice medicine and prescribe treatment as he saw fit, without his professional judgment being questioned.

Although these claims were debatable, one thing the model did do was to assure the creation of the two-tiered system. Because physicians operated on a for-profit basis, and because fee-for-service was an extremely expensive method of paying for care, physicians needed to locate in areas with the population density necessary to make solo practice economically viable. Thus, cities became "doctor magnets." The same was also true with hospitals. This situation was also impacted by the fact that cities had the service infrastructure, such as water, sewage, and waste treatment, that some rural areas lacked.¹⁵

The coal industry, in particular, suffered from the ways in which the health care system developed. As an extractive industry working primarily in a rural setting, it put up with all of the deficiencies outlined above and more. Central to that industry's health care delivery problems was the company doctor system. Using a prepaid method, physicians contracted with coal companies to provide health services to the miners. Operators paid these physicians via a wage deduction or "check-off" taken from the miners' salaries. While there is no question that some company doctors were dedicated and capable people, many company check-off practices had become a dumping ground for the incompetent and the addicted. The situation was so bad that an official from the coal miners' union, the UMWA, speaking at a conference on national health insurance held in 1938, characterized the system as the "bastard" of the American health system.¹⁶

It was clearly evident that something needed to be done. Although the federal government had done some work in this field during the Great Depression through the Farm Security Administration (FSA) and the United

States Public Health Service (USPHS), those programs were too small to meet the need. In addition, while people in the government had struggled to win passage of national health insurance, the AMA and the insurance industry were absolutely opposed to any such reform ever becoming law. Through their combined efforts, national health bills stalled in Congress, despite the fact that polls indicated a clear majority of Americans favored such legislation.[17] Thus, the only institution working in the Appalachian region that had the resources to effect change in this regard was the UMWA.

The UMWA Fills the Void

A number of factors thrust the UMWA into a leadership role on health care in Appalachia. By 1945, the UMWA had been under John L. Lewis's leadership for twenty-five years. In that time, the union had had its ups and downs, as did Lewis's reputation as a labor leader. Nevertheless, in that year the union reached its height in terms of economic and political influence. As for Lewis, his leadership was absolute. That did not mean, however, that the union's membership failed to impact his decision-making process. Prior to 1945, Lewis had not shown much interest in health reform other than making the occasional ritualized statement in support of national health insurance. Yet, during the middle to late 1930s, increasing numbers of UMWA local unions demanded that something be done about the company doctor system. These demands became so numerous and insistent that if Lewis had failed to act it could have exploded into a revolt against his leadership.[18] Ever pragmatic, Lewis decided that he needed to listen to these demands and act upon them. Therefore, instead of resisting this movement, he put himself at its forefront and began calling for the creation of a welfare fund during the 1945 national coal contract negotiations. The program Lewis envisioned would offer complete health coverage, as well as pensions, to all union miners and their families. Controlled by the union, the program would be financed by a royalty of five cents a ton assessed on all coal mined for sale or use.[19]

Although Lewis failed to win a fund in 1945, he made it a central feature in contract negotiations the following year. Threatened with a national strike, the federal government seized the mines under the War Labor Disputes Act, and negotiated with Lewis directly. On May 26, 1946, the union leadership ratified a new national contract and the United Mine Workers of America

Welfare & Retirement Fund officially came into existence.[20] While the fund was now a reality, it would be some time until its health care service got off the ground. Nevertheless, there was a great deal of interest. Through various private professional networks, word got out that the UMWA was putting a new health care program together.[21] This sparked the interest of a number of health care professionals who had been working in various New Deal social service agencies, such as the FSA and the Social Security Administration, as well as those working for the USPHS. Many of these people were involved since the 1930s in what one historian has characterized as "the health reform coalition" working toward national health insurance.[22]

By 1946, however, it appeared as if the proposal they were seeking to pass, the Murray-Wagner-Dingell Bill, was hopelessly mired in Congress. Disgusted, many of these people began leaving government service at this time but they were anxious to use their expertise relative to health care delivery in another venue. The fund provided that venue. The result was that between 1946 and 1948, the fund gathered a professional staff that was dedicated, capable, and willing to experiment.

Heading this staff were two exceptional individuals, Josephine Roche and Dr. Warren F. Draper. Roche had worked in public service for most of her adult life, and had served the Roosevelt Administration as Assistant Secretary of the Treasury. While there, she had worked as a co-author of the Social Security Act. Moreover, she had been the administration's go-to person on national health insurance, and was responsible for putting the proposal together that ultimately became the Murray-Wagner-Dingell Bill.[23] Dr. Warren F. Draper had served under Roche as chief deputy surgeon general of the United States. The organization for which Draper worked, the USPHS, was a branch of the Treasury Department. Although Roche and Draper would be responsible for creating the program, it was Draper who had the administrative vision and experience to make the fund's medical program a reality. Essentially, what Draper did was to borrow the USPHS's administrative model. Although headquartered in Washington, D.C., the service operated through a set of area offices placed in strategic cities throughout the nation. Through these offices, local medical administrators implemented national policy, and were given the latitude necessary in order to tailor those policies to fit local conditions.

Using this approach, the fund organized its medical program in the same way. It established a total of ten Area Medical Offices throughout the coalfields, each headed by a physician-administrator, who served as the Area Medical Officer (AMO). While broad policies were set at the fund's national office, each AMO had the autonomy necessary to best implement those policies in his service area.[24] By all accounts the system worked well, achieving several goals at once: improvement of treatment quality and cost containment, as well as uncovering incompetence and fraud.

The issues of fraud and incompetence proved to be very important early on. When the fund's medical program began operation, participation was open to all licensed physicians. However, as time went by, fund AMOs encountered two major problems: fraud and incompetent practice. In terms of fraud, the fund had to deal with over billing, billing for work never performed, double billing and the like.[25] As far as incompetence was concerned, in many instances physicians were doing work for which they simply were not qualified, especially surgery, with disastrous results. Even more disturbing was the fact that organized medicine, the AMA and its various county- and state-level affiliates, refused to do anything about the situation. Draper bent over backwards to work with the AMA on these matters. Indeed, he had been a member of the AMA's ruling body, the House of Delegates, for twenty years and sincerely believed that the association stood for quality care. However, between 1948 and 1957, Draper sought the AMA's assistance in fighting the abuses outlined above, but to no avail. Moreover, Draper and the fund encountered considerable hostility to the fund's efforts on the part of some AMA affiliates, most notably the Kentucky State Medical Association (KSMA). At one point, the KSMA's opposition was so vehement that it sought passage of a bill in the Kentucky state assembly that would have effectively banned the fund's medical program from operating in the state.[26]

Although relations with organized medicine varied from state to state, the conflict with the AMA's national office centered on the fund's desire to judge the quality of treatment physicians were rendering to fund beneficiaries. According to Dr. Draper, because the fund was paying for medical care, it had the right to pass judgment on whether or not the care provided to its beneficiaries met with acceptable professional standards. If those standards were not met, then the fund, or any other medical third party payer, had the right to drop the physicians in question from its participation list. From the AMA's perspective, this was unacceptable. While admitting that incompetence and fraud needed to be eliminated, the association insisted that physicians should only be policed by themselves. Medical third parties did not have this right, and the association insisted that insurance payers needed to regard any licensed physician as competent to practice medicine in all of its branches.[27]

Unfortunately, the fund's experience showed that physicians were not policing themselves. Rather, many "covered up" for their incompetent colleagues for the sake of the profession's appearance with the general public. This mentality was something reminiscent of the behavior described by A.J. Cronin in his novel *The Citadel*, which was an exposé about similar behavior among physicians in Great Britain.[28] Although it goes beyond the scope of this essay to describe the fund's struggle with the AMA over this matter in any detail, the fund did win the right to pass judgment on treatment quality. With this, the fund adopted a closed panel organizational model. A forerunner of the modern Health Maintenance Organization (HMO), closed panel plans were so named because they restricted participation to a panel of selected physicians.[29]

The AMA fought this method many times, claiming that it violated a patient's free choice of physician.[30] However, there were many instances prior to the fund's establishment where free choice of physician was routinely violated in the coalfields, with the AMA maintaining a conspicuous silence. Using the closed panel approach, the fund based participation primarily upon physician qualification.[31] AMA objections notwithstanding, the fund witnessed a number of benefits after the participation restrictions were put in place: the quality of treatment increased and the number of surgeries performed declined dramatically, as did instances of fraud and medical incompetence.[32] Although viewed as a victory for the fund, the real winners were its beneficiaries.

In tandem with these efforts, the fund also directed its attention to the problem of health care facilities in its operations area. At the time of the fund's creation in 1946, the federal government through the Department of the Interior conducted a survey of health and living conditions in the bituminous coal mining regions. The project was overseen by Deputy Surgeon General of the Navy, Rear Admiral Joel T. Boone, and the project's final report was named for him as a result. According to *The Boone Report*, bituminous miners living in close proximity to cities such as Pittsburgh, Wheeling, and Morgantown had access to excellent hospitals and other facilities. Those residing in more remote locations did not. In many cases the only facilities available were the small, substandard proprietary hospitals mentioned above.[33] Accordingly, the fund saw a need and attempted to fill it. From the fund's perspective, this need was two-fold. First, good quality hospitals had to be established in south-central Appalachia, where the need was the greatest. Second, more physicians had to be brought to the region, particularly with an eye towards eliminating company doctors. It was hoped that this could be achieved by creating new health care facilities, not only hospitals, but clinics as well.[34]

Established originally as the Miners' Memorial Hospital Association, the fund created a chain of ten hospitals and one free standing outpatient clinic that covered a 250 mile distance from Cumberland, Maryland, to Hazard, Kentucky.[35] This chain incorporated many of the theories that had been advanced by health reformers on how to best organize and distribute hospital services.[36] Sadly, the program eventually proved to be too much for the fund to handle, due primarily to the effects of the Eisenhower recession of the late 1950s, when coal demand fell to its lowest level since the Great Depression. Because of this, the fund eventually sold the hospital chain to the Board of National Missions of the Presbyterian Church, USA. The program's impact was and is on-going. Most importantly, it represented the first time that a private organization sought to bring urban-style hospital care to a medically underserved rural area.

The establishment of clinics proved to have an even more extensive legacy than the hospital program. Health reformers had advocated the virtues of group practice clinics for most of the twentieth century. Not only did they offer the patient a more sophisticated variety of care than what could be offered by a solo practitioner, the fees charged tended to be less expensive. Studies indicated that treatment in a clinic setting was far less expensive than similar treatment in a hospital.[37]

This point was significant. At the time, medical circles were debating over how to provide care in the most efficient manner possible. Central to this debate was the role of the hospital, an issue hotly contested within the fund's staff. According to those who advocated for clinics, America's health care delivery system was out of balance, since it tended to be overly hospital-centric. In many instances, American physicians tended to refer patients for hospital treatment for conditions that could have been effectively addressed either in a clinic, or in the physician's office. While the modern American hospital was a source of national pride, too much medical work was being done in that setting, and the overuse of hospital services was leading to chronic cost inflation.[38] Those who argued for clinics maintained that this problem could be avoided by performing much of the work done in hospitals in a clinic setting. Not only could high quality care be secured, clinics could provide it with greater cost containment.

The man making this argument on the fund's staff was Dr. John D. Winebrenner, AMO for Knoxville, Tennessee. Winebrenner's arguments had an impact. Because the coal industry was in decline, concerns about cost containment were always near the surface. Moreover, since many people on the fund's staff had come out of the health care reform movement, they were familiar with the arguments in favor of group practice, and thereby supported the creation of clinics.[39] Because of this, the fund's influence resulted in the establishment of a myriad of clinics throughout the bituminous coalfields, particularly in Appalachia. However, who established them varied. The fund created some through its various area offices; the UMWA local and district unions built others, while still others were created by coalitions consisting of social and fraternal organizations, civic bodies, and local professionals.[40]

To foster support, the founders of these institutions worked to rally the local community behind a clinic's establishment. The general tack used was to stress that the new facility was not simply a "miner's clinic" but a facility that would benefit all.[41] To that end, the groups that organized the creation of these facilities sought to promote as much involvement from the wider community as possible. This was usually done by seeking donations, either in the form of cash, equipment, or other usable items. However, what proved crucial was financial backing from the fund.[42] Clinics accomplished this in the form of a retainer, or fee-for-time arrangement. Originally developed to pay solo-practitioners, fee-for-time compensated physicians for the amount of time they spent treating fund beneficiaries as a percentage of their practice.[43] Negotiated between the fund and the physician, the retainer covered both overhead and income and thereby had a number of virtues. First, it was far less expensive than fee-for-service, but far more remunerative than capitation payment. Second, it did away with the paperwork associated with separate billing. Third, because it was negotiated by mutual consent, it was flexible and open to periodic renegotiation by either side. Fourth, because it guaranteed income, it attracted new physicians into areas that heretofore had been underserved.[44] With the guarantee of a fund retainer, many clinics got off the ground and thus served people whose only previous experience with medical care had been company doctors.

The Ill Winds of Change

By the early 1960s, the fund had become very much a part of Appalachia's existence. It was during this time that the region underwent a rediscovery. Central to that rediscovery was the apparent collapse of the coal industry and its side effects. As mentioned above, the severe recession of the late 1950s had taken a toll on the industry. While boom/bust cycles had been experienced in the past, what was happening at this time was different. In hindsight, this was the start of the long process of deindustrialization. Hitting the nation's extractive industries first, it would eventually affect America's manufacturing base.[45] Central hallmarks of this process were capital flight and high structural unemployment. However, at the time, it appeared as if this problem was specific only to Appalachia and the coal industry.

In response, the government created a myriad of programs, including VISTA, which originally was Volunteers in Service to Appalachia, but later became Volunteers in Service to America. In the area of health care, the USPHS and other groups developed scholarship programs whose purpose was to bring physicians to medically underserved areas. In addition, the federal government created Medicare and Medicaid to help provide medical attention to the elderly and the indigent. Although these efforts were needed and laudable, Medicare's creation opened a Pandora's box in terms of medical expenses. First, reimbursement was done on a fee-for-service basis, with the only limiter being a requirement that payment reflect an area's "usual and customary fee" for any given treatment. Central to this was a belief that physicians would act honorably in this matter and not overcharge for services. The fund's experience showed that such an approach would not work.

After just a few years of operation, Medicare was experiencing a cost inflation that needed to be contained.[46] Moreover, Medicare, following the general trend seen with the rest of nation's health care delivery system, tended to overhospitalize.

The result was an inflationary spiral of health care costs. Unfortunately, this spiral was a contributing factor in the demise of the fund's medical program.[47] The major coal operators looked upon that spiral with dismay. By this time, the industry had become highly mechanized, with the large companies such as Pittsburgh Consolidated employing comparatively few miners per ton of coal mined. Nevertheless, there remained a number of firms operating that were more labor-intensive. Owing to how the fund was financed, the highly capitalized firms were partially subsidizing the health insurance costs of their less-efficient competitors.[48] Dissatisfied with this situation, the leaders of the big firms decided to try and end the fund's medical program. They managed to do this, primarily due to the incompetence of the leadership that took over the UMWA, and by extension the fund, in 1972.

In that year, Arnold Miller was elected the UMWA's president, ending the corrupt administration of his predecessor, W. A. (Tony) Boyle. Viewing his election as a mandate for change, Miller and his supporters threw the baby out with the bathwater. With regard to the fund, this translated into the appointment of a new leadership that had no experience with either social policy or health care delivery.[49] The basic problem was the reformist mentality that the new leadership brought to its work. Prior experience was ignored and even denigrated in favor of a technocratic bias that sought to eliminate the personal factor from the fund's operations.[50] To achieve this end, a major effort was mounted to computerize all fund services. This proved to be a costly failure.[51] Even worse, in 1974, the fund was reorganized from one into four different trusts, separating the fund's medical system from its pension program. What this did was to give the nation's coal operators a veto power with regard to the fund's management that they had never before enjoyed. By law, the fund's trustee board was evenly divided between management, labor, and neutral opinion. Under the old system, assets could be transferred between accounts at will by the fund's director. Now, with the program broken into the four trusts, any such transfer required the trustee board's unanimous approval.[52] For the major operators, this represented a golden opportunity to end the fund's medical program. In addition to the structural inequity mentioned above, with regard to the more and less efficient firms, there was another problem. The coal companies resented the fact that the medical program gave their workers a considerable degree of independence from them as employers. From the coal companies' perspective, this was unacceptable.[53]

The denouncement came in 1977. Owing to how the fund's income was distributed among the four trusts, as well as other problems, such as the severe winter of that year, the fund's medical program went into a financial crisis.[54] Using their new-found authority, the nation's major coal operators forbade the asset transfers necessary to keep the medical program solvent. For their parts, Miller and his associates simply rolled over and accepted this action without protest, resulting in the health program being cut, and ultimately terminated.[55] This proved to be organized labor's first "giveback" contract and not only marked the demise of the New Deal Order, but heralded the dark years that followed, up to and including the present.

Tragic as the loss of the fund's program had been, the institutions that it had created, either directly or indirectly, continued. As a result, the disparities between urban and rural health care delivery, at least in Appalachia and the rest of the coalfields, had diminished. Considerable improvements had been made in a number of areas, including access to physicians, facilities, and the like. However, problems have remained. Although there are a greater number of physicians in rural areas than there had been before, disparities still exist.[56] In addition, there are continuing health problems in rural areas that are intertwined with matters relating to poverty, poor diet, and cultural attitudes. Included under this heading are such things as heart disease, poor oral health, and cancer.

Conclusions

Although Appalachia, and the rest of rural America for that matter, should not be defined by the presence of poverty, it would be a mistake to pretend that the problem does not exist. One only need look at the Appalachian Regional Commission's economic survey maps to see that many of the region's counties are economically distressed.[57] Poor health and economic deprivation go hand in hand. Although Medicaid is in place to provide health care services for the indigent, most people today in need of health services fall into a grey area. Broadly classified as "the working poor," these people make too much money to qualify for Medicaid, do not have health insurance from their employers, and yet make too little to afford health insurance on their own. Worse still, their numbers are growing. According to figures released in August of 2005 by the U.S. Census Bureau, the nation's poverty rate stands at 37 million people, 12.7 percent of America's population. The number of Americans currently without medical coverage stands at 45.8 million people, 15.7 percent of the nation's population.[58] These figures will only increase if things continue as they are now.

The problem is that the insured are currently subsidizing the uninsured. How it works is that someone who is uninsured defers treatment until it must be given under emergency conditions in a hospital. This is the most expensive form of care going. Since most hospitals are non-profit organizations, they are tax exempt. As a condition of that status, however, they cannot refuse care to someone in need who is unable to pay. The hospitals, however, do not write off such losses. Rather, they make it up by overcharging those with insurance, thus driving up costs.[59] Insurance carriers, in turn, increase premiums. This causes more employers to drop medical fringe benefits because the premiums are too expensive.

Thus, the number of insured is steadily shrinking while the uninsured population grows. With this, more and more people in the middle class are losing their coverage. Between 2003 and 2004, the number of people with health insurance fell from 60.4 percent of the population to 59.8.[60] If this continues, our health delivery system will eventually implode. The only way this can be stopped is for the creation of a comprehensive national health insurance system. The fund's experience, as well as the experience of other organizations in this area, certainly shows how such a program could be successfully operated.

With all of this, what must be remembered is that our current health insurance system is a hold-over from the old New Deal Order. That system was a patchwork quilt of a myriad of different insurance programs, which together stood in place of national health insurance. Pioneered by the labor movement, most Americans were covered by it, and its boosters claimed that it allowed this country to avoid the evils of "socialized medicine." In addition, because of the labor movement's strength, non-union employers were forced to offer their workers some form of health insurance, in order to keep unions out.

Since the late 1970s, the preconditions that made that system viable have ceased to exist. The labor movement has declined to a shadow of its former self and deindustrialization has taken a horrible toll. Because of this, the old method cannot be sustained. Thus, we need an entirely new way of doing things: specifically, a national health insurance plan that ensures fairness and access for all people. Such a plan is feasible. It could be operated through the USPHS and its various area offices. As far as service providers are concerned, the program could use the fee-for-time method that the fund pioneered. Although not as remunerative as fee-for-service, it would be a far more generous and fairer payment method than a flat capitation fee. At the same time, savings could be realized in other ways. For example, under the current system, with its myriad of health insurance payers, each payer has its own discrete and complex system of billing and reimbursement, which presents a mountain of paperwork. A single payer system would eliminate this.

Also, a single payer system could continue work on reducing those disparities between urban and rural health care that still exist. This could be done by the creation of group practice clinics in rural areas. Not only would such organizations offer sophisticated care, but they could do it in a cost-effective manner by providing treatment on an out-patient basis, and so avoid over-hospitalization. The methods are at our disposal, and the fund's experience more than shows the way. The time to act is now.

NOTES

1   U.S. Department of the Interior, *A Medical Survey of the Bituminous Coal Industry: Report of the Coal Mines Administrator, (The Boone Report)*, (Washington, DC: GPO, 1947), Bibliography, UMWA Health and Retirement Funds Papers, A&M 2769 (Hereafter referred to as "Fund Papers"), Box 1, West Virginia and Regional History Collection (WVRHC), West Virginia University, Morgantown, WV; Joel Halverson, *An Analysis of Disparities in Health Status and Accessibility to Health Care in the Appalachian Region* (draft, 2003); C. Horace Hamilton, "Health and Health Services," in *The Southern Appalachian Region: A Survey*, ed. Thomas R. Ford (Lexington:

University of Kentucky Press, 1962), 219–244; Interdepartmental Committee to Coordinated Health and Welfare Activities, Josephine Roche, Chairman, *Proceedings: National Health Conference, July 18, 19, 20, 1938* (Washington DC: GPO, 1938); Frederick W. Mott and Milton I. Roemer, *Rural Health and Medical Care*, (New York: McGraw-Hill Book Company, Inc., 1948); Richard P. Mulcahy, *A Social Contract for the Coal Fields: The Rise and Fall of the United Mine Workers of America Welfare and Retirement Fund*, (Knoxville: University of Tennessee Press, 2001); The President's Commission on the Health Needs of the Nation, *Building America's Health*, Five Volumes (Washington, DC: GPO, 1952); Jeffery Stensland, Curt Mueller, and Janet Sutton, *An Analysis of the Financial Conditions of Health Care Institutions in the Appalachian Region and Their Economic Impacts* (Bethesda: Project Hope: Center for Health Affairs, 2002); Paul Starr, *The Social Transformation of American Medicine: The Rise of a Sovereign Profession and the Making of a Vast Industry* (New York: Basic Books, 1982); White House Domestic Policy Council, *Health Security: The President's Report to the American People*, (Washington, DC: GPO, 1993).

2   Richard P. Mulcahy, Introductory Essay, "Health" Gary L. Burkett, Richard P. Mulcahy and Pamela M. Zahorik, section eds., *The Encyclopedia of Appalachia*, ed. Rudy Abramson and Jean Haskell (Knoxville: The University of Tennessee Press, 2006), 1631–1632; Stensland, Mueller, and Sutton, *Analysis of the Financial Conditions*, 3–22; Mott and Roemer, *Rural Health and Medical Care*, 50–57, 66–73, 151–185.

3   Janet E. Ploss, "A History of the Medical Care Program of the United Mine Workers of America Welfare and Retirement Fund" (master's thesis, Johns Hopkins University School of Hygiene and Public Health, 1981), 17–18; Dr. John D. Winebrenner, interview with the author, September 10, 1993. Dr. Winebrenner served as the Area Medical Administrator of Knoxville, Tennessee for twenty-five years. Hereafter referred to as "Winebrenner interview."

4   Stensland, Mueller and Sutton, *Analysis of the Financial Conditions*, 3–22.

5   Mott and Roemer, *Rural Health and Medical Care*, 151–160.

6   Starr, *Social Transformation of American Medicine*, 115.

7   Starr, *Social Transformation of American Medicine*, 112–115.

8   Starr, *Social Transformation of American Medicine*, 112–115.

9   Mulcahy, *Social Contract for the Coal Fields*, 62.

10   Mott and Roemer, *Rural Health and Medical Care*, 172.

11   Winebrenner interview.

12   Jere A. Wysong and Sherman R. Williams, "Health Services for Miners: Development and Evolution of the United Mine Workers Health Care Program," (Washington: National Center for Health Service Research, US Department of Health and Human Services, Office of Health Research, Statistics and Technology, 1981), 16.

13   United States Department of the Interior, *A Medical Survey of the Bituminous Coal Mining Industry*, 170–186, Fund Papers.

14   "Medical Hospitals in the Bituminous Coal Mining Areas of Kentucky – Tennessee – West Virginia, Preliminary Study to Survey Extent of Problems," 5–8, Series III, Subject Files, Box 7, Council on Medical Service, UMW Survey Report, Fund Papers, WVRHC.

15   Richard Mulcahy, "Health Care in the Coal Fields: The Miners Memorial Hospital Association," *The Historian*, 55, no. 4 (Summer, 1993): 646.

16   *Proceedings of the National Health Conference*, Interdepartmental Committee to Coordinate Health and Welfare Activities, Josephine Roche, Chairman (Washington: GPO, 1938), 68–70.

17   "An Address by Senator James E. Murray of Montana before the CIO Committee for Political Action," (New York, January 14, 1944), 2–3, Yellow Dot, Group Insurance, Box 78, Murray-Wagner-Dingell Bill, United Electrical, Radio, and Machine Workers of America Papers (hereafter referred to as UE Papers), Archive of Industrial Society, University of Pittsburgh, Pittsburgh, PA.

18   Alan Derickson, "Part of the Yellow Dog: US Coal Miner Opposition to the Company Doctor System, 1936–1946," *International Journal of Health Sciences* 19, no. 4 (1989): 714–718.

19   Melvyn Dubofsky and Warren Van Tine, *John L. Lewis: A Biography* (New York: Quadrangle Books/New York Times Book Co., 1977), 376.

20   Mulcahy, *Social Contract for the Coal Fields*, 8.

21   Dr. John Newdorp, interview with the author, December 10, 1992 and June 4, 1993; Dr. Newdorp joined the fund program in 1947 and originally headed its hospital program, the Miners' Memorial Hospital Association. After these institutions were sold to the Board of National Missions of the Presbyterian Church USA, Newdorp became the fund's deputy executive medical officer and replaced Dr. Draper as executive medical officer when he retired. Hereafter referred to as "Newdorp interviews." Mr. Henry C. Daniels, interview with the author, September 21, 1995. Mr. Daniels was a lay public health specialist and had worked for the Farm

Security Administration. He joined the fund's staff in 1947. Hereafter referred to as "Daniels interview."

22   Daniel M. Fox, *Health Policies, Health Politics: The British and American Experience, 1911–1965* (Princeton: Princeton University Press, 1986), 117–123.

23   See Richard Mulcahy, "Working Against The Odds: Josephine Roche, The New Deal, and the Drive for National Health Insurance," *The Maryland Historian* 25, no. 2 (Fall/Winter 1994): 1–21.

24   Dr. Lorin E. Kerr, interview with the author, May 30, 1988. Dr. Kerr joined the fund in 1947, having originally worked for the United States Public Health Service. Originally serving as the Fund's Area Medical Administrator for Morgantown, WV, he eventually joined the Fund's Washington staff and was a pioneer in winning recognition of Coal Workers' Pneumoconiosis (Black Lung) as a compensable, work-related disease afflicting workers in the bituminous coal industry. Newdorp interviews.

25   "Medical Hospitals in the Bituminous Coal Mining Areas of Kentucky – Tennessee – West Virginia, Preliminary Study to Survey Extent of Problems," 5. Series III, Box 7, Council on Medical Service, UMW Survey Report, Fund Papers.

26   "In the House of Delegates of the State of Kentucky," Regular Session, House Bill No. 343, Monday, February 17, 1958, Series IV, Subject Files, Box 6, A-II-6.1 Dr. Fred Zuspan vs. Floyd County, Fund Papers.

27   American Medical Association, *Suggested Guide to Relations Between State and County Medical Societies and the United Mine Workers of America Welfare and Retirement Fund* (1958), 6, Series III, Box 7, American Medical Association, Committee on Medical Care for Industrial Workers, 1957–1958–1960, Fund Papers.

28   See Archibald Joseph Cronin, *The Citadel* (Boston: Little, Brown, and Company, 1937). Cronin was in a position to do this, since, like his contemporary and fellow British novelist, W. Somerset Maugham, Cronin was a trained physician.

29   Richard P. Mulcahy, "Organized Medicine and the UMWA Welfare and Retirement Fund: An Appalachian Perspective on a National Conflict," *The Journal of the Appalachian Studies Association* 2 (1990): 92–93.

30   Mulcahy, "Organized Medicine," 92–93.

31   Mulcahy, *A Social Contract for the Coal Fields*, 118–119.

32   Mulcahy, *A Social Contract for the Coal Fields*, 118–119.

33   United States Department of the Interior, *A Medical Survey of the Bituminous Coal Mining Industry*, 170–186, Fund Papers.

34   Richard Mulcahy, "A New Deal for Coal Miners: The UMWA Welfare and Retirement Fund, and the Reorganization of Health Care in Appalachia," *The Journal of Appalachian Studies* 2, no. 1 (Spring 1996): 38–43; see also William A. Massie, *Medical Services for Rural Areas* (Cambridge: Harvard University Press, 1957), 11, 19, 21, 32.

35   "Remarks Given by Dr. Warren Draper at the Groundbreaking Ceremonies at Man Memorial Hospital, Man, West Virginia, October 31, 1953," Series IV, Construction/Administration, Box 2, MMHA Publicity, Publications, Fund Papers.

36   Rosemary Stevens, *American Medicine and the Public Interest* (New Haven: Yale University Press, 1971), 417–418; Fox, *Health Policies, Health Politics*, 16–20, 45–51; Richard Mulcahy "Health Care in the Coal Fields: The Miners Memorial Hospital Association," *The Historian* 55, no. 4 (Summer 1993): 645.

37   Starr, *Social Transformation of American Medicine*, 372.

38   John Winebrenner, "Medicine into the Sixties," *Tennessee State Medical Journal* 53, no. 10 (October 1960): 431–434; "The Reins of Progress" Paper presented at the Midwestern Regional Meeting, American Association of Medical Clinics (AAMC), Published in *Group Practice*, 14, no 9, September 1965.

39   Winebrenner, "Medicine into the Sixties," 431–434; "The Reins of Progress."

40   Massie, *Medical Services for Rural Areas*, 17ff.

41   Massie, *Medical Services for Rural Areas*, 17ff.

42   Allan Koplin, "Retainer Payment for Physician Services," 10–12, Box Addendum, Annual Reports, Misc. Medical Staff Publications, Gift, Yale University Library, Fund Papers.

43   Koplin, "Retainer Payment for Physician Services," 4–5, Fund Papers.

44   Koplin, "Retainer Payment for Physician Services," 3, 10–12, Fund Papers.

45   See Thomas Dublin and Walter Licht, *The Face of Decline: The Anthracite Region in the Twentieth Century* (Ithaca: Cornell University Press, 2005). What this study shows is that the anthracite coal industry was not so much a victim of decline, as it was the first victim of deindustrialization, with all of its attendant economic and social impacts. I highly recommended this study. A similar study of the bituminous industry is sorely needed.

46   Starr, *Social Transformation of American Medicine*, 383–388.

47   Wysong and Williams, "Health Services for Miners," 35.

48   Domanic Raino, interview with the author, September 8, 1993, hereafter

referred to as "Raino interview"; Albert Reese, *The Economics of Trade Unions*, 3rd ed. (Chicago: University of Chicago Press, 1989), 83–84. Mr. Raino worked as a member of the Fund's staff at its headquarters in Washington, D.C.

49 Ploss, "History of the Medical Care Program," 111–114.

50 Ploss, "History of the Medical Care Program," 133–134.

51 Mulcahy, *A Social Contract for the Coal Fields*, 174–175.

52 Mulcahy, *A Social Contract for the Coal Fields*, 167.

53 Thomas N. Bethell, "UMW in the Pits," *The New Republic*, April 1, 1978, 9; Ploss, "History of the Medical Care Program," 256–258.

54 Mulcahy, *A Social Contract for the Coal Fields*, 178.

55 Mulcahy, *A Social Contract for the Coal Fields*, 178–181.

56 Stensland, Mueller and Sutton, *Analysis of the Financial Conditions*, 4.

57 See Appalachian Regional Commission, "County Economic Status in Appalachia, FY 2007," map, 2006, http://www.arc.gov/index.do?nodeId=3130.

58 U.S. Census Bureau, Newsroom, "Income Stable, Poverty Rate Increases, Percentage of Americans Without Health Insurance Unchanged," August 30, 2005, http://www.census.gov/Press-Release/www/releases/archives/income_wealth/005647.html.

59 James Fallows, "A Triumph of Misinformation," *The Atlantic Monthly*, January 1995, 32.

60 U.S. Census Bureau, Newsroom, "Income Stable, Poverty Rate Increases."

1199 Comes to Appalachia:

Beginnings, 1970–1976

John Hennen

THE JUNE 1971 cover of *1199 News*, the monthly magazine of the National Union of Hospital and Health Care Employees, featured a photographic study that was out of the ordinary for the urban-based union.[1] Readers, who by and large were women and minority service workers in the rapidly changing health care industry, saw a pleasant family portrait of James and Mary Polk and two of their ten children. The Polks resided in rural Raleigh County, West Virginia, and readers learned that Mary Polk provided a critical source of income for the family, her husband having suffered a debilitating spinal injury in a coal tipple accident in 1966. James Polk, by virtue of his long membership in the United Mine Workers of America (UMWA), was guaranteed a disability check each month to supplement the family income. Mary's contribution as a $1.64-an-hour nurses' aid at Beckley Hospital, however, had been cut off since February 12. Mary was on strike for a union election by Local 1199 and had been one of thirty-two strikers arrested on March 9 for a sit-in at the hospital office of Administrator Albert Tieche, who had refused their request for an audience. "I'm 100 per cent behind my wife and all the others on the picket line," said James Polk. "Anyone who works should have a union." Mary agreed, affirming that "We'll stay on the line as long as we have to."

Local 1199, its militant leftist legacy forged in New York City's industrial union movement of the 1930s and nurtured by a commitment to civil rights, social justice, and the dignity of work, had only recently expanded its organizing mission to other areas of the country. In 1969 it had led a widely reported hospital workers strike in Charleston, South Carolina, which ended with increased wages and shop-floor power for workers, but fell short of formal recognition for the union. Now, as the headline above the Polks' cover photo proclaimed, 1199 had come to Appalachia.[2]

Will Appalachians Organize?

When 1199 organizer Larry Harless, a native of Pineville, West Virginia, returned to his home state from campaigns in New York and Baltimore to work on the Beckley campaign, Appalachian hospital and health care workers were historically underrepresented when it came to union recognition.[3] Some of them, such as the workers at Beckley Hospital, labored in so-called "proprietary," or privately-owned, facilities that were highly successful in controlling labor markets and influencing public opinion in Appalachian communities. Others worked in so-called "non-profit" hospitals—that is, facilities that were sponsored by a web of municipal, state, and federal subsidies and controlled by quasi-public boards of trustees and an increasingly professionalized administrative structure. These were not owned by stockholders, and therefore paid no dividends to private investors.[4] Workers in not-for-profit hospitals were not covered by the National Labor Relations Act (NLRA) (1935) and its Taft-Hartley (1947) and Landrum-Griffin (1959) amendments. Local 1199, an autonomous division of the Retail, Wholesale, Department Store Union (RWDSU), had built its base in New York without federal labor law protection, relying on a combination of strict union discipline within a structure of democratic practices, worker education, working-class coalitions, and seamless solidarity during several highly publicized strikes throughout the 1950s.[5]

A convergence of economic and social forces propelled 1199 to the forefront of Appalachian health care workers' consciousness in the late 1960s and early 1970s, long-term factors that convinced Larry Harless that the union would take hold in the region. Harless's conviction grew from his awareness of a history of labor militancy and success in the central Appalachian region, whose basic industries became heavily unionized with the emergence of the labor protections of New Deal-era labor law. West Virginia workers in particular had a rich background of organizational struggle in the coal, glass, and steel industries that contributed to a supportive environment for health care workers in many communities as they developed a conscious identity as industrial workers themselves. A maturing body of literature on Appalachian organizing that challenges the conventional wisdom that southerners and, more specifically, Appalachians, were psychologically "fatalistic" and skeptical of unionization, documents the extent of union support in the mountains. This history, forged in a twentieth-century context of explosive class and economic conflicts in the coalfields, shaped a collective memory and narrative favorable to organized labor in many Appalachian families and settings, a narrative undoubtedly familiar to Harless. The tangible benefits accrued through UMWA and the United Steelworkers of America contracts during and after World War II strengthened a widely held belief in unionization as a means to rectify the feudal structure of Appalachian industry in pre-Depression years.[6]

Moreover, structural changes in the American health care industry accelerated dramatically after World War II and contributed to the growing attractiveness of unionization to health care workers. Pressured by the United Mine Workers of America, the U.S. Department of the Interior issued the "Boone Report," an extensive survey of coal-mining regions. The report documented the scarcity of basic health services, high infant mortality, and shortages of physicians in rural Appalachian communities. These findings compelled Congress to pass the 1946 Hill-Burton Act, which was based on various schema of matching federal and local funding in order to subsidize the construction or expansion of modern hospitals in rural areas.[7] For example, Hill-Burton financed about 90% of a ten million dollar expansion of Fairmont General Hospital, a non-profit in Fairmont, West Virginia, in the mid 1960s.[8]

In *Hospital Administration*, a trade journal for health care management, Jonathon S. Rakich summarized the transformation of the private and public health care sectors, attributing the industry's growth to "three major factors": (1) Expanded third party hospitalization benefits, primarily through union contracts and the Great Society Medicare and Medicaid programs—by 1967, over eighty percent "of the civilian population had some form of hospitalization benefits"; (2) advances in medical technology, resulting in "increased specialization of services" that brought physicians into the hospital and away from general practice; and (3) increased disposable income combined with growth in the elderly and young demographics, sectors "more susceptible to illness." These growth factors, of course, "resulted in the need for more hospital employees," including employees with more formal training in their respective technical, service, and professional fields. Technological advances, said Rakich, required more highly skilled personnel "who, in turn, want correspondingly higher wages." The Fair Labor Standards Act of 1967 required hospitals to pay "premium wages for overtime," and other personnel costs (benefits, productivity, training) were also high on the agendas of what Rakich termed the "aggressive unionization" campaigns by 1199 and other health care unions. From 1961 to 1967, collective bargaining agreements in non-profits increased by 90 percent—this occurred when workers in not-for-profit facilities were yet to be covered by the NLRA. The wage and other fiscal implications of hospital unionization are obvious, but Rakich further cautioned that the "major consequence of a union is typically to lessen management's control" in the area of costs and work assignments.[9]

It is important to point out that 1199 sought to organize health care workers on an industrial union model. The organizing principle of the union was based on the general ideal of a community of all workers within what Barbara and John Ehrenreich called "the American health empire." Doctors, social workers, nurses, nurses' aides, laboratory technicians, food service workers, maintenance personnel, drivers, and laundry and janitorial staff all shared a common identity as industrial workers to 1199. This model was, not surprisingly, often problematic in workplace cultures and communities that were historically influenced by occupational and class hierarchies. Sometimes, highly professionalized physicians and nurses resisted the notion that they were, according to 1199, part of a working-class. On the other hand, some Appalachian health care professionals readily embraced the industrial analysis promoted by 1199, realizing that the modern hospital was roughly parallel to the "vast corporation" that had displaced the "small factory" at the turn of the century. These professionals, in all likelihood, accepted the spirit, if perhaps not the language, of the Ehrenreichs' assertion that independent physicians had "ceded the field to medical institutions." "The doctor's office," they said, "has been displaced from the center to the periphery of modern medical care; and the doctor himself has become a vassal to what was once just a rent-free workshop—the hospital."[10]

Whereas medical practitioners had traditionally enveloped themselves—or been expected to envelop themselves—in an "aura of selflessness and self-imposed poverty," and modern hospitals perpetuated the mythos as being places where "the few toil ceaselessly that the many might live," the new reality of rationalized, subsidized, and tax exempt "non-profit" hospitals was that by the 1960s the health industry was a big business.[11] Nursing, by virtue of the clear hierarchies in the office or hospital, had traditionally been free of labor-management disputes over wages and working conditions. Nursing was seen as "not merely a profession . . . but a ministry."[12] Hospital management, sometimes clumsily and sometimes expertly, fell back on the selflessness model of health care work (often called "Nightengalism") in every episode of emerging labor consciousness when 1199 came to Appalachia. "Health care executives," wrote William Rothman, argued that unionization and strikes "are intolerable because they endanger lives or delay the relief of suffering people." Actually, Rothman continued, hospital administrators were influenced primarily by "the repulsive thought that management will lose control of the operation of the institution."[13]

To maintain such control, hospital management redoubled the "rationalization" of the workplace by expanding scientific management. Scientific management, pioneered in the early twentieth century by industrial engineer Frederick W. Taylor, encompassed the division of work practices into identifiable pieces in order to quantify and enhance productivity and efficiency. Nurses and other workers "learned quickly that 'efficiency' was often synonymous with speed-up,"[14] and, as the increasingly systematized and lucrative health care structure matured, they were less reluctant to call management practices into question.

As an industrial union committed to the idea of workplace solidarity, 1199 continually confronted a culture of occupational status that segregated health care workers. Union staff and membership struggled to awaken and nourish a sense of common goals and vision that bridged these divisions, with mixed results. The rapid growth and increasing complexity of the modern hospital "required an explosive growth in the number of non-professional health workers—nurses aides, orderlies, ward clerks, therapy aides, community health and mental health aides, and so on," many drawn from minority and poor white communities who came into the industry skeptical of "professionalism." Conversely, it was not uncommon for highly-trained professionals, physicians especially, to disdain the notion of common identity with laundry or food service workers for purposes of collective bargaining. One 1199 nurse, a key activist during the 1975-1976 organizing campaign at Highlands Regional Medical Center (HRMC) in Prestonsburg, Kentucky, pinpointed some internal obstacles to solidarity:

> Surgery fought us harder than anybody. And the kitchen. Surgery and kitchen, probably, because they had a lady down in the kitchen, she was pretty hard on them and they were pretty scared. The people in the kitchen and the people in surgery. Now, the people in the kitchen were fearful, but the people in surgery, theirs was more "We're too good type of people to be involved in a union."[15]

Obviously, with rare (*exceedingly* rare) exceptions, hospital managers in the Appalachian industry were adamantly opposed to unionization and even more appalled and threatened by the idea of a strike. Their stated reasons for public consumption—some perhaps sincere, others manipulative—emphasized the exceptional character of health care as distinct from manufacturing, construction, mining, and other services, portraying the hospital as a "special institution where life and death are linked to a reliable and loyal workforce."[16] Kenneth Wood, executive director of Cabell-Huntington Hospital in Huntington, West Virginia, played this card beautifully during a 1977 work stoppage at the hospital. Wood had spearheaded the hospital's unsuccessful effort to keep 1199 from organizing service and maintenance workers and licensed practical nurses at Cabell-Huntington in 1975, employing textbook management tactics in what was commonly being referred to (euphemistically) as "union avoidance." When Wood and the hospital resisted serious bargaining for a contract renewal in 1977, the union struck. A surgical nurse chastised the strikers, and Wood spoke with reporters while standing in front of the hospital's newborn nursery. Jack Seamonds of the *Huntington Herald-Dispatch* documented their concerns:

In surgery, a tough-talking nurse stands with her hands firmly on her hips. "I was telling the doctors this morning that this was probably for the best," she said. "At least the professional people, the dedicated people, are here working. The doctors might get to like it."

Minutes later, executive director Wood is standing in front of the nursery. "We have 39 new babies in here right now," he says to no one in particular. He sighs. "This is what I can't understand. How the hell do you walk out on babies?"

And outside, the sky is still gray and the rain is still pouring down and the employees still are marching with signs demanding they be paid what they deserve.[17]

Indeed they were marching. But Seamonds suggested that hospital strikes, including this one at Cabell-Huntington, were strictly about wages. By doing so, consciously or not, his report reinforced a common theme that management sought to instill in the public, that striking health care workers were self-centered, grasping, even dangerous to the public health. True, 1199 members at Cabell-Huntington sought wage adjustments in contract renewal talks, but wages in hospital strikes could not be separated from other bargaining issues including staffing, job assignments, scheduling, training, grievance handling, job security, and, frankly, respect. Leon Davis often spoke of the important role that self-respect—as responsible, compassionate, valuable workers—played in the 1199 idea. Productivity increases, he said, "when a person thinks his job is important, and when the pay is important, then one of the major responses by workers is that they are 'somebodies' . . . . It means your job is important and you have to do it well. I don't know whether this approach performs miracles, but I think there is a much better feeling if you see yourself as a more useful person."[18]

Kenneth Wood undoubtedly understood that his reference to abandoned babies would elicit the management's desired response from some of the readers of the *Herald-Dispatch*. Contrary to the common management portrayal of 1199 as strike happy, however, the staff and members understood the seriousness and the risks to their public relations in strike situations. According to Barbara Melosh, "strikes posed a special dilemma for nurses":

If they relinquished the threat of withdrawing their labor, nurses negotiated from an untenable position. Yet hospital strikes raised serious ethical questions for many nurses, who agonized over refusing to care for the sick even if they defended their own rights as workers and fought for conditions that would improve patient care. Walkouts also created a tangle of tactical difficulties. Without a sustained public appeal, striking nurses were likely to find themselves isolated, as administrators won the support of an angry and uncomprehending community.[19]

Barbara Frame, a licensed practical nurse (LPN) of eighteen years' standing, embodied the nurse's dilemma Melosh articulated. Frame worked at Wetzel County Hospital in New Martinsville, West Virginia, when LPNs struck for union recognition by 1199 in the summer of 1979. The strike was an "absess," she said, "and if it lasts any longer the deeper the wound will be and the deeper the scar that will be left." She anguished over the immediate welfare of her patients, but believed that without the protection of a union contract she and other nurses would not have the staffing or the protection to compel management to provide the best care possible for the community. "A union will protect my job so I won't have to worry about speaking out about some of the small but significant problems this hospital has," she said. When the strike began, Frame stayed home in solidarity but kept away from the picket line. When she received a letter from management threatening her with firing, she joined the picket line, along with her sister Geraldine Wilcox. "I love my work and I sincerely don't want to see people hurt by this strike," said Frame. "We just want job security. The way it is now, if the hospital don't like the color of your eyes, they can fire you."[20]

Organizing Appalachia
The initial organizing campaigns for 1199 in Appalachia were in small proprietary hospitals and nursing homes. Larry Harless was confident that health care workers in the West Virginia coalfields, especially, would be sympathetic to unionization by virtue of the generally favorable impression most coalfield families had of the United Mine Workers. The first victory was at Madison General Hospital in Madison, West Virginia, described by *1199 News* as "a coal mining area where automation has produced extremely high

rates of unemployment." Shortly thereafter, Madison General shut down. The next win was at Doctors Memorial Hospital in Welch, a small proprietary owned by the Wisconsin Shoe Company of Kansas City, Missouri, where the employees were 40 percent black and 60 percent white. Staffers recall that the hospital sought to exploit racial differences, a common tactic faced by 1199, yet on February 2, 1971, Doctors Memorial LPNs, service workers, and maintenance employees voted 108-28 to be represented by 1199. Since Doctors was a private, for-profit facility, the National Labor Relations Board (NLRB) conducted the election, which then required the parties to begin collective bargaining negotiations. It should be noted here, however, that although the National Labor Relations Act requires that the employer *bargain* in good faith after a union win, it does not require that the bargaining ultimately guarantee a contract. The hospital therefore stalled contract talks, a common tactic by which management seeks to engender discontent and frustration with the newly elected union. In part because 1199 educated workers about such tactics and insisted that rank-and-file members carry much of the responsibility in the election campaign and contract negotiations, the hospital's strategy eventually failed. The new union members finally authorized a strike, which forced the hospital to agree on a first contract after twelve days. Some time later the hospital's clerical workers also voted to join 1199, with organizers Harless and Wendell Drake representing the union.[21]

A few months after the win at Doctors Memorial, Robert Muehlenkamp of the national union joined Harless to conduct three days of delegate training at the facility, a critical function of 1199's concept of democratic unionism. Delegates from the various job classifications represented in the bargaining unit—in the Doctors Memorial case, delegates included four LPNs, five nurses aides, one maid, three orderlies, and one laundry worker—were introduced to the time-consuming and often mundane life's blood of organization-building and contract administration. Muehlenkamp notified national organizing director Elliott Godoff that the training included running through

> several skits with the Delegates talking with New Hires about signing check-off cards and becoming Union members. . . . Generally, the Delegates are very smart and totally dedicated to the Union. I think that with the little classroom

training we were able to get in and the procedures we set up that they will be able to get along pretty well on their own most of the time.

I hope that we accomplished enough so Larry can be "freed" of that administrative burden. I think Larry feels better about it—feels he can leave Doctors' Memorial except for the regular meetings and emergencies. In fact, he and I went out to two other hospitals in the area and handed out a few cards. He is anxious to get things started again.[22]

Muehlenkamp had worked on a successful Johns Hopkins organizing campaign while Harless himself was working for 1199 in Baltimore; the tightly-knit and idealistic organizers for the union adopted the slogan "You organize or die," and Harless had earned a reputation among the staff and membership for personal courage and commitment. That reputation was enhanced when he "got things started again" in the form of the organizing drive at Beckley Hospital in Beckley, West Virginia, a campaign that was already underway when Harless and Muehlenkamp ran the delegates' training in Welch. In Beckley, Harless, Mary Polk, and the hospital's other pro-union workers faced a concerted resistance campaign led by hospital administration in conjunction with the town's mayor, John McCulloch.

McCulloch was part owner of the hospital and served as the facility's vice-president. The families of former West Virginia governor Hulett C. Smith and hospital boss Albert Tieche also partially owned Beckley Hospital, Inc., and commanded significant social, political, and economic power in Beckley. According to *1199 News*, the hospital board of trustees was dominated by "prominent banking, real estate and insurance interests." After six months, marked by multiple arrests of Harless, arrests of pro-union picketers, and at least one serious assault on Harless (the organizer was sucker-punched on the picket line on May 21 and suffered three broken bones in his face; his attacker was fined ten dollars in municipal court), exhausted 1199ers lost the NLRB representation election at the hospital.[23] In the wake of this crushing defeat, 1199's tiny staff of Drake and Harless organized a handful of small nursing homes in West Virginia and Kentucky. Then, in 1973, 1199 burst into the Appalachian consciousness with a dramatic campaign in Richlands, Virginia.

Harless could be excused if he was less than optimistic about 1199's chances of a win at the Clinch Valley Clinic Hospital in Richlands, which was owned by Bluefield Sanitarium, Inc. Virginia was (and is) a "right-to-work" state, meaning that even if the employees at Clinch Valley overcame significant management obstacles to win recognition for 1199, they would have to sustain the union not by contractual union security and automatic dues checkoff but by voluntary contributions by the membership. In spite of Virginia's right-to-work status, which was enshrined as a state legislative prerogative by Section 14(b) of the 1947 Taft-Hartley Act, service and maintenance workers at Clinch Valley voted for the union in an October 1972 election.[24] After thirteen weeks of "fruitless contract negotiations with hospital administrators," 150 Clinch Valley workers, comprising about 85 percent of the total work force, went on strike over wages, working conditions, and other contractual matters over which management resisted bargaining. Essentially, the hospital refused to recognize 1199 as the legitimate bargaining agent for the employees.[25]

Immediately, the hospital secured an anti-picketing injunction under a Virginia law prohibiting strikes against *not*-for-profit hospitals—which obviously did not include Clinch Valley. Harless and the strikers proclaimed the injunction illegal and set up picket lines, whereupon over sixty were promptly arrested. Before the U. S. District Court in Roanoke amended the injunction to allow limited picketing, Clinch Valley had witnessed the "largest mass arrests" in Tazewell County history. The herding of 140 workers, most of them women, mobilized the strikers and much of the public. "The big mistake," said one striker, "was when they arrested everyone's wives, mothers, and grandmothers . . . The miners didn't like that. The townspeople didn't like that."[26]

Most assuredly the local miners did not like the arrests. United Mine Workers of America President Arnold Miller joined strikers on the newly sanctioned picket lines. More dramatically, five thousand UMWA miners at eighteen operations staged a one-day walkout in solidarity with the Clinch Valley workers. Miller threatened to withdraw UMWA financial support for the hospital. Faced with snowballing resistance beyond the shop floor, and unable to control the public's perception of the strike, the hospital at last recognized the union at Clinch Valley and signed a contract.[27]

The Clinch Valley strike was enormously important in raising the profile of 1199 to health care workers in Appalachia, but beneath the euphoria of a watershed victory there were signs of tension between Larry Harless and the

National Union. One source of that tension emanated from a directive from Leon Davis to Harless that claimed Harless had "improperly used" some of the strike funds provided by the national to the Clinch Valley campaign. Harless had promised support for eight workers whom the hospital fired during the strike until the time they would be reinstated. The hospital adamantly refused to reinstate two of these workers (apparently for cause), whereupon Davis, who was already exercised that Harless had distributed the funds without prior approval, informed him that "I believe that the payments of monies that these strikers have been extended is already beyond a reasonable time and the workers ought to be told that no further money would be forthcoming." At that point, Davis's objections might be construed as simply a business matter, but Davis arguably pushed the matter. To the National Union president, Harless's actions represented "a gross violation of Union rules bordering on misappropriation and will not be tolerated."[28]

One can only speculate as to the impact such comments had on Harless, but it is reasonable to assume they were distressing. Harless had taken on a formidable task in working, together with organizer Wendell Drake, to prepare the ground for 1199 in Appalachia. He had been beaten and repeatedly threatened and, a year after the strike ended, served four months in the Wise Correctional Unit for violating the original injunction in the Clinch Valley strike, even though the injunction had been issued in violation of the law. When he finally resigned from 1199 in August of 1975, Harless, who was married with six children, was being paid a salary of $220 per week. Upon separating from 1199, Harless went on to earn a law degree and worked in public interest law in West Virginia.[29]

The Appeal of 1199

With the departure of Larry Harless, Wendell Drake worked with two organizers who had joined the West Virginia-based staff in the aftermath of the Clinch Valley strike. Danie Joe Stewart was about thirty years old and a native of Salt Rock, West Virginia. Tom Woodruff was a few years younger and a native of Charleston. As 1199 in Appalachia embarked on an ambitious organizing agenda, one which was propelled significantly by developments in federal labor law, these core activists worked out of a small garage on Charleston Avenue in Huntington. Stewart and Woodruff had each been involved in the campus antiwar movement at Marshall University and had been principle members of the Marshall chapter of the Students for a Democratic Society (SDS). The presence of SDS in Huntington had inflamed a group of conservative ministers and townspeople, culminating in a series of public meetings, letter writing campaigns, and demonstrations for and against recognition of the radical organization. In the end, Marshall President Roland Nelson, resisting intense public pressure, recognized the Marshall chapter of SDS. He noted that the organization's charter had nothing that "indicated advocacy of illegal acts or violence," even though local critics of the group were convinced that the violent disruptions and damage to property that marked SDS on some campuses was part of the organization's local agenda. No member of Marshall's SDS was singled out more in letters to Nelson than Stewart. One correspondent warned: "A member of my family who knew Stewart, the head of the SDS group at Marshall, says 'he will not give up.' Show this punk that you are stronger than he and his backers."[30]

Stewart and Woodruff built a working alliance during the tumultuous campus conflicts at Marshall, an alliance that continued in the labor movement until Stewart's death in 1997. Stewart arrived at Marshall as a freshman in 1963 and quickly gravitated toward the Civic Interest Progressives, a student-based civil rights organization on the campus. He originally worked to establish SDS at Marshall in 1965, but Nelson's predecessor, Stuart Smith, managed to deflect the attempt. In 1966, Stewart became part of a select class when he was drafted into the United States Marines. Conscription into the Marines affected only a handful of American men, and friends of Stewart have religiously attributed his February induction as retaliation by the local draft board for his emerging radicalism. When he returned to Marshall in 1968 after two stateside years in the service, Stewart joined Woodruff and others on a campus mobilized by the dramatic national events of 1968.[31]

It is not surprising that campus activists interested in working-class issues were attracted to 1199 ahead of other unions. Leon Davis had been working as a pharmacist and was active in the Trade Union Unity League (TUUL), the "labor organizing arm of the Communist Party," when he approached RWDSU for assistance in founding the Retail Drug Employees (the forerunner to 1199) in 1932. Davis's union was one of several Jewish and Communist-led New York unions in the eye of the organizing storm in the 1930s renouncing the historical separatism of clerks and pharmacists in favor of the industrial union model later institutionalized by the Congress of Industrial Organizations. Led by Davis and Godoff, a seasoned drug store organizer who joined Davis in 1957, 1199 insisted that a union—for ethical purposes and in order to survive—emphasize not only wages and benefits but social equality. Health care workers desiring representation from 1199 had to set aside differences based on education, class, race, sex, or ethnicity. According to long-time 1199 cultural officer Moe Foner, "1199's inclusiveness was both a tactical recognition of the need for unity in fighting the boss and a part of the American Left's active opposition to racism and ethnic prejudice." Solidarity across racial, ethnic, and craft lines was "not always easy," said Foner, but "became especially important when the union was faced with the highly diverse and stratified hospital industry."[32]

Davis acceded to the anti-Communist provisions of the 1947 Taft-Hartley Act, but 1199 continued to nurture the militant tactics, inclusive vision, and dogged commitment to political and economic education from the TUUL days. The union represented a break with the glacial bureaucratic unionism of the mainstream Cold War-era labor movement, providing a working class framework critical of corporate capitalism and embracing a "larger social agenda" than the mainstream unions. Members were predominately black, Latino, and female, "precisely those the big unions had failed or neglected to organize."[33]

Several factors solidified the union's appeal for campus activists. In March 1965, 1199 became the first national union to condemn the bombing of North Vietnam. Then, it won additional kudos from activists during the highly visible Charleston, South Carolina, hospital workers strike of 1969. Al Haber, the first president of SDS, had long advocated that the membership transcend the mainstream labor movement's poor reputation among student leftists and educate themselves and other students on the history and importance of unions. "Simultaneously," says Peter Levy, "Haber urged progressives within the labor movement to assert themselves."[34] It is likely that word of the hospital workers union reached Stewart and Woodruff on the Marshall campus, just as it had reached Bob Muehlenkamp at the University of Wisconsin at Madison. Muehlenkamp was a founder of the Wisconsin Teaching Assistants' Association (TAA) in 1967, which grew from the antiwar movement on campus.

The TAA became a "progressive force" within the state AFL-CIO and unsuccessfully lobbied that body to petition the national federation to break with the American Institute for Free Labor Development, a CIA front. The TAA was deeply involved with the Teamsters for Democracy movement, and "served as a training school or entrée into other unions."[35] It should be noted as well that while Muehlenkamp was beginning his career with 1199 in Baltimore, Tom Woodruff had served as an occasional contributing editor to *Mountain Life and Work*, the venerable monthly published by the Council on Southern Mountains that included coverage of labor struggles in each issue and featured several articles on the Clinch Valley strike not long before Woodruff went on 1199 staff.[36] Woodruff also helped operate a printing collective in Huntington called the Appalachian Movement Press. The press specialized in the reprint of pamphlets, broadsides, and poetry from Appalachian poet/agitator Don West and from the National Miners Union, the TUUL's radical alternative to the UMWA in the dual union period of the early 1930s.

Shortly after Woodruff and Stewart began working for 1199, the Democratic Congress passed and President Richard Nixon signed into law an amendment to the National Labor Relations Act that extended collective bargaining rights to employees at non-profit hospitals. This was a watershed development for 1199 and other health care unions, holding the promise of dramatic increases in membership, resources, and protection for health care workers. There were many reasons prompting the Congress to act—health care unions were militantly lobbying state legislatures to expand state protections for workers in the non-profits; hospital workers schooled in the civil rights, Black Power, women's rights, and antiwar movements were demanding representation and often ignoring the vague and increasingly meaningless nineteenth-century designation "not-for-profit" in an era of corporate health care; and some hospital administrators wanted the expansion of the NLRA so as to exercise some bureaucratic control over uncertain and potentially explosive labor-management relations.

Bob Muehlenkamp, contacted by workers from Highlands Regional Medical Center in Prestonsburg in the summer of 1975, welcomed their interest in 1199 and noted that "tens of thousands of hospital workers in twenty-two states" had capitalized on the new law and joined health care unions. At the second annual convention of the national union in December 1975, Leon

Davis reported that when Congress passed the 1974 NLRA amendment, "it was hailed as a new day for hospital and health care workers." The momentum for 1199 in Appalachia and elsewhere, however, was not uninterrupted. In the same report, Davis also told the delegates that "our expectations were unrealistic. What we didn't reckon with is that to pass a law is one thing, but to have it administered fairly and impartially is something altogether different." As it turned out, the "new day" held promise not only for health care workers, but for another occupational group—management consultants.[37]

When Bosses Organize

In the spring of 1975, some seven hundred service and maintenance workers and LPNs at Cabell-Huntington Hospital prepared to vote on whether or not to be represented by 1199. Woodruff was the chief organizer, and he and pro-union workers had petitioned the hospital's board of trustees three times in a series of meetings with them. The approval of the board was imperative, since as a publicly owned hospital Cabell-Huntington workers did not fall under the jurisdiction of the NLRB; therefore, the hospital board would have to consent to an election and would designate the bargaining units. On the fourth try, the trustees agreed to an election, but only if it covered all employees, including registered nurses and physicians who were skeptical of or opposed to unionization.

Using organizational and mobilization tactics typical of 1199, Cabell-Huntington union supporters demonstrated, leafleted, picketed, visited workers' homes, wrote songs and poetry with working-class themes tailored to their workplace, and held press briefings and rallies. They lined up solid support from the Huntington district AFL-CIO Labor Council; the local steelworkers', electrical workers', and chemical workers' unions; and the Huntington Policemen's Association. The hospital board, which included Mayor Harold Frankel, committed to managing a fair election, along with the city council and the Huntington Ministerial Association. The American Arbitration Association conducted the April 2 election, and 1199 won a close election (389-327) that marked a dramatic change in labor-management relations at Cabell-Huntington. When, after tough negotiations, the hospital and union signed a two-year collective bargaining contract, Cabell-Huntington became the largest organized hospital in the Appalachian region.[38]

The Cabell-Huntington campaign was significant not only as another important step for 1199 in Appalachia, but as an indication of the growing sophistication of anti-union strategies used nationally by health care administrators in the era of the industry's corporate restructuring. Determined to either avoid unionization in the first place, or to get rid of (decertify) existing unions at their facilities, management professionals adopted a systematic mixture of public relations, co-optation of workers, one-to-one (or "captive") boss-worker meetings, internal propaganda, and, at times, intimidation in order to keep hospitals "union-free." When Cabell-Huntington administrator Kenneth Wood referred to 1199 as an "outside third party who is not even a member of our family," warned that the hospital had the power to "permanently replace" workers striking over economic issues, or said that unionized workers "will lose their right to bargain individually with management," Wood was not the only administrator who thought that way.[39] Indeed, supervisors at Highlands Regional in Prestonsburg, also fighting 1199 in 1975 (where Stewart was the chief organizer), convened lengthy mandatory anti-union meetings that portrayed 1199 in Kentucky as the mouthpiece of a "nigger union" and a "communist union." Hospital workers also recalled that supervisors "tried to tell us that health care workers, you know, should never organize. What about the poor, sick people that we take care of?"[40]

These were not especially innovative or unusual tactics; these statements expressed the collective voice of most health administrators who, in the age of corporate health care, turned to a variant of hired guns to ensure control of workers and the workplace. Managers, human resource officers, and business organizations referred to these specialists as "management consultants." Ward Clerk Penny Burchett of Highlands Regional used the common term among workers: "They [HRMC management] also hired union busters." "Union busters is what we called them; let's just call it like it is, John," said Medical Technologist Larry Daniels. Daniels, along with Burchett and several others, worked with Stewart on the successful HRMC election and the contract bargaining committee in 1976. It is probably accurate to conclude that not all management consultants faced by 1199 were union busters, but all union busters identified themselves as management consultants.[41] In the Highlands Regional case, technicians, service and maintenance workers, housekeepers, and LPNs beat back the hospital's tactics and voted for 1199, 138–73—the first hospital workers' union win in Kentucky. "Despite the brainwashing and harassment we stuck together," said nursing technician Rebecca Wells. "The hospital really campaigned dirty, but we sent their union-busters packing."[42]

It was the proliferation of management consultant firms, combined with increased sympathy for management within the National Labor Relations Board bureaucracy, which led Leon Davis to conclude his union had unrealistic hopes. "While we expected that our efforts to organize the unorganized would meet with hostility from management," said Davis, "we did not anticipate the obstruction of the National Labor Relations Board."[43] Problems came particularly from the five-person board itself, which is appointed by the president of the United States and whose budget is drafted and administered under executive authority. Alan Kistler of the AFL-CIO identified consultants as "the new technicians of a new movement for totalitarian control. We estimate that today labor-management consultants are involved in two-thirds of all NLRB elections in planning strategy." Professional associations and universities, said Kestler, train management representatives to "exploit the fears of workers and convert those fears into a rejection of collective bargaining."[44] Administrators could also rely on an expansive pool of proscriptive literature in labor relations texts, trade journals, and manuals such as *How to De-Certify a Union: The Procedural and Practical Guidebook to Decertification*, by Memphis-based management lawyer Ted Yeiser.[45]

The principles and language of union avoidance begat a kind of secular liturgy that, over the years, linked management throughout the industry. For example, Kenneth Wood's sixteen "Facts You Should Want to Know" about the dangers of unionism, issued to employees during the Cabell-Huntington election campaign, summarize virtually intact the anti-union principles outlined by Yeiser in the foreword to *How to De-Certify a Union*, which was not published until 1979. Wood's practices during the campaign—letters, persuasion, "velvet hammer" intimidation, threats of layoffs or extended strikes— just as closely parallel an extended "Fact Sheet" and an advisory on the "Supervisor's Role in a Union Campaign"[46] that appeared in *Labor Relations in Hospitals* in 1984.[47]

Analysts of the decline of union density (that is, the percentage of union members in the private and public sectors, which has eroded from about one-third to one-tenth of the workforce in private-sector companies since the mid-

1960s) often debark from President Ronald Reagan's firing and permanent replacement of striking members of the Professional Air Traffic Controllers Organization (PATCO) in 1981. This episode was actually more of a catalyst than a commencement in terms of union decline. Employers determined to make American capitalism "union free" were certainly encouraged by Reagan's signal that union-busting "had support from the highest levels of government." But as evidenced by the experience of 1199 in Appalachia, and nationally, aggressive antiunionism was already being institutionalized.[48] These developments coincided with aggressive efforts by 1199 and other service sector unions to carry the social rights consciousness of the 1960s into the predominantly low-wage, minority, and women's health care industry. Former management consultant Martin Levitt, who made a fortune working for Modern Management Methods (3M), the most powerful management consultant firm in the country, bared his soul in the penitential *Confessions of a Union Buster* in 1993. He documented the alarm that permeated the health care industry:

> During the 1960s administrators of hospitals and nursing homes across the country watched with horror as their housekeepers, cooks, and nurse's aides took to the streets by the thousands in virulent protest of the poverty and indignity imposed on them by their employers. In 1974, with the blessing of the American Hospital Association, which wanted to end the devastating chaos within its industry, President Nixon signed the NLRA amendment into law, giving employees at the nonprofits the right to collective bargaining. Then the union-busting door burst open. Although 1199 and other unions had organized a great number of nonprofits before passage of the law, the amendment acted as a catalyst of fear on hospital administrators, particularly those in charge of small facilities where workers had felt they didn't have the strength in numbers to force union recognition, as did workers in huge big-city hospitals. . . . Local 1199 filed thousands of organizing petitions in the wake of the 1974 amendment. That, of course, meant thousands of jobs for anti-union consultants.[49]

Local 1199 encountered union avoidance strategies in every election, at every first contract negotiation, and at the expiration of every contract throughout its early years in Appalachia. Some administrators simply were tough bargainers and did not go beyond that into the realm of intimidation or violation of poorly enforced NLRB rules. Most times, however, the touch of union-busting sullied their efforts. Muehlenkamp, as organizing director of the National Union in the 1970s and 1980s, often saw 3M in action against 1199—although it appears that other companies, as well as local or regional lawyers from firms with specialists in union avoidance, prevailed in the Appalachian cases. It was not uncommon for them to advise supervisors to use "delay, intimidation, saturation propaganda campaigns, efforts to destroy rank-and-file leadership by transfer, promotion, firing, or character assassination; and efforts to control local mass media and politicians."[50]

Conclusion

When the newly unionized workers at Cabell-Huntington and Highlands Regional negotiated their first contracts, they proved that the first six years of 1199 in Appalachia had witnessed significant victories for health care workers and significant growth for the union. But, as forthcoming studies will show, 1199 faced powerful obstacles as it struggled to sustain its momentum and expand its power as a voice for labor. Management at Cabell-Huntington and Highlands negotiated with the union, but were resolved to decertify when the first contracts expired. The Huntington hospital essentially forced a strike in 1977, but 1199 held and won a new contract. At Highlands, management imported a powerful consulting firm, Southeastern Employers Service Corporation (SESCO) to break the union in 1978, but 1199 held and won a new contract. At Fairmont General in 1978, nurses led an organizing push that the hospital bitterly resisted; after a six-week strike, 1199 held and won a first contract. In 1980, 1,500 1199 members formed a new autonomous district, the seventh authorized by the National Union, to become 1199 WV/K. "Your history in 1199 has been one of fighting effectively and with determination," Leon Davis told the executive board of the new district as he swore them in on July 30, 1980, in Huntington. "Management, on the other hand, is stubbornly anti-union, so we have to renew our commitment to stay the course with 1199."[51]

Davis was right. While the new district took effect, 1199 was mired in a bitterly divisive and failing organizing battle at Wetzel County Hospital in

New Martinsville, West Virginia. Shortly thereafter, a crucial campaign at Charleston General Hospital failed, weakening 1199's resources and energy. Fairmont General launched a concerted effort to break the union in 1987, prompting a bitter strike. New management at Highlands Regional pushed a bruising strike and brought in a security force appropriately named the Nuckols Agency. Mary Schafer, a nurse at Fairmont General since 1969, who was one of the leaders of the 1978 recognition struggle and has sat on every contract bargaining committee since then, sums up 1199's task even where it has existed for nearly thirty years:

> I think people were so ready to be organized. They were so disgusted. We didn't get any overtime pay. We did not get any kind of holiday. We had to work holidays. We had to work stretches of, sometimes, thirteen, fourteen days in a row without a day off. They had this administrator who was like a southern gentleman and he didn't even like women, you know, at all—much less women telling him they're going to try to get a union. . . They have fought it ever since then. From the beginning till now. Till right now. They want rid of it.[52]

NOTES

1   The author thanks Justin Yelton and Kathryn Reeder, undergraduate fellows at Morehead State in 2006–2007, for their work on this project.

2   "1199 Comes to Appalachia," *1199 News* 6, no. 5 (May 1971): cover, 13–15.

3   There was a significant exception. Workers in the network of ten not-for-profit Miners Memorial Hospitals, established by the United Mine Workers of America Welfare and Retirement Fund in 1956, enjoyed collective bargaining contracts. When the UMWA sold the hospitals to the Board of National Missions of the United Presbyterian Church in 1963, they were renamed Appalachian Regional Healthcare. The ARH system expanded beyond the hospital bases and established clinics and primary-care centers throughout much of the region. Union contracts at the ARH hospitals were then serviced by the United Steelworkers of America, later joined by the Kentucky and West Virginia Nurses Association. According to Grant Crandall, now a UMWA lawyer who represented Local 1199 in the 1970s,

UMWA president Tony Boyle was happy to see the hospital rank-and-filers go, since many of them were part of the insurgency challenging his control. Grant Crandall interview with author, March 7, 2008. See Richard P. Mulcahy, Introductory Essay, "Health," *Encyclopedia of Appalachia*, ed. Rudy Abramson and Jean Haskell (Knoxville: University of Tennessee Press, 2006), 1634. Mulcahy is a historian of the UMWA postwar health system, chiefly in his *A Social Contract for the Coal Fields: The Rise and Fall of the United Mine Workers of America Welfare and Retirement Fund* (Knoxville: University of Tennessee Press, 2001).

4 Most hospitals in West Virginia were technically "non-profit" and exempt from taxation. Many did, however, operate in the black. For example, the *West Virginia AFL-CIO Observer* pointed out, "two West Virginia 'non-profit' hospitals during the fiscal year ending Sept. 30, 1975, showed profits most businessmen would like to have. The Charleston Area Medical Center reported a profit of 4 percent, or $1.9 million, and Thomas Memorial Hospital in South Charleston showed an 8.65 percent profit, or $835,353." *West Virginia AFL-CIO Observer*, July 1976, 4.

5 The essential history of the national union from its inception is Leon Fink and Brian Greenberg, *Upheaval in the Quiet Zone: A History of Hospital Workers' Union Local 1199* (Urbana and Chicago: University of Illinois Press, 1989). Chapters 1–5 cover the decades from the founding to the late 1960s. Chapters 3 and 4, "The Brewing Storm: Organizing from the Ground Up" and "The Battle of '59: Anatomy of a Hospital Strike," are especially helpful in establishing the philosophical and organizational framework of Local 1199 in New York, which was largely incorporated into the Appalachian model. As the authors explain, these two chapters "explore the takeoff of the hospital union through the event of the 1959 hospital strike, an unprecedented disruption of the traditional 'quiet zone' around hospital personnel relations. Through a twin focus on rank-and-file mobilization and the competing political strategies of union leaders, hospital officials, and government mediators, we seek to set political change in an appropriate interpretive context," xii, 44–90.

6 A sampling of the literature on Appalachian workers' mobilization in the pre-Depression era would include David Corbin, *Life, Work, and Rebellion in the Coalfields: The Southern West Virginia Miners, 1880–1922* (Urbana and Chicago: University of Illinois Press, 1981); Alan Banks, "Class Formation in the Southeastern Kentucky Coalfields, 1880–1920," in *Appalachia in the Making*, ed. Dwight Billings, Mary Beth Pudup, and Altina Waller (Chapel Hill: University of North Carolina, 1995), and Banks, "Miners Talk Back: Labor Activism in Southeastern Kentucky in

1922," in *Back Talk From Appalachia: Confronting Stereotypes*, ed. Dwight B. Billings, Gurney Norman, and Katherine Ledford (Lexington: University Press of Kentucky), 215–227; Ken Fones-Wolf, *Glass Towns: Industry, Labor, and Political Economy in Appalachia, 1890–1930s* (Urbana and Chicago: University of Illinois Press, 2007); and Ronald L. Lewis, *Black Coal Miners in America: Race, Class, and Community Conflict, 1780–1980* (Lexington: University Press of Kentucky, 1987). For post-World War I, pre-Depression, Depression, and pre-World War II, see Jerry Bruce Thomas, *An Appalachian New Deal: West Virginia in the Great Depression* (Lexington: University Press of Kentucky, 1998); John Hennen, "E. T. Weir, Employee Representation, and the Dimensions of Social Control, 1933–1937," *Labor Studies Journal* 26, no. 3 (Fall 2001): 25–49; and John Hevener, *Which Side Are You On?: The Harlan County Coal Miners, 1931–1939* (Urban and Chicago: University of Illinois Press, 1978). The essential guide to modern Appalachian community-based organizing and working class movements is Steve Fisher, ed., *Fighting Back in Appalachia: Traditions of Resistance and Change* (Philadelphia: Temple University Press, 1993). An important case study of the cultural factors in mobilization is Shaunna Scott, *Two Sides to Everything: The Cultural Construction of Class Consciousness in Harlan County, Kentucky* (Albany: State University of New York, 1995).

7   Mulcahy, Introductory Essay, "Health," 1634.

8   For more information on Fairmont General Hospital Inc., see https://www.fghi.com/PtPortal/OurMission/tabid/85/Default.aspx?PageContentID=20

9   Jonathon S. Rakich, "Hospital Unionization: Causes and Effects," *Hospital Administration* 18, no. 1 (Winter 1973): 7–10. Rakich's summary of the transformational factors in the health industry is largely reinforced by other analysts of contrasting interpretive positions, including Barbara Melosh, *'The Physician's Hand': Work Culture and Conflict in American Nursing* (Philadelphia: Temple University Press, 1982); John Ehrenreich, ed., *The Cultural Crisis of Modern Medicine* (New York and London: Monthly Review Press), 1978; Paul Starr, *The Social Transformation of American Medicine* (New York: Basic Books, Inc.), 1982; Arthur D. Rutkowski and Barbara Lang Rutkowski, *Labor Relations in Hospitals* (Rockville, MD: Aspen Systems Corporation), 1984; and Thomas Barocci, *Non-Profit Hospitals: Their Structure, Human Resources, and Economic Importance* (Boston: Auburn House Publishing Company), 1981.

10   Barbara and John Ehrenreich, *The American Health Empire: Power, Profits, and Politics* (New York: Random House, 1970), 30–31. For analysis of changing power

relationships in the pre-World War I industry, with specific analysis of Appalachian doctors and nurses, see Sandra Barney, *Authorized to Heal: Transformation of Medicine in Appalachia, 1880–1930* (Chapel Hill: University of North Carolina Press, 2000).

11   William A. Rothman, *Strikes in Health Care Organizations* (Owings Mills, MD: National Health Publishing, 1983), 65.

12   Melosh, *Physician's Hand*, 23

13   Rothman, *Strikes*, 63.

14   Melosh, *Physician's Hand*, 166

15   John Ehrenreich, *Cultural Crisis*, 7; Joanne Risner interview with author, September 5, 1998.

16   Barocci, *Non-Profit Hospitals*, 150.

17   *Huntington (WV) Herald-Dispatch*, October 2, 1977.

18   Al Nash, "An Interview with Leon Davis," *Hospital & Health Services Administration* 29, no. 6 (November/December 1984): 99.

19   Melosh, *Physician's Hand*, 201–202.

20   *Wetzel Chronicle*, August 16, 1979.

21   Al Bacon, *History of District 1199/SEIU*, unpublished typescript, nd., 5; "West Virginia," *1199 News* 5, no. 12 (December 1970): 3; "Doctors Memorial clerical workers vote union," *1199 News* 9, no. 4 (April 1974): 31. The third victory for 1199 was at another small proprietary hospital, Holden Hospital in Logan County, West Virginia, where service workers voted 24–4 for union representation; Bacon, *History*, 5; One of the practices of *1199 News* was to highlight the involvement of the rank-and-file in organizing and contract administration. In the Welch campaign, the journal noted, "The Doctors Hospital organizing committee included Joan Davis, Virginia Younger, Pat Williams, William Lindsey, Floyd Baker, Muriel Miles, Ruby Woods and LPNs Roger Roberts, Marian Terry and Naomi Box." See, "Two Wins in West Virginia," *1199 News* 6, no. 3 (March 1971): 22.

22   "West Virginia delegates complete course," *1199 News* 6, no.11 (November 1971): 19. Robert Muehlenkamp to Elliott Godoff, 13 September 1971, collection 5525, box 76, folder 1, Kheel Archives, Cornell University Library, Ithaca, New York. Hereafter designated as: "Kheel 5525, box 76, folder 1, " etc.

23   Fink and Greenberg, *Upheaval*, 186; "Hospital Strike stirs W.Va. town," *1199 News* 6, no. 4 (April 1971), 19; *Beckley Post-Herald*, March 1, 1971, August 13, 1971; *Raleigh Register*, August 12, 1971; and "Org. Harless recovering from beating," *1199 News* 6, no. 6 (June, 1971), 13.

24 Elliott Godoff, "Report on Organization," November 1973, in *Minutes: Founding Convention, National Union of Hospital and Health Care Employees RWDSU/AFL-CIO*, November 28–December 1, 1973, page 5. [Kheel Archives Library]

25 "Unions and Jobs: Richlands," *Mountain Life and Work* 49, no. 2 (February 1973), 11.

26 "Unions and Jobs: Richlands," *Mountain Life and Work* 49, no. 2 (February 1973), 11.

27 "Unions and Jobs: Richlands Wins!" *Mountain Life and Work* 49, no. 3 (March 1973), 15. The figure of 5,000 strikers was widely reported, including by Elliott Godoff, who visited the picket line along with Miller. The miners' walkout ended with an injunction. Godoff, *Minutes: Founding Convention*, 5.

28 Larry Harless to Elliott Godoff, July 21, 1973; Leon Davis to Larry Harless, August 7, 1973, Kheel 5651, box 7, folder 8.

29 Larry Harless to 1199 Payroll Department, August 11, 1975, Kheel 5651, box 7, folder 8; "Four Months in Jail for Peaceful Picketing in Virginia," *1199 News* 9, no. 6 (June 1974), 21; Larry Harless to Leon Davis, August 1, 1975, Kheel 5651, box 8, folder 7; "National Union Payroll---Officers, Organizers, Administrative & Clerical Staff As Of 4/30/75," Kheel 5651, box 5, folder [1].

30 Mary F. Loeser to Roland Nelson. Nelson papers, Marshall University. In John Hennen, "A Struggle for Recognition: Marshall University Students for a Democratic Society and the Red Scare in Huntington, 1965–1969," *West Virginia History* 52 (1993), 136.

31 George Q. Flynn, *The Draft: 1940–1973* (Lawrence: University Press of Kansas, 1993), 233; Thomas D. Morris, "Department of Defense Report on Study of the Draft, 1966," in Martin Anderson and Barbara Honegger, eds., *The Military Draft: Selected Readings on Conscription* (Stanford: Hoover Institution Press, 1982), 548; Transfer of Danie Joe Stewart (service number 2233373) from active duty to Marine Corps Reserve, January 17, 1968, Cabell County Court Discharge Records, Book 43, page 463. For some information on Stewart and Woodruff at Marshall I rely on interviews with: Jeanine Cawood Stewart, April 13, 1986; Tom Woodruff, April 5, 1986; Keith Peters, February 28, 1986; Philip Carter, July 6, 1999; Frank Helvey, July 9, 1999; and Roger Adkins, July 19, 1999. For insights on Stewart and Woodruff with 1199, I rely in part on interviews with Teresa Ball, December 30, 1998; the Philip Carter interview; Wayne Horman, July 12, 1999; and Robert Wages, June 29, 1999.

32   Fink and Greenberg, *Upheaval*, 19–20; Bacon, *History*, 3; Moe Foner, *Not for Bread Alone* (Ithaca: Cornell University Press, 2002), 33–35.

33   Fink and Greenberg, *Upheaval*, 19, 183–186; John Hennen, "Putting the 'You' in Union: The Highlands Regional Campaign of 1975–1976," *Journal of Appalachian Studies* 5, no. 2, (Fall 1999): 231–232.

34   Peter Levy, *The New Left and Labor in the 1960s* (Urbana and Chicago: University of Illinois Press, 1994), 12.

35   Peter Levy, *The New Left and Labor in the 1960s* (Urbana and Chicago: University of Illinois Press, 1994), 158–160.

36   To illustrate, Woodruff is listed as a contributing editor in volume 47, no. 2 (February 1971) and 47, no. 3 (March 1971).

37   Robert Muehlenkamp to Jo Ann Martin (Risner), July 30, 1975. Leon J. Davis, *Report From the President . . . to the Delegates of the Second National Convention*, December 10, 1975, 23.

38   *West Virginia [AFL-CIO] Observer* 8, no. 6 (February 1975), 1; *1199 News* 13, no. 1, historical calendar (January 1978); *1199 News* 12, no. 11 (November 1977), 5; *Huntington Herald-Dispatch*, numerous articles, January–June 1975; *1199 News* 10, no. 5, May 1975, 5; Harold Schlechtweg to author July 28, 2007.

39   Letter from Kenneth Wood to employees of Cabell-Huntington Hospital, March 31, 1975; Booklet, *Important Facts You Should Want to Know*, issued by management during the C-H election campaign. "Hospital Workers" file accessed at the West Virginia and Regional History Collection, West Virginia University.

40   Interview with Larry Daniels, HRMC, July 29, 1998; Interview with Penny Burchett, HRMC, August 5, 1998.

41   Interview with Larry Daniels, HRMC, July 29, 1998; Interview with Penny Burchett, HRMC, August 5, 1998.

42   "First 1199 win in Kentucky," *1199 News* 10, no. 12 (December 1975), 11.

43   Davis, *Report to the Delegates*, 23.

44   Statement of Alan Kistler in *Pressures in Today's Workplace*; "Overight Hearings Before the Subcommittee on Labor-Management Relations of the Committee on Education and Labor of the House of Representatives, 96th Congress, First Session, October 16–18, 1979. Volume I (Washington: U. S. Printing Office, 1979), 41, 53.

45   A sampling of textbooks and articles that collected "union-free" strategies in convenient formats includes Arthur Rutkowski and Barbara Rutkowski, *Labor*

Relations in Hospitals (Rockville, MD: Aspen Systems Corporation, 1984); Bruce Hatfield, president of Hospital Management Resources Corporation, "Quality Circles in Nursing," *Hospital Topics* 60, no. 1 (January–February 1982), 34, 40; Patrick Vaccaro and Doria Saletsky, "How to Preserve the Union-Free Status of Your Facility by Practicing Preventive Labor Relations" in *Hospital Topics* 60, no. 1 5–7; Charles Schanie, "Unionization in Hospitals: Causes, Effects, and Preventive Strategies, *Hospital & Health Services Administration* 29, no. 6 (November–December, 1984), 68–78; and Ronald Greenberg, "Union Avoidance for Hospitals," *Hospital and Health Services Administration* 28, no. 1 (January–February 1983), 24–29.

46   The foreword to Yeiser's primer is reprinted in *Pressures in Today's Workplace*, as part of materials supplied to the Subcommittee on Labor-Management relations by Nancy Mills of Local 880 of the Service Employees International Union, 1979, following page 121.

47   Rutkowski, *Labor Relations*, Exhibit 3-3, "First Letter on Hospital's Position Concerning Unions and Card Signing," 35; "Supervisor's Role," 40–45; Exhibit 3-5, "Disadvantages of a Union to Employees," 43.

48   Kate Bronfenbrenner, et. al. *Organizing to Win: New Research in Union Strategies* (Ithaca and London: ILR Press, an imprint of Cornell University Press, 1998), 4–5.

49   Martin Jay Levitt, with Terry Conroy, *Confessions of a Union Buster* (New York: Crown Publishers, 1993), 72–73.

50   Statement of Robert Muehlenkamp, *Pressures in Today's Workplace*, Oversight hearings, vol. 3, 1980, 172, 189; *1199 News* 14, no. 5 (May 1979), 9.

51   "Huntington Hospital Workers on Strike," *Mountain Life and Work* 53, no. 9 (October 1977), 6–9; "Unity wins contract gains at Highlands Regional," *1199 News* 13, no. 5 (May 1978), 21; *Fairmont Times-West Virginia*, passim, fall 1978; "Introducing 1199 WV/K," *1199 News* 15, no. 9 (September 1980), 22. The executive board included Woodruff and Stewart, and employees from Cabell-Huntington, Fairmont General, and Highlands Regional: Dee Bowyer, Joyce Lunsford, Maude Fay Adkins, Tanya Boggess, Becky Slaughter, Mary Bates, Bill Slone, Alberta Easley, Bonnie Ownes, Ellis Stevens, Jim Taylor, Ermel Cook, and Judy Siders. 1199 WV/K later merged with Ohio to create 1199 WV/KY/OH.

52   Interview with Mary Schafer, August 11, 2004.

Section III

Politics

Mining Reform After Monongah:

The Conservative Response to Mine Disasters

Jeffery Cook

ONE OF FAIRMONT COAL COMPANY'S most modern operations was located in Monongah, West Virginia. In fact, Monongah's interconnecting mines, No. 6 and 8, were considered the jewels of the Fairmont Field, safe and "up to date in every way."[1] The two mines spanned several hundred acres, and by best estimate, the company possessed over one hundred acres for development. By 1907, these operations led the state in machine-mined coal, producing 433,285 long tons with the aid of fourteen cutting machines.[2] Governor Aretas Brooks Fleming and his associates were so proud of the Monongah operations that visiting dignitaries often concluded tours of the Fairmont coalfield at the Monongah plants.[3] Moreover, the company was convinced that the Monongah operations were among the safest mines in West Virginia.[4] Monongah Superintendent A. J. Ruckman, for example, responded to a young woman's query about the safety of No. 8, opining that "some mines might be dangerous, but this one was as safe as her parlor at home."[5] In spite of Ruckman's optimism and the company's pride, in the blink of an eye the Monongah operation was transformed from a showplace into the worst coal mine disaster in American history.[6]

On a damp Friday morning, December 6, 1907, simultaneous explosions occurred in Fairmont Coal Company's No. 6 and 8 mines; fire and dust cascaded from both. A large cloud of smoke spewed from No. 6 and one hundred feet of the mountainside was completely blown out by the blast, leaving a giant crater where the No. 8 mine entrance had once been. The force of the explosion broke windows in houses, buckled the pavement, and produced tremors that were felt six miles away in the distant towns of Fairmont and Grafton. Monongah residents quickly converged on the mines and stood in a state of confusion, watching, praying, and waiting for some word on the status of their family members. As the pit mouth billowed what one observer described as "dirty white smoke," the situation looked hopeless. Of the nearly 400 men and boys who reported to work, only one man, Peter Urban, managed to escape through an air hole to safety.[7]

Fleming and his fellow operators were stunned by the Monongah coal mine disaster. If this was truly a safe operation, how could a disaster of this magnitude occur? Fleming and many contemporary mine owners failed to recognize the full implication of mine mechanization. Machinery introduced new dangers to the mine and the miner, and these hazards had to be addressed by an active ownership. Unfortunately, it took the December 6, 1907, Monongah mine disaster and the deaths of 361 men and boys to drive the point home.

Reluctant Reformer

As if the Monongah disaster and the wave of mine explosions that swept the nation were not sufficiently perplexing, coal operators encountered another challenging problem during the first decade of the twentieth century: the relationship of their conservatism to the changing political culture. Conservatives were beleaguered, and frequently repelled, by the country's movement away from free markets, laissez-faire, self reliance, and limited government toward the world of the centralized state. Conservatives had traditionally looked to Alexander Hamilton, Thomas Jefferson, and Andrew Jackson for strength, but now some embraced reform rather than resisting it. After the Monongah explosion, Fleming adapted to the modern industrial society by organizing a powerful group of coal operators, mine inspectors, and scientific experts to push for the establishment of the Bureau of Mines in 1910. His support for the bureau, however, was not so much a break from the past as an extension of the limited government doctrine.[8]

In terms of politics, finance, and the law, there was no detail too insignificant to escape Fleming's notice, so why did the Monongah disaster occur? Was it simply due to a greedy capitalist who failed to maintain a safe workplace? A historian of coal mining safety in the Progressive era, William Graebner, explained mine safety through the competitive framework.[9]

Graebner dismisses corporate greed and asserts that "industry economics ensured that operators would view safety not as efficiency, but as an expense to be avoided." Therefore, only the largest coal corporations could afford to comply with safety standards.[10] Graebner's competitive model is reinforced by the work of Anthony F. C. Wallace. Wallace provides a thought-provoking analysis of the nineteenth-century coal operator. Even though Wallace's theory focused on the Pennsylvania anthracite coalfield, it is applicable to Fleming and the bituminous coal operators. Wallace argued that coal operators "avoided the sure financial loss attendant upon taking all of the safety precautions and opted for the more risky, and ultimately more costly alternative of neglecting safety and gambling they could get away with it" because of market pressures and the lack of external regulations.[11] In short, coal operators avoided the sure financial investment in making the mines safer and risked an explosion instead. The Monongah mine disaster demonstrated that coal mine explosions were costly in terms of human life and dollars.[12] To permit productive workers to die and interrupt production was an unintelligent way to conduct business.[13]

Therefore, long-term plans for dealing with mine disasters were among Fleming's primary considerations following the positive verdict of the Marion County Coroner's jury. However, the proximate cause of the disaster pathology was open to speculation. Historian Ronald D. Eller has asserted that the "causes of most of the mine explosions, generally the accumulation of gas and coal dust, were widely known," but Eller is mistaken.[14] The coal community was divided over the importance of coal dust and its role in the mine explosions. The essence of the controversy hinged on whether coal dust could ignite and propel an explosion without the presence of methane gas. Fleming and fellow West Virginia coal operator Justus Collins took the dust theory seriously, but other operators were hesitant.[15] Seven years earlier, coal operator William Nelson Page dismissed the possibility that a coal mine disaster at one of his collieries had been caused by coal dust.[16] Other operators joined Page in rejecting the coal dust theory. One Pennsylvania coal operator asserted that coal dust could not "take fire. Don't see how it can take fire."[17] The operators failed to comprehend the changes that occurred with the introduction of machinery. As it turned out, Fleming was right and Page was wrong.

Coal mining disaster and death rates began rising in the United States at the turn of the century when the secretary of the interior first began to gather figures to "investigate the nature and extent of mine disasters." The number of mine deaths slowly rose from 1889 to 1900 before shooting upward to 2,061 men between 1900 and 1906.[18] The data, one could argue, showed that a disastrous trend began to emerge at the same time that the industry began to mechanize. A simple examination of the four major disasters in December of 1907 revealed that mechanization was the common denominator.[19] The *Engineering and Mining Journal* applauded Fairmont Coal for trying to operate a safe colliery, but noted there was room for advancement.

The journal opined, "Many improvements might however, be adopted and result in providing greater safety. No matter how patriotic we desire to be, it must be acknowledged that we are far behind the coal operators on the continent."[20]

The frequency of these disasters presented a problem for West Virginia's bituminous coal operators. Early reports indicated that West Virginia Governor William M. O. Dawson was set to convene a special session of the West Virginia legislature to revise the state mining laws in 1908.[21] West Virginia's coal operators, including Fleming, had consistently resisted reform legislation, and in this case their refusal may have cost the loss of life and property.[22] Responding to the threat of mine safety legislation, coal operator George Caperton turned to Fleming, stressing that "there is to [sic] much law now, making the matter of mining a burden and that no further laws could be enacted to help out the situation."[23] The opinionated William N. Page agreed "that undigested legislation may follow which will affect every mine in the state more or less."[24] Journals of the coal trade voiced similar concerns, editorializing that "a cloud of drastic and unnecessary mining laws" seemed to emerge on the horizon, but hoped it would be "dispersed by the wind of sound judgment."[25]

West Virginia's operators were afraid that the state legislature would react to the wave of disasters and impose an arbitrary safety standard upon all working mines, thereby raising production costs, market prices, and jeopardizing West Virginia's competitive advantage.[26] Operators questioned the ability of the state to enforce safety regulations in a uniform manner and thought any variance in standards would result in lost coal sales for those who were held to a higher standard.[27] Pennsylvania trade journals balked at a Pennsylvania law that one editor argued would increase production costs and cut the industry out of the competitive market.[28] In practice, the factors that influenced competitive advantage embraced more than mine safety, but West Virginia's coal operators were worried.[29]

These concerns were compounded by the actions of West Virginia's Chief Mine Inspector James W. Paul. A coal dust advocate and outspoken critic of the state's mine laws, Paul had gone public in an attempt to pressure Republican Governor Dawson and his GOP-controlled legislature to revise the mining laws and to expand the authority of the State Department of Mines.[30] Once considered an ally who had applied for a job as an engineer with Fairmont Coal Company, Paul became the operators' chief antagonist.[31] The combined fear of Paul and restrictive legislation, along with the smoldering resentment between liberal reformers and business aroused during the 1903–1905 tax debates, made it unlikely the coal operators could control the outcome of the legislative debate.[32] Within the framework of legislative politics, mine safety legislation provided a serious issue around which liberal reformers, the state mine inspectors, and labor leaders could rally. Such a powerful coalition could gain concessions at the bargaining table and have some confidence of future gains.[33]

It was the fear of radical action by the West Virginia legislature that aroused the West Virginia Board of Trade to action. The board was a conservative business organization boasting three thousand members in twenty-seven constituent bodies, which provided a voice for the state's business community.[34] Each business organization maintained its individual autonomy but banded together in the larger body for the "advancement and upbuilding of the state."[35] The board elected officers and rotated them in and out of office to give old and young an opportunity to lead the organization.[36] The active membership included West Virginia's chief political and economic leaders, notably Fleming, Henry G. Davis, and Stephen B. Elkins.[37] Coincidentally, Fleming replaced Henry G. Davis as president in October 1907 and was serving in that capacity at the time of the Monongah disaster.[38] The emergence of the West Virginia Board of Trade also reflected the growing strength of progressive reforms within West Virginia's two major political parties. Businessmen found the two major parties too diverse to meet their needs so they united with other businessmen to gain concessions at the expense of their rivals.[39] As a political entity, the West Virginia Board of Trade found interest-group politics very effective in its efforts to stamp out corruption in city government, and in lobbying the West Virginia legislature to shape and direct the legislative agenda.[40]

In late December 1907, Roy B. Naylor, secretary of the West Virginia Board of Trade, formulated a plan to sway the legislative process and sent it to Fleming. Naylor proposed the appointment of a special committee of five to seven men, including mining experts Professors Israel C. White and Henry Payne of West Virginia University, and a few men not closely identified with the coal industry.[41] Naylor believed that the special committee, the governor, and West Virginia lawmakers together could find a marketplace solution that would not damage the coal industry.[42] The events of 1907 had proven to Naylor that the state's mine operators could not solve the problem when left to their own devices. Only through a cooperative effort between the state and the private sector could the business community direct the legislation along conservative lines.[43]

Naylor's plan was too late, and negotiating with the state legislature was not an option, so Fleming rejected the proposal. Coal operator J. C. McKinley made one attempt at persuading Fleming to accept the Naylor plan, pointing out that there would be "an attempt at radical legislation in Pennsylvania, Ohio, and West Virginia."[44] Fleming understood the connection between the disasters and the industry's technological advances, but did not think new technology required new laws. Nevertheless, Fleming considered what he must sacrifice from a practical perspective. The most fundamental considerations were: Who would make the decisions, and what would be the total cost of the remedy?

Recognizing that McKinley and Naylor spoke for large numbers of businessmen whose support for the coal industry was vital, Fleming felt compelled to reply to their proposal. He informed McKinley that he was convinced that dust contributed to mine explosions. He stressed this point in a confidential letter: "mines should be classified by some competent authority and those dangerous from dust and gas be operated with greater expense than those not so dangerous." Fleming understood the full import of mine mechanization, asserting that "any coal mine using coal mining machines might be termed dusty." He embraced other changes as well, including hiring only trained shot firers at all gassy and dusty operations, and requiring the miners to use only permissible powders. Fleming argued that most coal operators would implement these controls because they would make the mines safer.[45]

By early January, with the West Virginia legislature scheduled to convene on January 28, Fleming left Fairmont for a special meeting of the West Virginia Mining Association.[46] The meeting of the operators was held in Washington, D.C., on January 8, 1908, at the National Metropolitan Bank Building.[47] Fleming was joined by operators Jere Wheelwright, William Ohley, William N. Page, Neil Robinson, several operators from Pennsylvania, and members of the United States Geological Survey to discuss the recent explosions.[48] As president of the West Virginia Mining Association, William N. Page asserted the need to determine the cause of the explosions and to avoid them. Page stated that a "single individual can accomplish nothing" in the effort to determine the cause of the explosions. "We stand to-day in the position of a sick man, and we have the Legislature and the Congress as doctors," he observed. "They want to prescribe medicine when they have not diagnosed the disease." While the operators discussed a variety of topics, including ventilation and the hiring of trained shot firers, those present passed three official resolutions: they agreed to support appropriations for research into the causes of mine disaster; they expressed their conviction that the disaster pathology was unknown; and they concluded that the federal government "should take the necessary steps to determine the causes before any attempt is made to apply legislative remedies." Since the resolutions did not vest too much power in the government, Fleming and the Fairmont Coal Company supported the resolutions as did many of West Virginia's bituminous coal operators.[49]

In the meantime, in spite of some reasonable assurances that Governor Dawson would not request harsh mine safety legislation until the memories of the Monongah victims had faded, several southern West Virginia coal operators were concerned that Fleming was becoming a bit of a maverick.[50]

Coal lobbyist, long-time business associate, and close friend William Ohley tried to dissuade Fleming from hiring trained shot firers to detonate charges at Fleming's coal collieries.[51] Ohley feared that if Fairmont Coal implemented such a procedure, the West Virginia legislature would recommend that all operators follow Fleming's lead. Coal operator Neil Robinson, who was also secretary of the West Virginia Coal Operators Association, voiced similar concerns.[52] Recognizing the political influence of Fleming and the Fairmont Coal Company, Ohley and Robinson could only register their concerns. Fleming disregarded their pleas and quietly hired some experienced coal men to detonate the charges at Fairmont Coal Company's Monongah operations, and he also required the use of permissible powders at most of his mines.[53]

Fleming took further steps in studying mining conditions and improvements in mine safety by opening the Monongah mines to the scientific community. He cooperated with Joseph A. Holmes and his staff at the Technological Branch of the United States Geological Survey, a federal agency empowered to explore the causes of coal mine disasters. Holmes had examined the Monongah mines after the explosion and was apparently convinced that dust was the culprit.[54] In any event, Fleming became Holmes's closest ally when the coal operator opened the Monongah mines to Holmes and his staff to study "the effect of exhaust steam on the dust."[55] Fleming was confidant in Holmes and adapted many of his safety measures at the Fairmont mines. Holmes also appreciated Fleming's efforts, and later praised the coal operator and the Fairmont Coal Company for earnestly attempting to make the mines safer.[56]

The United States Bureau of Mines

While Holmes, Fleming, and the bituminous coal industry joined forces to push for the passage of the Bureau of Mines, the UMWA pushed for mine safety reform where they could. For instance, at the state level, the UMWA pushed for safety reforms they could control, while on the national level, the organization failed to exert any pressure on Congress to appropriate funds to study mine safety. The president of the United Mine Workers of America, John Mitchell, expressed his support to Joseph G. Cannon, speaker of the House of Representatives, but organized labor generally appears to have been content to wait until the Geological Survey had proved beneficial.[57]

The near silence of the UMWA was eclipsed by the unusual lack of interest among the mainstream press. Reform had gained its most dramatic push in the early twentieth century from journalists who exposed the underside of human endeavors and made them public. The number of these newspapers and magazines and their circulation grew at a faster pace than the corporations. In the period between 1880 and 1910, the number of newspapers jumped from 1,000 to 2,600 in terms of titles published daily.[58] In the main, the press became far more effective than the political parties in exposing and curbing corporate wrongdoing and the antisocial extremes of capitalism. The press, in fact, drew attention to a number of abuses, notably government corruption, child labor, and the abuses in the meat-packing industry.[59] With the mounting loss of life in America's coal mines, one would naturally assume that mine safety reform would have attracted these investigative journalists.

Nevertheless, in the months following the Monongah explosion, America's national magazines were lethargic. According to Louis Filler, *Leslie's Weekly* gave "the accident issue enough momentum and drama to make it national," but *Leslie's* carried nothing on its December 19, 1907, opinion page, and only a picture story of the Monongah catastrophe appeared in its December 26, and January 2, 1908, issues.[60] Apart from the photograph, *Leslie's* did little that differed from its competitors. The mine safety issue was all but marginalized by the other magazines, notably *Collier's,* which reported the loss of life, but failed to discuss mine safety legislation for several months. Nor were *Leslie's* and *Colliers* alone in ignoring the mine safety issue. *Arena,* no friend of corporate America and the print source most effective in dealing with deaths and railroad regulation, wrote only one story on industrial accidents of any kind between 1907 and 1908. These journals, therefore, failed to display the moral outrage that had become commonplace among magazines during the early twentieth century. Only Edgar Allen Forbes's *World's Work* dared to offer some suggestions for eliminating the problems. Forbes outlined a series of broad reforms, including the hiring of trained shot firers to detonate all charges, and he assailed ignorant miners and inefficient mine inspection. Forbes concluded that the best course was the proposed Bureau of Mines, which would serve as a clearinghouse for research information.[61]

Newspapers, on the other hand, responded quickly to the accidents and called for federal action. On December 23, 1907, a *Pittsburgh Chronicle* editorial recommended an investigation by a special committee. The *Washington Star* (D.C.) claimed that the federal government had no jurisdiction over the affairs since mine safety was controlled by the states, but it did stress the need for investigation. The *New York Times* concurred with the *Chronicle* and the *Star* and helpfully listed for its readers an itemized account of seven reforms proposed by Joseph Holmes. "Dr. Holmes points out the obvious remedy— official investigation in each coal mine by expert commissions," and "with the data supplied by one of these boards of experts courts and juries would know how to fix responsibility for each accident, and mine owners would take suitable precautions for their own protection."[62]

The rise of independent newspapers and magazines with the financial resources to conduct thorough investigations into corporate abuses and antisocial behavior had some impact on the issue of mine safety reform. While the press did print sensational stories, reporters did not discover the growing dangers in coal mining, nor did they reveal any unknown practices. According to William Graebner, "Muckraking in the sense of interpretive and investigative reporting simply was not present in the coal fields."[63]

Why did the political weeklies remain lethargic and fail to investigate the wave of death and destruction sweeping America's coal mines? Generally speaking, most targets of the investigative journalists were located in urban settings and the abuses they exposed affected greater numbers of people. The coal mines, on the other hand, were located in remote areas of the country, in little-known towns, with a heavy foreign population, and with whom the urban dwellers had little connection. While many Americans were probably aware of the coal mine disasters that swept the country in December 1907, Fleming's fight for the establishment of the Bureau of Mines may have preempted this issue, and national politics was actually ahead of the press.[64]

Fleming's apparent willingness to support the establishment of the United States Bureau of Mines took him into the halls of Congress. This was familiar territory for one of the industry's most effective lobbyists, but this time Fleming was called to testify before a congressional subcommittee. The congressional hearings began in 1908 to consider the establishment of the Bureau of Mines. On the morning of March 9, 1908, a broad range of witnesses, including Fleming, some UMWA representatives, and a spokesman from the American Mining Congress appeared before the House Committee on

Mines and Mining.[65] Fleming informed the federal law makers that the bituminous coal operators supported the bureau because the "cost would seem insignificant compared to the good it would do." In particular, the bureau would provide the "information" that "would enable the operators to make the mines more safe, thus saving life and property."[66] Fleming concluded his testimony with a note of urgency, asserting "that if these explosions are going to continue we are going to have to go out of business. We can not stand it. Another such explosion as we had would kill us all."[67] Fleming's nostrums called for some government direction. "It is the province of the Government to do it, when you consider this great industry, second to none in the country," he observed, "not only to conserve the coal and prevent its waste and the destruction of property, but above all to prevent the loss of life."[68]

Fleming's plea was consistent with his ideas of limited government. Previously, he had made exceptions to the doctrine of limited government and lobbied for subsidies in the form of high duties on coal. For Fleming, protecting the bituminous coal trade reaped personal financial reward, but also bolstered the economy, and hence benefited the people of West Virginia.[69] In addition, there was a large gap between Fleming's passionate laissez-faire rhetoric and the reality. He was not a pure laissez-faire advocate; he supported governmental action that established the rules of law and maintained the structure of the market economy.[70] He went far beyond these measures and became a member of the West Virginia Shippers Association to break up the perceived monopolistic practices of the Baltimore and Ohio Railroad.[71] A government agency created to determine the cause of mine explosions could thwart a hysterical campaign against the coal industry and, therefore, fit the protective model. Moreover, support for the bureau did not interfere with the constitutional doctrine of federalism.

Following the congressional hearings, Fleming joined forces with the American Mining Congress (AMC), an organization composed of mine operators, owners, and western miners that had been formed for "the purpose of advancing the mining and metallurgical industries in all their branches." As a professional organization, the AMC had considerable influence with the United States Congress, and it made sense for Fleming to ally with the organization to pursue their common goal of the creation of the Bureau of Mining under the Department of the Interior.[72]

Fleming participated as a delegate from West Virginia at the Eleventh Annual Session of the American Mining Congress held in Pittsburgh, Pennsylvania.[73] He questioned the speakers on the issues of coal dust and the proper speed of ventilation fans.[74] Fleming's emphasis on dust and ventilation indicated that the two were related in his mind. Some experts believed that sufficient volumes of air prevented the accumulation of methane gas, but too much air dried the environment, creating more dust. Some of Fleming's queries remained unanswered, and he took the podium himself to call for the creation of a bureau to study mining and make scientific recommendations.[75]

The most important way that Fleming aided the AMC and Holmes was by fostering favorable sentiment toward the establishment of the bureau among West Virginia's congressional delegation through regular correspondence with the state's representatives in Washington, D.C. Holmes kept Fleming informed with the latest scientific data, and Fleming used the information to influence West Virginia's representatives and to urge their support for the Bureau of Mines.[76]

Fleming also received favorable support from Republican Senator Nathan B. Scott of West Virginia.[77] In 1909, Fleming clearly outlined the benefits of the proposed bureau in a letter to Scott. "There is nothing perhaps which at this time would do more to establish confidence among the miners and with the public generally," he wrote "or would do more to prevent injurious mining legislation in different States, than the prompt establishment of this bureau of mines."[78] Fleming understood the importance of public opinion, and he recognized the industry's support for reform would serve to deflate the public perception that mine operators were willing to sacrifice human life in the blind pursuit of profit.

Moreover, Fleming also acknowledged the growing political power of organized labor.[79] He did not go so far as coal operator Justus Collins, who viewed mine safety within the context of the UMWA's efforts to unionize West Virginia. Collins was convinced that some radical miners entered the coal mines and intentionally set off several hundred pounds of high-percentage dynamite to create mine explosions. After an explosion, the union could use the mine disaster as a vehicle to extend their influence and pressure state legislative bodies to enact restrictive mine safety legislation.[80] Fleming did not harbor these conspiratorial theories himself, but he knew that the establishment of the bureau would take the issue of mine safety away from the United Mine Workers of America.

The promotion of the bureau bill by Fleming and the Fairmont Coal Company, the AMC, and the scientific community finally forced the bill to a vote. The bill, which eventually became law, was crafted by a coal operator then serving as chairman of the House Committee on Mines and Mining, Congressman George Huff of Pennsylvania.[81] After much planning, debating, and lobbying by the coal operators, the Committee on Mines and Mining forwarded the bureau bill to the full House during the final days of the session in 1908.[82] The House ratified the bill and transmitted it to the Senate, but the bill was detained in the upper chamber and failed to gain approval before Congress adjourned.[83] Nevertheless, supporters of the bureau were confident that the bill would be passed during Congress's next regular session.[84]

The impression Fleming gave of himself, quite deliberately, was that of a devoted reformer, yet he demonstrated some inconsistencies in his commitment to mine safety. Aware that one of his properties had failed to comply with West Virginia safety regulations, Fleming wrote kinsman and general manager of Consolidated Coal Company, George T. Watson. Fleming informed Watson that "the company itself could be indicted for misdemeanor and fined," but "in the event we should have some serious accident at one of these mines," and it was determined that "no special rules had ever been adopted," the company "might get in very serious trouble." Moreover, Fleming concluded that if "one or more persons should be killed," the company "might be held liable for manslaughter."[85] The letter to Watson demonstrated the sharp divide between Fleming's reform-minded public face and the private reality. Fleming came to seem all the more calculating on the outside, all the more ready to play the role of corporate attorney and operator behind the scenes.

As anticipated, however, the mine safety bill was introduced to the sixty-first Congress on January 22, 1909. After some initial discussions, the Congress passed the measure on May 16, 1910, establishing the United States Bureau of Mines under the direction of the Department of the Interior.[86] As enacted, the United States Bureau of Mines was limited to studying mining methods, research, education, and training; mine inspection was outside its jurisdictional scope.[87] While some modern critics focus on the narrow scope of the bill, operators were delighted that it did not violate the constitutional doctrine of federalism and kept the mines under their control.[88]

Even though Fleming supported the creation of the Bureau of Mines, this was not a sign of his conversion to liberalism. Nor was he a man trying to stabilize and enhance his social status in an emerging urban society.[89] Fleming was trying to balance limited government without compromising individual freedom within a system of competitive free enterprise. Fleming's conservative political ideology revolved around the fundamental belief that government, in all of its actions, should be bound by rules and fixed beforehand. Under such a principle, tariffs, workmen's compensation, and the newly established Bureau of Mines were permissible.[90] His doctrine of limited government respected the autonomy of the individual, freed citizens from government coercion, permitted freedom of choice, and most importantly protected the autonomy of private property. For Fleming and many of his contemporaries, the ownership of private property was the cornerstone of American civilization.[91]

Control of private property enabled the operators to direct the decision-making process. Fleming and other operators argued that they had to be free to develop and discover the best ways to deal with their own individual operations. For instance, Fleming's Fairmont Coal Company mined the rich Pittsburgh coal seam, a highly volatile coal that releases large amounts of methane gas when mined.[92] Measures adopted by Fleming and operators in the Fairmont Field made little or no sense to initiate in a coal mine that did not use cutting machines.

An equally significant and related issue to the Bureau of Mines was the appointment of a director of the new agency.[93] From the conservative operators' perspective, it was important to appoint a person who was not a bureaucrat.[94] Bureaucrats outside the coal industry were unable to deal with industry problems and would have been of little or no use to the coal operators. More importantly, they could create new demands for their skills, by passing a complex web of rules to regulate the coal industry.[95]

As it happened, Joseph Holmes had excellent credentials, including the qualifications most desired by Fleming and the operators.[96] He was not a reactionary: in fact, he moved slowly, conducting experiments to determine the exact cause of these explosions. Moreover, Holmes thought mine safety fell within the jurisdiction of the individual states and rejected the notion that the federal government should regulate America's coal mines.[97] Fleming was comfortable with Holmes, a man whom Fleming described as a "great friend of ours," and the two men stood together to promote the bureau.[98]

Hoping to keep the bureau in the hands of an ally, Fleming and his Fairmont Coal Company associates turned to West Virginia's congressional delegation and applied equal pressure on Maryland's congressional envoys.[99] Fleming informed United States Senator Stephen B. Elkins that Joseph Holmes was "the very best person that can be found for the head of the new department."[100] Fleming also wrote to President William Howard Taft. Fleming and Taft were not close friends, nor were they completely aligned ideologically, but the two men had met in court while Taft was serving as the United States Assistant Attorney General.[101] Fleming assured the president that Holmes's "appointment to the head of the Bureau will, I believe, give great satisfaction" to the industry and America's bituminous coal operators.[102] In effect, Fleming was thoroughly committed to Holmes and was willing to make his requests known to the highest level of power, the president.

Fleming followed the correspondence with action. He led a coalition of supporters to Washington, D.C. in early spring, and again in late June 1910 to plan a proper course of action to press Holmes's nomination.[103] Fleming and his group of supporters gained an audience with President Taft. Upon entrance into Taft's office, Fleming's delegation appealed for the appointment of their candidate, Joseph Holmes.[104] While Fleming and the coal operators bolstered Holmes's candidacy from the start, the former president of the United Mine Workers of America, John Mitchell, also lent support to Holmes.[105] The combined advocacy of the operators, Senators Stephen B. Elkins, Nathan Scott, and John Mitchell was substantial.[106]

With the bituminous coal operators supporting Holmes's appointment, it was natural that there would be opposition. The miners and the UMWA desired a man "who had swung a pick in the coal mine [rather] than a technically educated man."[107] The vector for a practical mining man not only drew support from the rank and file miners, but also from the State Mine Inspectors' Institutes in the states of Ohio, Pennsylvania, and Illinois.[108]

Opposition to Holmes opened the door for an alternative candidate, David Ross of Illinois. Ross had served as secretary of the Bureau of Labor Statistics, a former UMWA organizer, and a member of the Miners' Protection Association.[109] Ross's supporters tried to push his candidacy at the Mine Inspectors' Convention held in Chicago in the summer of 1910, but the movement was stopped by a combination of Pennsylvania and West Virginia delegates led by West Virginia coal operator and West Virginia Chief Mine Inspector John Laing.[110]

Fleming could never settle for Ross as director of the bureau and continued to lobby for Holmes.[111] Ross's union leanings aroused the former governor, and Fleming applauded Chief Inspector Laing for blocking Ross's nomination.[112] Fleming bubbled over with appreciation in a letter to Laing. "I cannot let the matter pass without congratulating you upon the stand you took," he wrote to the chief inspector, and "the credit you have done your state, in being able to lead in so important a fight against the iniquitous propositions." Confronted with the urgency that the office of the director had to be in friendly hands, Fleming noted that "it would have been exceedingly unfortunate to have placed the mine inspection in the control of politicians, or rather so that it would be controlled by politics."[113]

Fleming's political maneuverings worked, however, and the alternative candidates were pushed aside for Joseph Holmes, who President Taft appointed to the directorship with an annual salary of $6,000.[114] Critics saw Taft's decision as a favor to the coal operators. Once again Fleming and his allies had their way.[115] After two years of driving the political machine, Fleming and the Fairmont Coal Company were relieved that the operators' friend had been appointed.[116] The American press likewise praised the decision, and some looked to a brighter future. Holmes may "perform a great public service" and supply "the operators of the mines with information needed to work them more economically and humanely," wrote an editorialist for the *New York Times*.[117] Fleming probably hoped the editorial was correct. The man that Fleming had supported and defended against all opponents was now in place and ready to oversee the Bureau of Mines.[118]

Conclusion

The Wallace-Graebner explanation figures prominently in understanding Fleming's view of mine safety prior to the Monongah explosion. Very little capital was invested in the early coal ventures so the operators thought mine safety was the exclusive responsibility of the individual miner. However, as Fleming opened larger operations and installed modern cutting machines in these collieries, it was no longer feasible to gamble on mine safety by allowing the miner to assume responsibility for his individual safety as well as the safety of his fellow workers. The operator needed to be more proactive and work to keep his mines safe and productive.

In Fleming's scheme of things, with production high and costs low, the profits would take care of themselves. The Monongah disaster killed productive workers, interrupted production, and destroyed machinery. Most importantly, however, the Monongah explosion dramatically marked the end of an era in American coal history, an era made distinct by constructive achievements, but blemished by the terrible month of December 1907.

Fleming's response was to take the initiative by becoming one of the first bituminous coal operators to view safety as a reform movement that made economic sense. He led the charge for the establishment of the United States Bureau of Mines in 1910, an agency that could provide tax-supported scientific knowledge to the coal operators but not regulate the coal mines. Fleming, the coal operator turned reformer, served as a link between the operator of the past and the bituminous producers who led the "Safety First" campaign in the 1920s.[119] Historian William Graebner is certainly correct in writing that Fleming and the coal operators provided tangible support to the establishment of the Bureau of Mines.[120] Fleming played the Pied Piper that led the coal operators along the safe path while outflanking the more dangerous state legislation. Fleming's acceptance and support of the new agency was by no means a new direction, but rather an evolution of the limited government approach. Fleming permitted the federal government to establish the rules of law that maintained the market economy and high protective duties on coal. This new agency was no different; it provided scientific research to Fleming and the coal operators while its narrow aim did not interfere with the fundamentals of conservatism and private property.

NOTES

1   A. B. Fleming to G. A. Grasselli, 14 December 1907, General Correspondence, A. B. Fleming Papers, West Virginia and Regional History Collection, West

Virginia University. The Fleming collection at West Virginia University, traces the political and economic development of West Virginia through the eyes of Governor Aretas Brooks Fleming. The collection, although somewhat sanitized, is a rich source and is divided into three parts: General Correspondence, Personal Papers, and Miscellaneous. Each part is arranged chronologically. Hereafter, all manuscript collections cited will be from the West Virginia and Regional History Collection unless otherwise noted.

2   Carlton Jackson, *The Dreadful Month* (Bowling Green: Bowling Green University, 1982), 23. The availability of work served as a virtual magnet as men migrated to Monongah for employment; West Virginia Department of Mines, *Annual Report of the Department of the Mines, 1907*, 36–45, 47.

3   "Governor Warfield and Others Talk of What They Found," *Fairmont (WV) Times*, 9 August 1907, 1. For more on Governor Fleming, see Jeffery B. Cook, "The Ambassador of Development: Aretas Brooks Fleming, West Virginia's Political Entrepreneur, 1839–1923" (PhD diss., West Virginia University, 1998).

4   The Monongah plants were probably the safest mines in West Virginia. However, by comparison, they fell short of the safety standards maintained in Ohio.

5   This story is recounted in a newspaper column by editor, C. E. "Ned" Smith in "Good Morning!" *Fairmont Times-West Virginian*, 22 April 1934, 2.

6   In reality, mines No. 6 and 8 were a blend of the traditional and the modern. The company used mechanical undercutting and haulage equipment, and watered three times per day, while two large ventilation fans pumped air into the operation. During the time these modern practices were employed, ponies were still used and the men performed most work by hand.

7   See, for example, Davitt McAteer, *Monongah: The Tragic Story of the 1907 Monongah Mine Disaster, The Worst Industrial Accident in US History* (Morgantown: West Virginia University Press, 2007), 114–170; and Cook, "The Ambassador of Development," 263–328.

8   Concepts for this paragraph were extracted from Robert Green McCloskey, *American Conservatism in the Age of Enterprise: A Study of William Graham Sumner, Stephen J. Field, and Andrew Carnegie* (Cambridge: Harvard University Press, 1951).

9   See for a discussion of the competitive nature of the bituminous coal industry Michael E. Workman, "Political Culture and the Coal Economy in the Upper Monongahela Region: 1774–1933" (PhD diss., West Virginia University, 1995); Glenn F. Massay, "Coal Consolidation: Profile of the Fairmont Field of Northern

West Virginia, 1852–1903" (PhD diss., West Virginia University, 1970); William Graebner, "Great Expectations: The Search for Order in Bituminous Coal, 1890–1917," *Business History Review* 48 (1974): 49–72; and Dix, *Work Relations in the Coal Industry: The Handloading Era, 1880–1930* (Morgantown: Institute for Labor Studies, 1977), 81.

10   William Graebner, *Coal-Mining Safety in the Progressive Period: The Political Economy of Reform* (Lexington: University Press of Kentucky, 1976), 154. For a shorter version of this work see William Graebner, "The Coal Mine Operator and Safety: A Study of Business Reform in the Progressive Period," *Labor History* 14 (1973): 483–505.

11   Anthony F. C. Wallace, *St. Clair: A Nineteenth-Century Coal Town's Experience with a Disaster-Prone Industry* (Ithaca: Cornell University Press, 1987), 450.

12   There is some extensive literature on the social costs of disaster; see, for example, Kai T. Erikson, *Everything In Its Path: Destruction of Community in the Buffalo Creek Flood* (New York: Simon and Schuster, 1976), 186–245; Davitt McAteer, *Coal Mine Health and Safety: The Case of West Virginia* (New York: Praeger, 1970), 19; Charles E. Fritz, "Disaster," in *Contemporary Social Problems*, ed. Robert K. Merton and Robert A. Nisbet (New York: Harcourt and Brace, 1961); Martha Wolfenstein, *Disaster: A Psychological Essay* (Glencoe: Free Press, 1957); and Allen H. Barton, *Communities in Disaster* (Garden City: Doubleday, 1969).

13   Production was interrupted as the Monongah No. 8 remained idle until January 27, 1908, while No. 6 resumed production in February, "Work Resumed in No. Eight Monongah Mine Today," *Fairmont Times*, January 20, 1908, 1; "Number 6 Will Start in a Week," *Fairmont Times*, January 28, 1908, 1.

14   Ronald D. Eller, *Miners, Millhands, and Mountaineers: Industrialization of the Appalachian South, 1880–1930* (Knoxville: University of Tennessee Press, 1982), 180.

15   A. B. Fleming to J. C. McKinney, 3 January 1908, General Correspondence, A. B. Fleming Papers; Justus Collins to Jairus Collins, 13 January 1909, Justus Collins Papers; Justus Collins to A. M. Herndon, superintendent of Winding Gulf Colliery Co., 20 November 1911, Winding Gulf Colliery Company Papers.

16   After the bodies of the miners had been recovered from the Red Ash Colliery, Page entered the operation. He located the origin of the explosion, mapped the path of the blast and concluded that "firedamp" and not dust caused the disaster.

Page thought dust played an "important part in all explosions, when the gas is once ignited," Louis L. Athey, "William Nelson Page: Traditionalist Entrepreneur and the Virginias," *West Virginia History* 46 (1987): 11.

17   Graebner, *Coal-Mining Safety*, 45.

18   "United States Mines Kill Three to One in Europe," *Fairmont Times*, December 21, 1907, 3; "Prevention of Coal Mining Accidents," *Fairmont Free Press*, January 9, 1908, 1.

19   For a brief list of the major disasters that struck the Appalachian coalfields see Eller, *Miners*, 180. 1907 was by far the worst year in the history of industry: 3,241 miners lost their lives in America's mines, see Braithwaite, *To Punish or Persuade*, 1.

20   Floyd W. Parsons, "Disaster at Monongah Coal Mines Nos. 6 and 8," *The Engineering and Mining Journal* 84 (14 December 1907): 1123.

21   G. H. Caperton to A. B. Fleming, 17 December 1907, General Correspondence, A. B. Fleming Papers.

22   Glenn F. Massay, "Legislators, Lobbyists, and Loopholes: Coal Mining Legislation in West Virginia, 1875–1901," *West Virginia History* 32 (April 1971); and Massay, "Coal Consolidation," 126–192.

23   George Caperton to A. B. Fleming, 17 December 1907, General Correspondence, A. B. Fleming Papers.

24   William Page to A. B. Fleming, 14 December 1907, General Correspondence, A. B. Fleming Papers.

25   "Editorial," *Coal Trade Bulletin* 20 (February 15, 1909): 22.

26   Neil Robinson to C. W. Watson, 30 December 1907, A. B. Fleming Papers; J. C. McKinney to A. B. Fleming, 2 January 1908, General Correspondence, A. B. Fleming Papers; William Page to A. B. Fleming, 14 December 1907, General Correspondence, A. B. Fleming Papers.

27   George Caperton to A. B. Fleming, 18 October 1907, General Correspondence, A. B. Fleming Papers.

28   "Editorial," *Coal Trade Bulletin*, 20 (1 March 1909), 22; "Editorial," *Coal and Coke Operator* 8 (25 February 1909), 142.

29   The two most important factors in terms of production costs were freight rates and labor. See Workman, "Political Culture and the Coal Economy," 171–180; Massay, "Coal Consolidation"; and Graebner, *Coal-Mining Safety*, 102–103.

30   William Page to A. B. Fleming, 14 December 1907, General Correspondence, A. B. Fleming Papers; "Mining Laws are Defective," *Fairmont Times*, March 9,

1906, 6; and "Mine Inspector J. W. Paul Makes Recommendations," *Fairmont West Virginian*, February 6, 1908, 1. Other West Virginia mine inspectors questioned the state's mine laws as well, see, "Mining Laws Are Too Inadequate," *Fairmont Times*, February 16, 1906, 4.

31   James W. Paul to A. B. Fleming, 14 October 1901, General Correspondence, A. B. Fleming Papers. Paul also manufactured O. V. tobacco and sold the product at Fairmont Coal's company stores, see James W. Paul to A. B. Fleming, 27 August 1901, General Correspondence, A. B. Fleming Papers. The *United Mine Workers Journal* was hard on Chief Inspector Paul, and charged him with incompetence, especially in connection with the Monongah disaster. The *Journal* asserted that the "administration of his office is marked by one long bloody trail of human slaughter, caused by negligence, inefficiency, by wanton nullification of every mining law in the state," quoted in Graebner, *Coal-Mining Safety*, 16.

32   After 1905, each session of the West Virginia legislature was a battleground over liberal reforms, including pure food and drug laws, severance taxes and various tax reform measures, primary election laws, and antitrust laws. See John Alexander Williams, *West Virginia and the Captains of Industry* (Morgantown: West Virginia University Library, 1976), 196–256; and John Alexander Williams, *West Virginia: A History* (New York: W. W. Norton and Company, 1984), 153–154.

33   Workman, "Political Culture and the Coal Economy," 173–175. Robert F. Wesser argues that business's primary role is to oppose liberal reformers, see "Conflict and Compromise: The Workers' Compensation Movement in New York, 1890s–1913," *Labor History* 11 (Summer 1971): 345–372.

34   R. B. Naylor to A. B. Fleming, 26 September 1907, General Correspondence, A. B. Fleming Papers.

35   For an example of the board's political activism and emerging role in West Virginia politics see "State Board of Trade Writes to Members of the Legislature," *Fairmont West Virginian*, December 23, 1906, 6.

36   It was a common practice for these commercial organizations to rotate their leaders; see, for instance, Don H. Doyle, *New Men, New Cities, New South: Atlanta, Nashville, Charleston, Mobile, 1860–1900* (Chapel Hill: The University of North Carolina Press, 1990), 165–167.

37   *Fairmont Times*, September 5, 1907, 5.

38   Letterhead for the West Virginia Board of Trade has Henry G. Davis as president, R. B. Naylor to A. B. Fleming, 4 October 1907, General Correspondence,

A. B. Fleming Papers; "Governor Fleming is a Delegate," *Fairmont Times*, January 9, 1907, 8; "Governor Fleming New President of State Board of Trade" *Fairmont Times*, October 10, 1907, 1. To gain some insight into the rise and importance of political interest groups, see Samuel P. Hays, *Response to Industrialism, 1885–1914* (Chicago: The University of Chicago Press, 1957); Robert H. Wiebe, *The Search for Order, 1877–1920* (New York: Hill and Wang, 1967); and Richard L. McCormick, *From Realignment to Reform: Political Change in New York State, 1893–1910* (London: Cornell University Press, 1981).

39   This was a national trend that swept through both parties; see Arthur S. Link and Richard L. McCormick, *Progressivism* (Arlington Heights, Ill: Harlan Davidson, Inc., 1983), 48–49.

40   Williams, *Captains*, 210; R. B. Naylor to A. B. Fleming, 2 July 1908, General Correspondence, A. B. Fleming Papers; R. B. Naylor to A. B. Fleming, 12 January 1909, General Correspondence, A. B. Fleming Papers.

41   R. B. Naylor to A. B. Fleming, 28 December 1907, General Correspondence, A. B. Fleming Papers. Morgantown was a center for geology and mining education, and Dr. Israel C. White was primarily responsible for that reputation; see "Morgantown's Wealth of Fuel," *Black Diamond* 71 (August 11, 1923). For a short biography of White, see Lloyd D. Brown, "The Life of Dr. Israel C. White," *West Virginia History* 5 (October 1946): 5–104. Dr. Henry Payne also assisted Fairmont Coal Company during the 1907 Monongah Mine disaster, and Fleming later recommended Payne for West Virginia's chief inspector of mines; A. B. Fleming to W. S. Kuhn, 16 December 1909, General Correspondence, A. B. Fleming Papers. For more on Naylor see John W. Kirk, "Roy B. Naylor," in *Progressive West Virginians 1923* (Wheeling: The *Wheeling Intelligencer*, 1923), 305.

42   In addition to White and Payne, Naylor suggested twenty-two possible representatives and the list was dominated by members of the coal community, including J. C. McKinney, Stephen B. Elkins, A. B. Fleming, Samuel Dixon, M. T. Davis, Justus Collins, John Tierney, Isaac Mann, G. H. Caperton, C. W. Watson, L. L. Malone, E. W. Knight, Neil Robinson, William Page, and Joseph Holmes, head of the Geological Survey; see R. B. Naylor to A. B. Fleming, 28 December 1907, General Correspondence, A. B. Fleming Papers.

43   R. B. Naylor to A. B. Fleming, General Correspondence, 28 December 1907, A. B. Fleming Papers.

44   J. C. McKinley to A. B. Fleming, 2 January 1908, General Correspondence, A. B. Fleming Papers.

45   A. B. Fleming to J. C. McKinley, 3 January 1908, General Correspondence, A. B. Fleming Papers.

46   "Extra Session Will Begin January 28," *Fairmont Times*, December 20, 1907, 8; "Extra Session January 28," *Fairmont West Virginian*, December 19, 1907, 1; "Gov. Dawson Sends Out Call for Extra Session of the Legislature," *Fairmont West Virginian*, January 8, 1908, 1; "Number 6 Will Start In a Week," *Fairmont Times*, January 28, 1908, 1.

47   William Page, Neil Robinson to A. B. Fleming, 30 December 1907, General Correspondence, A. B. Fleming Papers; William Page to A. B. Fleming, 2 January 1908, General Correspondence, A. B. Fleming Papers; A. B. Fleming to J. C. McKinley, 3 January 1908, General Correspondence, A. B. Fleming Papers; William Page to Neil Robinson, 4 January 1908, General Correspondence, A. B. Fleming Papers.

48   William Ohley to A. B. Fleming, 16 January 1908, General Correspondence, A. B. Fleming Papers; William Page to Neil Robinson, 4 January 1908, General Correspondence, A. B. Fleming Papers; "Coal Operators for Better Laws," *Fairmont Times*, January 6, 1908, 2; "West Virginia Mining Association Will be Formed," *Fairmont West Virginian*, January 9, 1908, 1.

49   William A. Ohley to A. B. Fleming, 18 January 1908, General Correspondence, A. B. Fleming Papers; D. E. Llewellyn to A. B. Fleming, 2 March 1908, General Correspondence, A. B. Fleming Papers; and Graebner, *Coal-Mining Safety*, 19–20. In the midst of the meeting in Washington, D.C., the United States Geological Survey issued their report on the condition of America's coal mines, "Prevention of Coal Mining Accidents," *Fairmont Free Press*, January 9, 1908, 1.

50   William Ohley to A. B. Fleming, 11 January 1908, General Correspondence, A. B. Fleming Papers; William Ohley to A. B. Fleming, 15 January 1908, General Correspondence, A. B. Fleming Papers; "The Governor's Reasons for Calling the Extra Session,"*Fairmont West Virginian*, January 9, 1908, 1; Williams, *Captains*, 251; and William Ohley to A. B. Fleming, 19 January 1908, General Correspondence, A. B. Fleming Papers.

51   William Ohley to A. B. Fleming, 16 January 1908, General Correspondence, A. B. Fleming Papers.

52   Neil Robinson to A. B. Fleming, 20 January 1908, General Correspondence, A. B. Fleming Papers.

53   William Ohley to A. B. Fleming, 16 January 1908, General Correspondence, A. B. Fleming Papers; West Virginia Department of the Mines, *Annual Report of the Department of Mines, 1907–1908* (Charleston: Tribune Printing Company, 1908), 303. These practices were continued in the subsequent years as well, see West Virginia Department of Mines, *Annual Report For the Year Ending June 30, 1909* (Charleston: News Mail Company, 1910), 321–322; and West Virginia Department of Mines, *Annual Report of the Department of Mines for the Year Ending, June 30, 1910* (Charleston: The News Mail Company, 1911), Sec. 3, 8–9.

54   William Page to A. B. Fleming, 14 December 1907, General Correspondence, A. B. Fleming Papers.

55   J. A. Holmes to A. B. Fleming, 12 February 1908, General Correspondence, A. B. Fleming Papers; J. A. Holmes to A. B. Fleming, 19 March 1909, General Correspondence, A. B. Fleming Papers.

56   J. A. Holmes to A. B. Fleming, 19 March 1909, General Correspondence, A. B. Fleming Papers; J. A. Holmes to A. B. Fleming, 24 December 1909, General Correspondence, A. B. Fleming Papers.

57   Graebner, *Coal-Mining Safety*, 19.

58   Paul Johnson, *A History of the American People* (New York: Harper Collins, 1998), 601; and refer to table "Number of Daily Newspapers in the US, 1770–1990," in Eric Foner and John Garraty, *The Readers Companion to American History* (Boston: Houghton-Mifflin, 1991), 691.

59   Fred J. Cook, *The Muckrakers: Crusading Journalists Who Changed America* (Garden City: Doubleday and Company, Inc., 1972), 181.

60   Louis Filler, *The Muckrakers: New and Enlarged Edition of Crusades for American Liberalism* (University Park: The Pennsylvania State University Press, 1976), 212.

61   Graebner, *Coal-Mining Safety*, 14, 16–17.

62   Graebner, *Coal-Mining Safety*, 17; "Preventable Deaths in Mines," *New York Times*, 22 December 1907, sec. 2, 8. Some of the local newspapers followed the pattern established by the national press. The *Fayette Journal*, a newspaper that consistently supported Fayette County's coal operators, reflected the need for some reform, remarking that the "frightful annual slaughter of coal miners in this country has got on the nation's nerves," and it was sad that America had fallen a "generation behind

Europe in the study of safety regulations," see "The Coal Mine Slaughter," *Fayette Journal*, February 13, 1908, 4.

63   Graebner, *Coal-Mining Safety*, 17.

64   The inaction of the national media, according to William Graebner, was due to the fact that most coal mine deaths were the result of the unspectacular roof falls, and the American public thought railroading was by far the more dangerous occupation, Graebner, *Coal-Mining Safety*, 14.

65   The following men testified before the subcommittee: Dr. Israel C. White, West Virginia state geologist; J. H. Walker, UMWA district president of Springfield, Illinois; W. D. Van Horn, UMWA district president of Indiana; J.M. Craigo, UMWA district president of West Virginia; F. J. Drum, UMWA district president of Maryland; James Purcell, UMWA district president of Clearfield, Pennsylvania; and J. F. Galbreath, secretary of the American Mining Congress. House Committee on Mines and Mining, *Hearings to Consider the Question of the Establishment of a Bureau of Mines. 60th Cong., 1st sess.,* 1908, 3.

66   House Committee, *Bureau of Mines*, 15–16.

67   House Committee, *Bureau of Mines*, 23.

68   House Committee, *Bureau of Mines*, 24.

69   For an explication of the importance of regionalism, see Workman, "Political Culture and the Coal Economy."

70   Fleming, like most American businessmen, did not accept the doctrine of laissez faire in its classical sense. See, for instance, Thomas Richard Ross, *Henry Gassaway Davis: An Old-Fashioned Biography* (Parsons: McClain Printing, 1994); and Joseph Frazier Wall, *Andrew Carnegie* (Pittsburgh: University of Pittsburgh Press, 1989), 391; and for the nuances of the laissez-faire approach to economic devel-opment see Ronald L. Lewis, *Transforming the Appalachian Countryside: Railroad, Deforestation, and Social Change in West Virginia, 1880–1920* (Chapel Hill: The University of North Carolina Press, 1998), 103–129.

71   For a discussion of the genesis of the West Virginia Coal Shippers Association, see Workman, "Political Culture and the Coal Economy," 176–180.

72   J. F. Callbreath to A. B. Fleming, 24 March 1908, General Correspondence, A. B. Fleming Papers. For more on the American Mining Congress, see Graebner, *Coal-Mining Safety*, 25–27.

73   E. L. Boggs to A. B. Fleming, 13 November 1908, General Correspondence, A. B. Fleming Papers.

74   T. J. Garman to A. B. Fleming, 12 November 1910, General Correspondence, A. B. Fleming Papers.

75   T. J. Garman to A. B. Fleming, 12 November 1910, General Correspondence, A. B. Fleming Papers.

76   J. F. Callbreath to A. B. Fleming, 23 July 1908, General Correspondence, A. B. Fleming Papers; J. A. Holmes to A. B. Fleming, 4 February 1909, General Correspondence, A. B. Fleming Papers; A. B. Fleming to Jere Wheelwright, 23 May 1908, General Correspondence, A. B. Fleming Papers.

77   A. B. Fleming to Nathan Scott, [n.d.], 1909, General Correspondence, A. B. Fleming Papers; Nathan B. Scott to A. B. Fleming, 16 March 1910, General Correspondence, A. B. Fleming Papers. There are numerous letters on this subject between August and December 1909, in General Correspondence, A. B. Fleming Papers.

78   A. B. Fleming to Nathan Scott, [n.d.], 1909, General Correspondence, A. B. Fleming Papers.

79   A. B. Fleming to William M. O. Dawson, 17 September 1907, General Correspondence, A. B. Fleming Papers.

80   Graebner, *Coal-Mining Safety*, 144.

81   Graebner, *Coal-Mining Safety*, 33.

82   William B. Wilson to A. B. Fleming, 25 May 1908, General Correspondence, A. B. Fleming Papers. Pennsylvania Congressman William B. Wilson, was considered pro labor and spoke for the interests of the state's coal miners, see Paul Prichard, "William B. Wilson, Master Workman," *Pennsylvania History* 12 (April 1945): 81–108. Wilson was in accord with Fleming, the bituminous coal operators, and others in regard to the bureau. Wilson told the House Committee on Mines and Mining in 1908, "I am firmly now of the opinion it should be in the Department of the Interior. I fear that if it is placed in the Department of Commerce and Labor it will simply be used for the purpose and the extension of trade and the promotion of trade rather than for the protection of life, limb, and property, as it really should be," House Committee on Mines and Mining, *Hearings on the Bureau of Mines*, 50–51.

83   William B. Wilson to A. B. Fleming, 24 May 1908, General Correspondence, A. B. Fleming Papers; William B. Wilson to A. B. Fleming, 2 June 1908, General Correspondence, A. B. Fleming Papers.

84   William B. Wilson to A. B. Fleming, 2 June 1908, General Correspondence, A. B. Fleming Papers; J. F. Callbreath to A. B. Fleming, 23 July 1908, General

Correspondence, A. B. Fleming Papers; American Mining Congress to A. B. Fleming, 20 November 1908, General Correspondence, A. B. Fleming Papers.

85 A. B. Fleming to George W. Watson, 22 November 1909, General Correspondence, A. B. Fleming Papers.

86 Graebner, *Coal-Mining Safety*, 30–34.

87 Improved training can reduce mining accidents. For a discussion of training and its role in mine safety see Braithwaite, *To Punish or Persuade*.

88 Sociologist Daniel J. Curran also contends that mine safety was being neglected by the coal operators and the bureau in favor of production, see Daniel J. Curran, *Dead Laws for Dead Men: The Politics of Federal Coal Mine Health and Safety Legislation* (Pittsburgh: University of Pittsburgh Press, 1993), 67–69. Curren argues for government regulation, but government regulation can fail as well.

89 This thesis has been presented by Robert Wiebe. Progressivism, according to Wiebe, was the effort of a new middle class, who was trying to stabilize and enhance its position in American society; Wiebe, *The Search for Order*.

90 The federal government enacted the nation's first full workman's compensation law in May 1908 to protect workers killed or injured while working in dangerous occupations for the United States. Many of the states followed suit, and while Fleming and the Fairmont Coal Company had opposed most state mining reforms, they learned to live with the workmen's compensation law that was enacted by the state in October 1913. Graebner, *Coal-Mining Safety*, 149; and J. C. McKinley to A. B. Fleming, 8 January 1913, General Correspondence, A. B. Fleming Papers. Fleming and company's initial opposition to workman's compensation to acceptance comports with a pattern in two state-centered studies see Stanley P. Caine's, *The Myth of a Progressive Reform: Railroad Regulation in Wisconsin, 1903-1910* (Madison: Historical Society of Wisconsin, 1970); and Richard L. McCormick, *From Realignment to Reform: Political Change in New York State, 1893-1910* (London: Cornell University Press, 1981).

91 Keith Dix also notes the importance of private property rights and its affect on the mine safety debate, see Dix, *Work Relations*, 80. For a discussion of conservatives in this period and their values, see McCloskey, *American Conservatism in the Age of Enterprise*.

92 Michael E. Workman, "The Fairmont Field," in *Northern West Virginia Coal Fields: Historical Context* (Morgantown: Institute for the History of Technology and Industrial Archaeology Technical Report Number 10, West Virginia University, 1994), 25.

93   This view is expressed by Civil Engineer Sam Taylor to A. B. Fleming, 17 May 1910, General Correspondence, A. B. Fleming Papers.

94   While Holmes was a member of the Technological Branch of the Geological Survey, he was a bit of a maverick. After being ordered by his superior to stop lobbying for the passage of the Bureau of Mines, Holmes ignored the request and continued to press for the passage of the bill. He was demoted to subordinate duties, and his superior pushed another candidate that was loyal to the bureaucracy for the directorship, see Graebner, *Coal-Mining Safety*, 38–39.

95   Historian William Graebner asserts that Fleming and his allies would have tolerated the appointment of a bureaucrat as director. This may be, but they certainly worked through the only channel that was open to them, and pushed their candidate for the directorship; see Graebner, *Coal-Mining Safety*, 40.

96   Samuel A. Taylor to A. B. Fleming, 17 May 1910, General Correspondence, A. B. Fleming Papers. Holmes graduated from Cornell with a bachelor's degree in agriculture and served as a professor of geology and natural history at the University of North Carolina. Holmes became director and state geologist for North Carolina in 1891, and later took charge of the Technological Branch of Geological Survey in 1907. For more on Holmes see Graebner, *Coal-Mining Safety*, 35–37.

97   Graebner, *Coal-Mining Safety*, 103.

98   A. B. Fleming to C. W. Watson, 18 May 1910, General Correspondence, A. B. Fleming Papers; A. B. Fleming to The President [Taft], 18 May 1910, General Correspondence, A. B. Fleming Papers.

99   C. F. Snyder to A. B. Fleming, 19 May 1910, General Correspondence, A. B. Fleming Papers; Jere Wheelwright to A. B. Fleming, 19 May 1910, General Correspondence, A. B Fleming Papers.

100   A. B. Fleming to Stephen Elkins, 18 May 1910, General Correspondence, A. B. Fleming Papers.

101   A. B. Fleming to Joseph Holmes, 18 May 1910, General Correspondence, A. B. Fleming Papers; A. B. Fleming to The President [Taft] 18 May 1910, General Correspondence, A. B. Fleming Papers.

102   A. B. Fleming to The President [Taft], 18 May 1910, General Correspondence, A. B. Fleming Papers.

103   Jere Wheelwright to A. B. Fleming, 19 May 1910, General Correspondence, A. B. Fleming Papers.

104   A. B. Fleming to John Laing, 2 July 1910, General Correspondence, A. B. Fleming Papers.

105   Graebner, *Coal-Mining Safety*, 54.

106   A. B. Fleming to C. W. Watson, 18 May 1910, General Correspondence, A. B. Fleming Papers; G. F. Snyder to A. B. Fleming, 19 May 1910, General Correspondence, A. B. Fleming Papers.

107   Andrew Roy to John Mitchell, 3 June 1908, Mitchell Papers, Wise Library, West Virginia University.

108   Graebner, *Coal-Mining Safety*, 37.

109   Graebner, *Coal-Mining Safety*, 37.

110   A. B. Fleming to John Laing, 2 July 1910, General Correspondence, A. B. Fleming Papers; Frank Haas to A. B. Fleming, 2 July 1910, General Correspondence, A. B. Fleming Papers; John Laing to A. B. Fleming, 5 July 1910, General Correspondence, A. B. Fleming Papers. While Laing served as chief mine inspector, he also was president and general manager of the Wyatt Coal Company, see "The Rise of a Call Boy," *West Virginia Review* 6 (March 1929): 206.

111   John Laing to A. B. Fleming, 5 July 1910, General Correspondence, A. B. Fleming Papers.

112   John Laing to A. B. Fleming, 5 July 1910, General Correspondence, A. B. Fleming Papers.

113   A. B. Fleming to John Laing, 2 July 1910, General Correspondence, A. B. Fleming Papers. Nearly three years after this event, Fleming wrote that "We consider the present incumbent, Mr. Laing, the best Chief Mine Inspector the State ever had," A. B. Fleming to His Excellency, Henry D. Hatfield, 15 August 1913, General Correspondence, A. B. Fleming Papers.

114   Curran, *Dead Laws for Dead Men*, 67.

115   One disgruntled coal miner, Charles Tinlin, informed the *United Mine Workers Journal* that appointment of Holmes as director was a "big joke," Jackson, *The Dreadful Month*, 136.

116   A. B. Fleming to C. W. Watson, 18 May 1910, General Correspondence, A. B. Fleming Papers; J. H. Wheelwright, vice president of Consolidation Coal Company, to A. B. Fleming, 19 May 1910, General Correspondence, A. B. Fleming Papers.

117   "The Director of Mines," *New York Times*, September 5, 1910, 6.

118   For a public interview with Dr. Joseph Holmes ("Fitness Not Politics Should Govern Choice of Mine Inspectors," *Fairmont Times*, March 28, 1911, 2). During Holmes's term as director, he maintained a working relationship with Fleming and his associates at Fairmont Coal and Consol. Holmes briefed the group on pending federal legislation and offered assurances that the lines of communication would remain open, Frank Haas to A. B. Fleming, 7 March 1913, General Correspondence, A. B. Fleming Papers.

119   Dianne Bennett and William Graebner, "Safety First: Slogan and Symbol of the Industrial Safety Movement," *Journal of the Illinois States Historical Society* 68 (June 1975): 243–56.

120   Graebner, *Coal-Mining Safety.*

Depression, Recovery, Instability:

The NRA and the McDowell County, West Virginia, Coal Industry, 1920–1938

Mark Myers

MCDOWELL COUNTY, the southernmost county in West Virginia, depended on the coal industry for its economic livelihood throughout the twentieth century. During the latter half of the century, however, many of the coal operations in the county closed and left the region destitute. Mechanization within the coal industry, along with other factors, including competition from alternative fuels, caused the process of deindustrialization in McDowell.[1] The Great Depression, more than any other event, can be seen as a turning point in the coal industry of the county. During the 1920s, coal companies faced an uncertain future as the industry suffered through a decade-long downturn. The economic collapse of 1929 contributed to the virtual collapse of the industry in McDowell County. By 1932, conditions in McDowell County, and the entire Central Appalachian region, were among the worst in the nation.

The 1932 election of Franklin D. Roosevelt as president, and the implementation of his New Deal, resulted in significant socioeconomic changes in McDowell County. The passage of the National Industrial Recovery Act (NIRA) and the authority given to the National Recovery Administration (NRA) in promulgating industrial codes of fair competition resulted in changes in mining methods and in the relationship between miners and employers. Although it seemed that the NRA succeeded in bringing recovery to the coal industry immediately after the implementation of the coal codes, the agency ultimately failed to achieve the stabilization of the industry. The production and employment figures for three representative mines in McDowell County—Olga No. 1 in Coalwood, United States Coal and Coke No. 6 in Gary, and Peerless Coal and Coke in Vivian—show that the coal industry went through another downturn in the late 1930s.[2] Furthermore, the NIRA, with its support for collective bargaining, led to increased labor costs and encouraged companies to implement mechanization in an effort to increase profits. Although the mines of Central Appalachia did not mechanize to the extent that mines in the organized Central Competitive Field did, NRA policy encouraged the adoption of machinery, which was one of the factors leading to deindustrialization, eventually displacing many miners in McDowell County.

Depression in Coal, 1920–1933

By the mid-1920s, the McDowell County coal industry held an important place in the markets of the world. Migrants who poured into the region to find work in the coal mines settled into the company towns dotting the landscape. Coal mining had become a way of life. On the surface it appeared that the economy of the county was strong, but industry analysts knew that the coal industry was in trouble. A depression in the bituminous coal industry began in 1919. The industry recovered from the downturn, yet remained unstable. The instability resulted from World War I, the very cause of coal mining's expansion in McDowell County. The demands of war industries led to the opening of new, marginal mines. As war industries downsized at the end of the conflict, the nation's coal mines overproduced. As a result, many mines closed and coal prices fluctuated. Because of the importance of mining to West Virginia, the problems in the coal industry set the stage for the Great Depression years before 1929.[3]

Although wartime expansion was the primary cause, other factors contributed to the industry's instability. The operators in the Pocahontas coalfield, which included McDowell County, defeated postwar organizing efforts by the United Mine Workers of America (UMWA). Widespread success in southern West Virginia evaded the UMWA during the 1920s in general.

Southern West Virginia coal was still popular in commercial markets because of its low-sulfur content and smokeless burning. Both the lack of union success and the quality of the coal encouraged the opening of new mines, which in turn inundated the market with more coal than could be sold.[4]

The problems facing the national coal industry during the 1920s were very complex. The supply of coal increased during the decade as demand fell. Although downsizing of wartime industries constituted the major reason for the decline in demand, there were other reasons for the drop as well. Increased output and productivity of the oil and natural gas industries made these alternative fuels cheaper than coal in many regions of the country. The invention of the diesel engine, for instance, led to oil replacing coal as the preferred fuel for the railroads. Because of the decline in demand, it became difficult for mines to sell coal at a price that covered production costs.[5]

National and statewide trends in the coal industry foreshadowed hard times to come. Labor strikes, external competition, and the inability to achieve industry-wide unity led to chronic losses for the operators and frequent mine failures. The labor strife that resulted from continued UMWA efforts to organize non-union fields added to the problems facing the coal industry in the 1920s. Production declines, which occurred during 1919 because of the national strike conducted by the UMWA, caused shortages and drove up prices. The rise in prices encouraged the opening of new mines. However, when prices began to drop in 1923, operators suffered tremendously. To prevent bankruptcy, operators reduced wages or used labor-saving machinery. The reduction of wages led to a near-starvation existence for many of the miners. Many smaller mines closed and left a multitude of miners out of work. The mining community, therefore, faced the negative consequences of overproduction, which plagued the industry throughout West Virginia; statewide, coal employment and production declined during the 1920s. Employment peaked at 121,000 in 1923, but fell to 107,000 by 1929. Production peaked at 144 million tons in 1927 and then began to fall.[6]

West Virginia's economic future seemed bleak. When West Virginia became the leading coal producer in the nation in 1928, the state's newspapers cheered the news, despite the obvious downturn in the entire industry. The *Charleston Gazette* took an optimistic approach when it declared that downward trends in the industry "cannot mean anything but that the coal business is readjusting itself and lopping away the deadwood."[7] The *Clarksburg Exponent*, on the other hand, took a realistic approach when it noted that in 1928, production declined, companies failed, and the numbers employed decreased. It was clear to the editors of the *Exponent* that coal was a sick industry and that the economy of West Virginia was on tenuous footing.[8]

In McDowell County, the coal industry's production and employment trends followed the national patterns. In 1927, the county's production totaled almost 22 million tons. In 1929, countywide production declined by almost one million tons.[9] Employment declined as well, from 20,109 in 1926 to 18,558 in 1928.[10] By the eve of the economic collapse in 1929, the county's coal industry seemed to rebound from the industry-wide depression of the 1920s. Employment increased to 19,424 and production increased to almost 38 million tons.[11] Individual companies also saw an improvement. Although the *Annual Report of the Department of Mines* for West Virginia does not give employment figures for individual companies between 1929 and 1934, the increase in production numbers for both Peerless and U.S. Coal and Coke's No. 6 mine between 1928 and 1929 suggests that, contrary to national trends, the state of the coal industry in McDowell improved during the late 1920s.[12] Outside economic factors, however, destroyed the gains made by the industry and seriously harmed the lives of the miners in McDowell County.

The stock market crash on October 29, 1929, ushered in a period of economic hardship unknown before in the United States, a period later called the "Great Depression." The economic collapse hit the entire country; banks closed, men and women lost their jobs, and a general sense of hopelessness among Americans pervaded. In West Virginia, the depression struck the struggling coal industry hardest. The economic slump during the 1920s had depressed the industry enough to make collapse inevitable after October 1929. Coal production fell 40 percent from 1929 to 1933, matched by a corresponding drop in market demand. As a result of the drop in both demand and production, wages fell significantly. Coal families faced unprecedented economic hardships. The Depression particularly damaged McDowell County. Because of the importance of the coal industry to the people of the county, the continued downturn in coal, along with the overall economic malaise, created a sense of despair among the county's miners and their families. During the Depression, thirty of the county's ninety mines closed; almost 3,000 of McDowell's miners lost their jobs, and the remaining 16,000 worked only a few days per week. At best, a mining family could barely survive on the reduced wages that miners earned. At worst, a family faced the prospect of starvation and homelessness, because the dominance of coal in McDowell did not allow laid-off miners an opportunity to find other work. Migration was not a viable option because jobs were scarce throughout the nation. Mining families had to find other ways to survive their predicament.[13]

As the Depression deepened in West Virginia, conditions developed statewide that were among the worst in the United States. Some 33,000 coal jobs disappeared statewide as coal production fell from 144 million tons in 1927 to 84 million tons in 1932.[14] Throughout the state, coal families lost their homes and became trapped in a culture of hopelessness. Relief for the suffering families became a top priority for local government officials. Local governments and relief agencies exerted much effort to find aid for the suffering citizens of West Virginia, yet budget problems and the scope of the crisis rendered them helpless in the effort to provide relief. Each county dealt with the crisis in its own way because fiscal conservatives who controlled the state legislature refused to aid the counties in their relief efforts. In McDowell, the county court met in November 1930 to discuss the burgeoning problem. After the meeting, the court asked the coal operators of the county to divide available work among each family head so that everyone had an opportunity to work each week. The commissioners further argued that "it should be the responsibility of the large coal companies to prevent suffering in their camps whenever possible," and urged the companies to cooperate with charitable organizations.[15]

Some government officials believed that relief efforts were not necessary, exhibiting agreement with American conservatives who insisted that state and federal assistance would create economic dependency. Conservative Americans argued that people, if given aid of any kind, would not work and contribute to society. Every relief organization struggled with the problem of identifying those who were only interested in a "free ride." In McDowell, the county court sought assistance from the coal companies in identifying those worthy of assistance. The coal operators' identification of needy applicants, who were willing but unable to find work, resulted in the denial of aid to "loafers" so that assistance would go to those whom they deemed "truly worthy."[16]

As the economic crisis deepened in McDowell County, it virtually eliminated the modest gains in coal mining employment in the late 1920s. The lack of available markets for coal required companies to layoff thousands of miners. In McDowell, coal mining employment decreased from 19,424 in 1929 to 16,098 in 1932.[17] Peerless and U.S. Coal and Coke felt the economic crunch as well. Production at Peerless declined from 358,149 tons in 1929 to 146,841 tons during the height of the Depression in 1932.[18] Although the miners employed in the Peerless mines at Vivian suffered tremendously from the Depression, the economic crisis of the 1930s affected the miners working at U.S. Coal and Coke's No. 6 mine even more. Production at No. 6 fell from 904,580 tons in 1929 to 278,638 tons in 1931. By 1932, U.S. Coal and Coke decided to close the No. 6 mine, resulting in the displacement of hundreds of miners.[19]

The mass poverty caused by unexpected layoffs hindered relief initiatives. County and private relief efforts could not help everyone in need. Many of the people who usually contributed to relief efforts quickly found themselves on the relief rolls. Conditions deteriorated so much that county officials petitioned the state for funding to cover growing assistance needs. The problems in McDowell affected many areas of life. Six thousand children in the county could not attend school due to a lack of clothing and books. Sanitation and nutrition diminished for those who still lived in coal camps. An alarming growth in the incidences of typhoid, diphtheria, and dysentery struck McDowell County coal towns.[20]

As the election of 1932 approached, there appeared to be no end in sight to the economic crisis. The Republican administration of Herbert Hoover supported traditional relief programs, those provided by private individuals and charities. It was clear that the approach of the administration was not working. In 1932, the voters of West Virginia supported the Democratic presidential nominee, Governor Franklin D. Roosevelt of New York; the election of Democrats indicated that West Virginians realized the Republican approach was not working and government relief programs were necessary. Riding a tidal wave of discontent, Roosevelt defeated Hoover and immediately began to change the way in which the country helped those affected by the economic collapse. The New Deal, as Roosevelt's program came to be called, was a set of initiatives designed to stabilize the economy, aid in relief efforts, and stabilize the industry.[21]

The Search for Stability, 1933–1935

The passage of the National Industrial Recovery Act in 1933 influenced industry in several different ways. The NIRA set forth a general procedure for the formation of industrial wage and price codes. The establishment of the industrial codes fell to the National Recovery Administration. The NIRA did not, however, provide guidelines as to the type of provisions to be found in the codes. The only specifics required to be included in the codes were those dealing with labor. Labor and industry representatives negotiated the rest of the industrial codes. Section 7 of the NIRA stipulated that the codes would set minimum wages, maximum hours, and appropriate working conditions for workers. Section 7A outlawed yellow-dog contracts and guaranteed to labor collective bargaining rights.[22] The NIRA gave operators privileges as well. Businesses that accepted the industrial code agreements would be exempt from antitrust laws. Ideally, widespread acceptance of the codes would stabilize the coal industry.[23]

The NIRA was significant because it was the first peacetime attempt by the national government to regulate industry. During the latter part of 1933, representatives of various industries promulgated codes of fair practice under the direction of the NRA. However, the negotiations were tense in most industries because business and labor representatives had different ideas concerning the solutions to the nation's economic problems. Business leaders wanted to protect business, government officials wanted to thwart cutthroat competition, and organized labor wanted to protect the interests of workers. The NRA ultimately failed for several reasons. First, the NRA tried to accommodate too many contradictory interests. Second, the federal government did not have the resources to regulate business adequately. They left that to the trade associations. Thus, unenthusiastic government support of the NRA doomed it to failure.[24]

The coal industry found the task of enacting the industrial codes difficult because mining and wage standards varied in all thirty-three mining states. The differences between the coal mining states with regard to unionization impeded the future negotiations of the coal code. In an effort to protect their interests, many coal companies submitted versions of the coal code for their region that were sympathetic to them. The administration believed that the best way to resurrect industry in the United States was through national coordination; as a result, the federal government thwarted all efforts to submit regional codes. Another problem was the lack of knowledge General Hugh Johnson, head of the NRA, had of the coal industry. Johnson's ignorance retarded the debates on a code of fair competition rather than aiding them. The coal code negotiations went well until Johnson, on August 22, 1933, stated that he would clarify section 7A. The non-union southern operators believed that Johnson threatened their interests and immediately refused to bargain further. Consequently, negotiations came to a halt. There was also dissension among union and non-union operators. Primarily non-union southern operators wanted regional codes in order to protect their interests. Northern operators, especially in Indiana and Illinois, and the union argued for a single national code in order to achieve parity and unionism in the South.[25]

General Johnson worked diligently to coerce the industry representatives to sign the code, but the debates dragged on through the summer. The UMWA believed that the delayed negotiations were part of the coal operators' effort to continue the system of paternalistic control that hindered unionization. In the *United Mine Workers Journal* (*UMWJ*) dated September 15, 1933, the union argued that the negotiated code was fine and that the objection of the operators was nothing more than an effort to continue exploiting the miners. The final code would, in the UMWA's opinion, stabilize the coal industry. Finally, President Roosevelt became involved in the process. On September 14, 1933, the president told representatives that if they did not come to an agreement within twenty-four hours, he would impose one. The committee reached agreement one hour before the deadline, but the pact would prove to be a very fragile one.[26]

The code was a compromise between operators of union fields and operators of non-union fields because it required both a national board, pursuant to the wishes of the union operators, and regional boards, which the non-union operators advocated. The code provided for five regional divisions and fifteen wage districts. A national board governed the industry and each division had its own labor board to handle disputes. Together with the code agreements, the committee also signed the Appalachian Agreement on September 21, 1933. The Appalachian Agreement was a labor compact that gave the code national recognition. The agreement provided for an eight-hour day and forty-hour week, the right of the miners to choose their own checkweighmen, the end of compulsory purchases at the company store, the abolition of scrip, the end of required housing, the institution of a minimum age of seventeen to work in the mines, and the right to collective bargaining.[27] The right to collective bargaining in the Appalachian Agreement was significant because it was the first time that the non-union operators agreed to recognize this right.[28]

The NIRA aided the UMWA in its efforts to organize the southern West Virginia coalfields. Section 7A gave a languishing organization needed momentum to conduct a largely successful organizational drive in southern West Virginia. The success of the membership drive did not derive only from the changes instituted by the NIRA. Even before the passage of section 7A, John L. Lewis, president of the union, began planning the most extensive unionization movement in West Virginia history. It is clear, however, that the passage of section 7A was critical to the success of such a movement. During the drive, many organizers told the crowds, "the president wants you to join the union."[29] The tactic of invoking the president seemed to make joining the union synonymous with patriotism, yet when asked, organizers admitted that they meant the president of the union. Before the organization drive by the UMWA, wages in southern West Virginia were as low as $1.50 a day. After the Appalachian Agreement was signed, wages were about $4.20 a day, only about fifty cents below the minimum for the organized Central Competitive Field.[30]

Furthermore, the UMWA saw the code as the final solution to the problems facing miners and their families. In the first *United Mine Workers Journal* after the code was signed, the union stated, "This gigantic achievement gives to the bituminous coal industry a code of fair competition and sound business and industrial principles that should prove to be the salvation of the industry and all who are interested in it."[31] Not only did the new contract cover every mining field in which the UMWA enjoyed some success in previous years, it also included non-union fields such as McDowell County, where the union had previously been forbidden to enter. Therefore, changes resulting from the NRA produced a higher degree of labor stability in the coal industry, or so it seemed.[32]

Not every mine came under the jurisdiction of the NRA. Originally, the NRA did not affect captive mines, such as U.S. Coal and Coke, because they did not compete in the commercial coal market. All of the coal mined at U.S. Coal and Coke mines went to its parent company, U.S. Steel. Miners employed at captive mines wanted access to a fair and even contract as well. After a wildcat strike at captive mines in western Pennsylvania, most of the steel companies agreed to follow the provisions of the bituminous coal code in their mines. However, U.S. Coal and Coke resisted the code more than other captive mines. In December 1933, two months after most of the steel industry agreed to follow the code, the UMWA published a letter that was sent to the miners at the Gary mines of U.S. Coal and Coke in District 17. The UMWA called on the miners to make a decision about the union without fear of reprisal from the company and promised them that the federal government, under the auspices of the NRA, would ensure that the company would allow the miners to organize. The company continued using force to keep the union out of Gary, however. In April 1934, a letter from W. O. Stuart, president of the UMWA Local at Gary, appeared in the *UMWJ*. Stuart noted that the organizing process was extremely difficult at Gary. He stated, "two union men were beaten with blackjacks by company thugs."[33] Stuart lamented about the lack of federal support for the unionization movement at Gary, yet ended the letter by restating the union men's resolve to organize the mines at Gary.[34]

The workers at U.S. Coal and Coke eventually succeeded in gaining recognition for their local union without a long work stoppage. Miners at other mines in McDowell were not so lucky. After the signing of the bituminous coal code and the Appalachian Agreement, miners at the Fordson mine at Twin Branch organized under the auspices of the UMWA. The management at Fordson refused to negotiate with the union; rather than allowing the UMWA to gain a foothold at their mine, management of the Twin Branch mine closed their operation in January 1934. The company asked every miner to turn in all equipment owned by the company and report for final settlement of wages. The mine stayed closed for at least six months. In July, a letter appeared in the *UMWJ* from the recording secretary of the Local at Twin Branch. The letter insinuated that the company was considering reopening the mine. It is unclear, however, whether the company agreed to negotiate with the union miners of Twin Branch.[35]

At the state level, the NRA had many economic, social, and political consequences. In keeping with the spirit of the NRA, the state government of West Virginia struck several blows to the old order. In October 1934, Governor Herman Kump ordered Sheriff Maginnis Hatfield of McDowell County to disarm and disband the 195 deputies who had public authority, but were paid and controlled by the operators. In 1935, the state legislature abolished the mine-guard system, approved a prevailing wage-rate law, amended workmen's compensation, and provided compensation for victims of silicosis. Within a five-year period, it seemed that the miners in West Virginia were freed from the yolk of oppression by the operators.[36]

Despite the apparently harmonious relationship between labor and industry in West Virginia, defenders of the old order did not disappear. At the twenty-seventh annual meeting of the West Virginia Coal Mining Institute, for instance, William Beury, vice-president of the Algoma Coal Company, defended practices condemned by the NRA. Beury argued that it was necessary for coal operators to be patriarchs for their miners. Companies built camps because the areas where coal could be found were so desolate that adequate housing did not exist. After the coal industry expanded, the camps kept the workers satisfied. Beury argued that payment in scrip protected families because wages could be used for liquor or labor racketeers (i.e. union organizers). Beury continued the tradition of "paternalism" by arguing that miners would not or could not provide for their families. Beury also believed that mine guards were superior to elected officials because they were controlled by the company.[37] Other operators agreed with Beury's assessment; some decided to use traditional methods in an effort to ensure the survival of the open-shop. A miner from Welch told the UMWA that although organization of McDowell County was nearly complete in November 1933, the Baldwin-Felts agents continued to operate in the county.[38] The "gun-men," as the miner called them, tried to confuse the workers of McDowell by circulating false contracts. The operators hoped that the confusion over the false contract would result in miners striking against the true UMWA contract and breaking up their local unions. Although some miners struck against the contract, union efforts to explain the situation kept most workers on the job.[39]

Was the coal code successful in relieving unemployment and stabilizing the industry? Some argued that the code did little to help workers; the Progressive Miners of America, for instance, objected to the code because they believed that it did little to help the unemployed. They felt that the code should allow work-sharing between employed and unemployed miners.[40] Although the Progressive Miners had little influence in the creation of public industrial policy, their opposition served as a benchmark for further and future opposition to the NRA and the coal code.

Within a year after the promulgation of the coal code, the agreement reached in 1933 began to fall apart. The differences in local conditions that hindered the original negotiations recurred with efforts to modify the original code in order to allow parity within the industry. For example, in 1934 operators from Alabama, western Kentucky, and Tennessee sought relief from the minimum day rates included within the code. They felt that only by amending the code could operators from regions where local conditions increased production costs compete in the market. The NRA agreed to pass some amendments, which allowed for increases in tonnage and piece work rates. While this action, like the original code, aided some portions of the coal industry, operators in other regions, such as West Virginia, believed that the government unfairly discriminated against them and did not allow for fair competition. In public comment on the proposed amendments, H. R. Hawthorne, representing the Smokeless Operators Association of southern West Virginia, argued that increasing piece work rates was illegal and inimical to the precepts of the NIRA and would "secure the maintenance of the existing lack of parity in certain districts."[41] Whatever the issue, local considerations hindered the federal government's efforts to ensure fairness in industry.[42]

Even the UMWA eventually turned against the NRA. By early 1935, the UMWA began to argue that the NRA had failed because it did not provide adequate relief for the unemployed. Both UMWA President John L. Lewis and American Federation of Labor President William Green argued that the only solution to the unemployment problem was a thirty-hour week. Under the codes, there were no limits on the number of hours employees could work. The codes only retained the provisions for an average forty-hour week over a one year period. Without some method to share available work, organized labor believed that the effort to remove workers from relief rolls and place them back on the job fell far short of its goals.[43]

Instability, 1935–1938

The NRA failed to address a number of problems common to the coal industry. Overproduction was the most serious of the industry's ailments. Price increases exacerbated overproduction by giving the small operators incentive to open marginal mines. However, the reduction in hours by the NRA increased hourly wages, leading to a large increase in the cost of labor. As a result, many small operators realized that their marginal mines could not produce the amount of capital required to meet production costs. Smaller operations that relied on hand mining (mining coal without any mechanical aid) faced an uncertain future. Because of the higher labor costs, larger operators began to strip mine and to use mechanized loaders in the underground mines. Mechanization, once begun, intensified until machines replaced many underground miners by 1960.[44]

What machines affected coal miners in the 1930s? Similar to most factors of coal production, the type of machinery varied by mine location. Some of the unionized fields, such as those found in Illinois and other Midwestern states, instituted mechanical coal loaders during the economic crisis in the 1920s. Mechanical loading machines varied by design, but most included gathering arms that swept the loose coal onto a conveyor. Prior to unionization, the labor cost in the southern West Virginia coalfields were so low that coal operators could profitably mine coal with less-advanced methods. The most common machinery used in McDowell County mines were undercutters. Undercutters used a chain blade to notch the coal at the base of the seam, thus relieving the miner of the painstaking job of cutting the coal at the base by hand. Although the use of undercutters increased during the 1920s, the majority of McDowell mines did not take advantage of the machinery available on the market. For most of the 1920s, Peerless mined the majority of its coal by hand. U.S. Coal and Coke, in contrast, used undercutters to produce nearly all of their coal during the decade.

There were two reasons for the more widespread use of machinery in U.S. Coal and Coke mines during the 1920s. First, U.S. Coal and Coke, more than any other operation in the county, led the way in research and development. In the early 1920s, Colonel Edward O'Toole spent considerable time working on building a machine that would improve efficiency and increase profits for the company. O'Toole claimed that he built the first machine that combined both the cutting and loading actions in one motion. Although the No. 6 mine did not use the O'Toole machine, the company, as a result of O'Toole's passion for coal machinery, implemented undercutters at No. 6.[45]

A more significant reason for U.S. Coal and Coke's use of machinery was that by the mid-1930s, the UMWA entered into a national contract with the coal industry and established minimum labor costs. If a company wanted to lower its prices, it had to reduce costs of production, not by lowering wages, but by employing cost-efficient machines. New Deal labor policies alone, however, did not explain increased mechanization in the 1930s. Low interest rates and capital costs, an increase in the demand for coal in 1934 and 1935, the concentration of large-scale production in the larger mines, and the selection of machinery offered by the mine supply industry all aided the spread of mechanization. The decrease in interest rates and the increase in the demand for coal influenced operators to invest in the new machinery developed by the mine supply industry. Because production was concentrated in the larger mines, which had the resources to implement new machine technology, mechanization affected more miners in the 1930s than ever before.[46]

Although the onset of mechanization inevitably led to the loss of jobs, UMWA leaders, including John L. Lewis, supported mechanization because they believed that the evolution of the coal industry required mechanization. The rank and file of the UMWA disagreed with their assessment. At the UMWA national convention in January 1934, a delegate proposed that the union oppose all mechanized mining. The delegate thought it was the union's place to protect the jobs of all members of the union. In response to the motion, Lewis shouted in protest, "You can't turn back the clock!"[47] The motion died as a result of Lewis's opposition. Lewis favored mechanization because he believed that it would stabilize industry in the long run. He had promoted mechanization for many years prior to 1934. In his book, *The Miners' Fight for American Standards*, Lewis elaborated on his view that the values of the UMWA necessitated mechanization. "Fair wages and American standards of living are inextricably bound up with the progressive substitution of mechanical for human power. It is no accident that fair wages and machinery will walk hand in hand. . . . The policy of those who seek a disruption of the existing wage structure would only postpone mechanization of the industry and perpetuate obsolete methods."[48] Lewis probably was correct in his assessment of the influence mechanization would have on the coal industry, yet the miners who lost their unskilled jobs to machines had every right to be dismayed with the progress of mechanization.[49]

By looking at the numbers, can historians argue that the NRA succeeded in stabilizing the industry and improving the lives of workers? The immediate aftermath of the institution of the coal code showed remarkable improvement in both statewide and countywide production. State coal production increased from 84 million tons in 1932 to 98 million tons in 1934. McDowell County production showed just as impressive an increase; county production jumped from 13 million tons in 1932 to 17 million tons in 1934.[50] Individual companies also showed a significant improvement in business after the NIRA. The Olga No. 1 mine (the new name of Consolidation Coal Company's Coalwood mine after its acquisition by the Carter Coal Company) increased production in its first two years under Carter from 462,314 tons in 1933 to over one million tons a year later. Peerless improved its output to 311,863 tons and U.S. Coal and Coke reopened its No. 6 mine and produced 413,773 tons in 1934.[51] Therefore, by 1935, most indicators showed the coal industry lifting out of the depression and achieving some stability.

As the 1930s progressed, however, it became clear that the efforts of the Roosevelt administration to stabilize industry had mixed results. The coal industry in McDowell County reached its highest production level of the decade in 1936, when county mines produced more than 23 million tons of coal. After a slight decrease in production in 1937, the bottom dropped out again in 1938, when county production fell to 16 million tons.[52] Similar to the economic collapse during the early part of the decade, miners suffered the consequences of another crisis within the coal industry. McDowell employment declined in one year, between 1937 and 1938, from 21,521 to 18,606.[53] The impact on miners, however, weighed most heavily on those employed by small, marginal mines. All three mines represented in this study showed increased employment between 1937 and 1938. Olga No. 1 employed 952 miners in 1938, up from 922 the year before. Peerless had an increase of fifteen miners, while U.S. Coal and Coke employed ninety-four more miners in 1938.[54]

Why did only the smaller mines of McDowell County lay off workers during the late 1930s? The smaller mines did not have the resources necessary to compete with more prosperous companies. The realities of the coal industry during the late 1930s required mines to mechanize in order to produce enough coal to pay miners the higher wage scales that resulted from the NRA. Olga No. 1 used undercutters to mine all of its production since Carter acquired the mine in 1934. Peerless, by 1938, undercut a majority of its coal: 227,898 of its total production of 424,772 tons.[55] The reason the companies that survived the economic decline of the late 1930s could hire new miners was because of the increased productivity and cost effectiveness of the undercutting machine. Under the room and pillar system of mining, prevalent in the southern West Virginia coalfields during the early twentieth century, miners controlled every aspect of the mining process for a certain area, or "room." Before the implementation of the undercutter, the miner had to undercut the coal seam, blast the coal from the wall, load the coal, and make all necessary safety precautions. Because undercutters did not displace any miners, the machines allowed for an increase in both production and employment.[56]

In long-organized coalfields, the combination of the support for unionization by the federal government that began with the NRA and the support for mechanization by the union contributed to the continued use of mechanical loading machines. Even before the creation of the NRA, operators in the Illinois coalfields worked with union officials to negotiate an agreement that allowed operators to mechanize without fear of union reprisal. The UMWA agreed to the proposal and, after ratification of the contract in 1928, Illinois mines began to mechanize their mines so that by 1933 machines loaded 58.9 percent of the underground coal in Illinois. By contrast, West Virginia companies only loaded 1.2 percent of their underground coal by machine in the same year.[57] The increased wage rates mandated in the coal code increased reliance on loading technology. Even after the Supreme Court ruled that the

NIRA was unconstitutional in 1935 and revoked the authority of the industry codes, federal influence continued to prevent operators from breaking agreements made with the UMWA under the auspices of the NIRA. As a result, price competition resumed after the Supreme Court ruling, yet the decision did not cause wage reduction.

Because operators were reluctant to break their contracts with the UMWA and needed to stay competitive in an uncertain coal market, operators increasingly looked to machinery in an effort to increase profits.[58] Interestingly enough, the McDowell County coal industry did not institute mechanical loaders during the latter half of the 1930s. Because of its non-union tradition, it was not until World War II and the resulting manpower shortage that McDowell County mines looked to mechanical loaders in an effort to meet the increased demand.

Conclusion

Although the NRA failed to fully stabilize industry, it produced many changes in McDowell County. Traditional paternalistic institutions collapsed in the coal towns of the county. The administration's support of unionism, however clandestine, led to union contracts in West Virginia that increased labor costs. Mechanization was the answer for operators seeking to maximize its profits and decrease its costs. Although the mechanization that resulted from the NRA did not lead directly to the implementation of mechanical loaders, which would have displaced numerous miners, it set the precedent for the future use of machinery in the mines of McDowell. A vicious cycle thus began. During World War II, McDowell mines relied increasingly on mechanical loaders to meet the rising demand for coal. However, by the 1950s, a decline in demand caused by increased competition from alternative fuels led the companies to implement continuous mining machines, which combined the mining and loading processes in one step and represented a major component of deindustrialization. In addition to the continued reliance of McDowell County on the coal industry, the use of the continuous miner led to a massive displacement of West Virginia's workers, and in combination with the continued problems plaguing the industry nationally, led to a significant deterioration of McDowell County's economy. Throughout the nation and the Central

Appalachian region, the story was the same. The process of deindustrialization resulted in the unemployment of thousands of miners nationally and, for West Virginia and Central Appalachian coal miners in particular, contributed to endemic poverty.

NOTES

1   Deindustrialization is a decline in the country's productive capacity. Although commonly used in the context of Midwestern industrial capacity, deindustrialization can be used to describe the decline in production and employment in the West Virginia coal industry after 1950. For further information, see: Barry Bluestone and Bennett Harrison, *The Deindustrialization of America: Plant Closings, Community Abandonment, and the Dismantling of Basic Industry* (New York: Basic Books, 1982).

2   The three primary mines used in this study represent different size operations. The figures for Olga No. 1, which only begin in 1934, suggest that Olga was a large operation. Peerless was a smaller, yet significant contributor to the McDowell economy. U.S. Coal and Coke's No. 6 mine was a subsidiary and captive mine for United States Steel and was, therefore, an entirely different example.

3   *Charleston Gazette* (WV), September 16, 1929; Thomas Longin, "Coal, Congress, and the Courts: The Bituminous Coal Industry and the New Deal," *West Virginia History* 34 (January 1974): 101; Jerry B. Thomas, *An Appalachian New Deal: West Virginia in the Great Depression* (Lexington: University Press of Kentucky, 1998), 8.

4   Thomas, *An Appalachian New Deal*, 8.

5   Harold Barger and Sam H. Schurr, *The Mining Industries, 1899–1939: A Study of Output, Employment, and Productivity* (New York: National Bureau of Economic Research, 1944; repr., New York: Arno Press, 1972), 78 (page citations are to the reprint edition); James P. Johnson, *The Politics of Soft Coal: The Bituminous Industry from World War I through the New Deal* (Urbana: University of Illinois Press, 1979), 122; Longin, "Coal, Congress, and the Courts," 102.

6   State of West Virginia, *Annual Report of the Department of Mines* (Charleston, 1927), 102; State of West Virginia, *Annual Report of the Department of Mines* (Charleston, 1929), 119.

7   *Charleston Gazette*, September 26, 1929.

8   *Charleston Gazette*, September 26, 1929; Ellis W. Hawley, *The New Deal and the Problem of Monopoly: A Study in Economic Ambivalence* (Princeton, NJ: Princeton University Press, 1966), 205–206; Stanley Vittoz, *New Deal Labor Policy and the American Economy* (Chapel Hill: University of North Carolina Press, 1987), 51–52; Thomas, *An Appalachian New Deal*, 8–9.

9   State of West Virginia, *Annual Report* (1927), 101; State of West Virginia, *Annual Report* (1929), 84.

10   State of West Virginia, *Annual Report of the Department of Mines* (Charleston, 1926), 192; State of West Virginia, *Annual Report of the Department of Mines* (Charleston, 1928), 170.

11   State of West Virginia, *Annual Report* (1929), 84, 95.

12   State of West Virginia, *Annual Report* (1928), 74–75; State of West Virginia, *Annual Report* (1929), 69–70.

13   State of West Virginia, *Annual Report* (1929), 85, 118; State of West Virginia, *Annual Report of the Department of Mines* (Charleston, 1933), 75, 112; James S. Olson, "The Depths of the Great Depression: Economic Collapse in West Virginia, 1932–1933," *West Virginia History* 38 (April 1977): 223.

14   State of West Virginia, *Annual Report* (1927), 102; State of West Virginia, *Annual Report of the Department of Mines* (Charleston, 1932), 89; West Virginia Commissioner of Labor, *21st Biennial Report of the Department of Labor* (Charleston, 1933), 6.

15   *Welch Daily News* (WV), January 6, 1931.

16   *Bluefield Daily Telegraph* (WV), November 9, 1930; James T. Patterson, *The New Deal and the States: Federalism in Transition* (Princeton, NJ: Princeton University Press, 1969), 26–27.

17   State of West Virginia, *Annual Report* (1929), 118; State of West Virginia, *Annual Report* (1932), 120.

18   State of West Virginia, *Annual Report* (1929), 69; State of West Virginia, *Annual Report* (1932), 71.

19   State of West Virginia, *Annual Report* (1929), 70; State of West Virginia, *Annual Report of the Department of Mines* (Charleston, 1931), 70.

20   Thomas, *An Appalachian New Deal*, 115–116.

21   Thomas, *An Appalachian New Deal*, 69.

22   There were many types of yellow-dog contracts, but the most commonly used one, which companies required employees to sign, stated that an employer

would not employ a member of a union and that the employee would neither join a union nor aid in the organization of a union.

23 Robert F. Himmelberg, *The Origins of the National Recovery Administration: Business, Government, and the Trade Association Issue, 1921–1933* (New York: Fordham University Press, 1976), 207.

24 James A. Hodges, *New Deal Labor Policy and the Southern Cotton Textile Industry, 1933–1941* (Knoxville: University of Tennessee Press, 1986), 5; James P. Johnson, "Drafting the NRA Code of Fair Competition for the Bituminous Coal Industry," *Journal of American History* 52 (December 1966): 521–540; Hugh S. Johnson, *The Blue Eagle from Egg to Earth* (New York: Doubleday, 1935; repr., New York: Greenwood, 1968), 240–243; Irving Bernstein, *Turbulent Years: A History of the American Workers, 1933-1941* (Boston: Houghton Mifflin, 1970), 40–50; Longin, "Coal, Congress, and the Courts," 105.

25 In this instance, the term "southern operators" describes the operators in the southern United States, primarily in Appalachia.

26 K. M. Simpson to S. A. Wender, 9 September 1933, Records of the National Recovery Administration (NRA), Record Group 9, Box 12, National Archives, College Park, MD; *Charleston Gazette*, September 10, 1933; *United Mine Workers Journal*, 15 September 1933; *New York Times*, June 18, July–11, 1933; Johnson, *The Blue Eagle*, 150; Bernstein, *Turbulent Years*, 44.

27 Prior to the 1930s, companies paid miners by the ton. Checkweighmen observed the weighing of the coal to ensure that the companies paid miners accurately for the amount of coal mines. Scrip was a form of pay that companies often gave miners in lieu of legal tender. Scrip could only be redeemed at the company store, giving companies the opportunity to charge higher prices and increase profits.

28 "Appalachian Agreement," Records of the NRA, Box 13; *New York Times* September 8, 1933; *Coal Age* 38 (September 1933): 317; Paul Salstrom, *Appalachia's Path to Dependency: Rethinking a Region's Economic History* (Lexington: University Press of Kentucky, 1994), 87–88; Johnson, *The Blue Eagle*, 159–160.

29 Thomas, *An Appalachian New Deal*, 93.

30 Melvyn Dubovsky and Warren Van Tine, *John L. Lewis: A Biography* (New York: Quadrangle, 1977), 184–186; Salstrom, *Appalachia's Path to Dependency*, 87–88. The Central Competitive Field encompassed the bituminous coalfields in Illinois, Indiana, Ohio, and Pennsylvania.

31 *United Mine Workers Journal*, 1 October 1933.

32   *United Mine Workers Journal*, 1 October 1933, Dubovsky and Van Tine, *John L. Lewis*, 184–186; Bernstein, *Turbulent Years*, 41; Longin, "Coal, Congress, and the Courts," 106; Salstrom, *Appalachia's Path to Dependency*, 87–88.

33   *United Mine Workers Journal*, 15 April 1934.

34   *United Mine Workers Journal*, 15 April 1934, *United Mine Workers Journal*, 15 October 1933, 15 December 1933.

35   *United Mine Workers Journal*, 15 February 1934, 15 July 1934.

36   *United Mine Workers Journal*, 11 November 1933.

37   The companies hired private security forces, known as the mine guard system, in an effort to keep the peace and block the union.

38   The Baldwin-Felts company provided mine guards for many coal companies in southern West Virginia.

39   William Beury, "The Social Aspects of Coal Mines," in *Proceedings of the West Virginia Coal Mining Institute, Twenty Seventh Annual Meeting, Bluefield, West Virginia, December 5–6, 1933* (Morgantown: West Virginia Coal Mining Institute, 1933), 63–77.

40   "Objections by the Progressive Miners of America to Code," Records of the NRA, Box 14.

41   H. R. Hawthorne, "Comments on Proposed Amendment to Bituminous Coal Code for Increasing Piece Work Rates," Records of the NRA, Box 13.

42   "Hearing for the Modification of Bituminous Coal Code, April 9–11, 1934," Records of the NRA, Box 12; Smokeless Operators Reports to the NRA, Records of the NRA, Box 12.

43   *United Mine Workers Journal*, 15 February 1935.

44   Winco-Block Coal Company to Wayne P. Ellis, 24 March 1934, Records of the NRA, Box 13; Johnson, *Politics of Soft Coal*, 191.

45   C. E. Lawall, I. A. Given, and H. G. Kennedy, *Mining Methods in West Virginia* (Morgantown: West Virginia University, 1929), 36–39; *Coal Age* 27 (28 May 1925): 783–787; Keith Dix, *What's a Coal Miner to Do? The Mechanization of Coal Mining* (Pittsburgh: University of Pittsburgh Press, 1988), 33; Curtis Seltzer, *Fire in the Hole: Miners and Managers in the American Coal Industry* (Lexington: University Press of Kentucky, 1985), 12–13; Crandall A. Shifflett, *Coal Towns: Life, Work, and Culture in Company Towns of Southern Appalachia, 1880–1960* (Knoxville: University of Tennessee Press, 1991), 203.

46   Dix, *What's a Coal Miner to Do?*, 198–199.

47   *Coal Age* 39 (February 1934): 83; *New York Times*, 25 January, 26 January 1934.

48   John L. Lewis, *The Miners' Fight for American Standards* (Indianapolis: Bell Publishing, 1925), 108.

49   United Mine Workers of America, *Proceedings, 1927 Convention* (Indianapolis: Cornelius Printing Company, 1927), 215; Dix, *What's a Coal Miner to Do?*, 161.

50   State of West Virginia, *Annual Report* (1932), 86–87; State of West Virginia, *Annual Report of the Department of Mines* (Charleston, 1934), 94–95.

51   State of West Virginia, *Annual Report* (1933), 58; State of West Virginia, *Annual Report* (1934), 77–78.

52   State of West Virginia, *Annual Report of the Department of Mines* (Charleston, 1936), 93; State of West Virginia, *Annual Report of the Department of Mines* (Charleston, 1938), 95.

53   State of West Virginia, *Annual Report of the Department of Mines* (Charleston, 1937), 95; State of West Virginia, *Annual Report* (1938), 98.

54   State of West Virginia, *Annual Report* (1937), 73–75; State of West Virginia, *Annual Report* (1938), 75–77.

55   State of West Virginia, *Annual Report* (1938), 33–35.

56   Dix, *What's a Coal Miner to Do?*, 81–82, 183.

57   Dix, *What's a Coal Miner to Do?*, 182–183.

58   Dix, *What's a Coal Miner to Do?*, 198.

To Dance with the Devil:

The Social Impact of Mountaintop Removal Surface Coal Mining

Shirley Stewart Burns

"There are two roads in life, a right one and a wrong one.
There is no in-between path to take."
—Pauline Canterberry, resident of Sylvester, West Virginia

AS THE NINETEENTH CENTURY gave way to the twentieth, West Virginia's rural backcounties experienced a fundamental transformation. Natural-resource speculators pervaded the area; chief among them were the coal and timber industries, along with their handmaiden, the railroad industry. Throughout West Virginia, beautiful hardwood forests came crashing down until, by the 1920s, nearly all of them were gone.[1] Railroads penetrated the rugged countryside to whisk the natural treasures of timber and coal away from the state and into the large cities beyond. Older agricultural communities were soon joined by new industrial towns that dotted the landscape for the express purpose of providing a home for workers and their families. The repercussions of this rapid-fire change resonated throughout the southern region. Subsistence farmers accustomed to bartering soon disappeared, replaced by wage-earning laborers who toiled in the mines rather than in the fields.

Then, as the industrial age shifted to the information age, coal miners found themselves struggling for their economic lives. Technology had rendered them nearly obsolete. Underground miners saw their ranks slashed as the continuous miner and longwall machinery replaced tens of thousands of men. Surface workers witnessed the introduction of twenty-story draglines that performed the work previously requiring hundreds of workers. The amount of surface-mined acreage has continually increased since 1982, and surface-mining production has been on the rise since 1991.[2] This is largely attributable to the newest surface-mining machinery, such as that used in mountaintop removal (MTR).[3] MTR is a coal extraction process wherein the tops of mountains are removed in order to expose underlying coal seams near the surface. The resulting overburden (the soil and rock that comprised the mountain) is shoved into adjacent valleys where it often covers headwater streams. Since the introduction of the twenty-story dragline—instrumental in the MTR process—in the 1980s, coal-mining employment has plummeted from 59,700 in 1980 to 15,200 in 2004.[4] Coal production has increased, but the bottom line for companies has vastly improved since the highest cost of operating—labor—has been virtually eliminated.

Coupled with increased demand for "cheap" electricity and the desire of the coal industry to cut labor costs, the MTR method of coal extraction became the latest stopover in the trajectory of strip mining history. Companies soon found the fastest and least expensive way to meet the insatiable demand for cheap energy while still meeting federal air quality standards was to utilize low-sulfur coal reserves that are found in the southern West Virginia coalfields. Also needed was the very land the communities inhabited.

These changes in mining methods dramatically affected coalfield residents. Few alternative economic opportunities were available in these areas where coal had been in power for more than one hundred years and where shortsighted politicians had done little to advance economic diversification. While many individuals migrated from southern West Virginia, others stayed because of personal ties to their families and communities. Those who refused to migrate found themselves in the most precarious position of all, caught between dwindling coal jobs and the desire to protect their own homes and families from what they deemed a very unpromising future. Many of those who did not leave remained loyal to the coal industry, but now they found themselves in a predicament where the coal companies needed the mountains and valleys in order to meet the increasing demand for coal. The very land these residents had already sacrificed so much to live on was now itself being obliterated. Quite a few residents flatly refused to leave the land their families had settled more than two hundred years ago, regardless of the conditions or pressures exerted on them by the companies and their agents. These social complexities still pervade the southern West Virginia communities facing depopulation by the encroaching MTR operations.

More than a Nuisance

Pauline Canterberry and Mary Miller live in Sylvester, Boone County. Sylvester is a small, incorporated town of some 195 people on the outskirts of Whitesville.[5] Sylvester was founded in 1952 in hopes that it would be a haven for those who did not want to live inside a coal camp while they made their living by mining coal.[6] The town has seen a number of coal companies on its fringes during its more than five decades of existence. Both Miller and Canterberry note that previous underground coal companies had bosses and supervisors who lived in the town, which ensured that the companies had a vested interest in the health and safety of the community.[7] Then came Massey Energy Company, one of the largest coal operators in southern West Virginia, along with their underground and huge MTR operations. Initially, nothing much changed as coal trucks carried out vast amounts of coal just as the other companies had done. All of that changed in 1997 when Massey opened its Elk Run Coal Company preparation plant just outside the Sylvester city limits. This preparation plant cleaned coal derived from some of Massey's underground mines at Elk Run as well as the surface-mined coal from its Progress mine.[8] With Massey's new preparation plant fully operational, the town and its residents began to experience significant problems.

The first significant change directly related to MTR is blasting. Once blasting commences, the effects can be felt for miles around. Often these blasts disturb properties, separating walls and floors from each other and from the foundation. Blasting can also hurl boulder debris known as flyrock from the MTR site into residential lawns, cause damage to private property, and ruin water wells. Russell Elkins from Rawl, Mingo County, saw the windows fall out of his house immediately after a Massey subsidiary blasted nearby. Elkins estimated that nine out of ten homes in the hollow were affected by the blasting in some way, but he claimed that their owners were afraid to come forward because they or their loved ones were employed by

Massey. Dickie Judy from Foster Hollow, Boone County, experienced damage to his home as well, both inside and out, during a blast on a nearby MTR site. His foundation split, and his walls shook so hard that pictures fell off. Larry Brown, also from Rawl, observed many kinds of blast-related damage in his church and in other structures in the town—cracked foundations, split windows, ruined wells. Summarizing the problems, he noted, "It's destroying property and the state. The beauty of our state is being cut out . . . torn away from us."[9]

Those far away from the blasting, in some instances miles away, are able to hear the distant rumbling as the dynamite explodes. Those closer to the blasting may experience tremors in their houses and flyrock on their property. An MTR site close to Carlos Gore's home in Blair produced flyrock the size of softballs that pelted his house and landed in his front yard. Emphasizing the danger of such flyrock, Gore commented to regulators from the West Virginia Department of Environmental Protection (DEP), "If a rock this big hits you or your car or your house, you're going to have more than a headache. It's going to ruin your whole week, because there's going to be a funeral."[10] Gore's family is one of fewer than thirty still remaining in the small, historic community of Blair. As the MTR permits have increased, the small community has been dismantled house-by-house, hollow-by-hollow.[11] Similar experiences have occurred and continue to occur throughout the southern West Virginia coalfields.

Federal studies refuse to acknowledge residents' assertions of the problems resulting from blasting and MTR. One such document, the Final Programmatic Environmental Impact Statement (FPEIS) on MTR, declared that:

> [t]he existing regulatory controls provide adequate protections from coal mining-related blasting impacts on public safety and structures including wells [and that] the existing regulatory programs are intended to ensure public safety and prevent damage rather than eliminate nuisances from coal mine blasting activities. Some blasting within legal limits may still constitute a nuisance to people in the general area. As with all nuisances, the affected persons may have legal recourse regarding blasting nuisances through civil action. Consequently, blasting is not considered a "significant issue" and no actions are considered in this [Environmental Impact Statement].[12]

Coalfield residents tell a different story. Noise, dust, and property damage associated with blasting has been a frequent complaint.[13] In areas where blasting occurs on a regular basis, cracked foundations, loss of wells, and blown-out windows are commonplace. Unlike traditional contour strip-mining where blasting would last from weeks to months, MTR blasts can last (and affect close neighbors) for years.[14] While legal recourse in civil courts is an option, the residents must prove that damage was caused directly by blasting and not through faulty construction. To do this, they must have an independent assessment of their homes that details all findings and current damages. Residents within one-half mile of the permit area can request a blasting survey, but those further out are on their own.[15] These surveys can be costly, especially in a depressed region where people have problems even meeting basic survival needs. Thus, many outside the one-half-mile radius do not have surveys completed.

The noise and dust created from the constant coal-truck haulage also pose certain problems for communities. In 2003, the legal weight for hauling coal in fifteen southern West Virginia coal counties increased to 120,000 pounds from the previous 65,000–80,000 pound limit. The roads in many of these areas are very narrow, and even at the previous low rate of 65,000-pounds, these behemoth machines got into accidents with coal-community residents on numerous occasions. Sometimes the accidents were fatal.[16]

Many individuals living in the fifteen counties opposed raising the weight limits, fearing for their safety. They knew the roads and bridges in their neighborhoods were not meant for so much weight, and they had seen the results of illegal overweight trucks running through their neighborhoods: demolished roads that were rarely, if ever, repaired, and unnecessary deaths.[17] Protests against the tonnage increase went unheard, and West Virginia's 2003 legislature increased the amount. This should have surprised no one familiar with the history of overweight coal trucks in the southern coalfields. The trucking industry indicated that they were only competitive for contracts if they hauled over the legal limits. Otherwise they risked being underbid.[18]

It is common knowledge in the industry and coal communities that these trucks had run illegally for at least twenty years while authorities turned a blind eye.[19] This inaction resulted in the purchasing of even larger trucks while the state continued to ignore their illegal activity. When the state tried to address the situation in late 2001 through increased fines on overweight trucks, effectively shutting down operations for several days, truckers in the southern coalfields complained bitterly.[20] Bickering between the two factions continued for months. In March 2002, about a dozen coal-truck drivers rallied around the state capitol, blowing their horns in support of new, higher weight limits. At the same time, a group of close to one hundred legislators, community activists, and union members gathered at the capitol in a rally focused on keeping and enforcing the current weight limits.[21] Proponents of the increase saw livelihoods at stake; opponents cited public safety. Caving under pressure, the state continued its watered-down enforcement efforts and, in the end, legalized an activity that had been occurring illegally for decades.

Perhaps the best example of the consequences of the coal dust associated with processing as well as the hauling of coal in overweight trucks is the town of Sylvester. The town experienced firsthand the noise, dust, and other problems caused by huge coal trucks. So troublesome was the dust in Sylvester that in 2001, 154 of the town's residents filed a lawsuit against Elk Run, a subsidiary of Massey. Mary Miller stated, "You're a prisoner in your own home, breathing this coal dust twenty-four hours a day."[22] In what was for many an amazing turn of events, Boone Circuit Judge E. Lee Schlaegel Jr. ruled against Massey and declared that the coal company must contain the dust that was polluting the town or cease operations. Massey originally asserted that the pine trees they had planted in front of the plant were enough to keep the dust to a minimum, but the trees were small and in no way could have cleansed the area. Eventually, the company complied with the order by erecting a huge nylon dome over their preparation plant. The dome worked better, but has ripped twice since its construction and has had to be replaced.[23] The extent of the dust problem is evident by a quick drive through the town: coal dust blankets almost everything, including patio furniture that is protected by a cover. Even spraying homes with power hoses becomes futile, considering that a few weeks later, they will once again be coated with coal dust. Residents remain concerned about the possible health hazards the coal dust presents.[24]

The process of cleaning coal is a dirty one. The plant washes the coal to remove the ash, leaving a thick, gooey substance known as slurry, which is contained by an impoundment that holds vast amounts of the impurities and wastewater remaining after the washing.[25] The holding capacity for a coal-slurry impoundment can range from millions to even billions of gallons of slurry. In addition, slurry impoundments are frequently hundreds of feet deep and encompass several acres. One such impoundment is the Brushy Fork impoundment, built about five miles from the Sylvester-Whitesville neighborhood. This impoundment, owned by Massey, is nine hundred feet high and will hold 8.166 billion gallons of slurry once it is completed. It will be the largest impoundment in the nation.[26]

One of the main fears voiced by Sylvester residents is the exit route on file at the DEP in case of a break has the community leaving *toward* the flow of the impoundment break. Hydrogeologist Rick Eades performed a survey of the Brushy Fork impoundment. He, too, was alarmed that the evacuation plan prepared by the Massey subsidiary, Marfork Coal Company, which owns the impoundment, instructed the citizens of Sylvester to travel four miles into the path of any sludge release. He called for a new emergency evacuation plan to be constructed, one that would not have inhabitants driving into the danger.[27] Mary Miller noted, "They are . . . trapping us down here in these valleys with no hope of escape."[28]

In October 2000, a Massey slurry impoundment in Martin County, Kentucky, broke through the underground mine it rested above. From there, the 300 million gallons of slurry poured through local creeks, eventually making its way to the Tug Fork of the Big Sandy River, on the border between West Virginia and Kentucky, as well as into the Ohio River.[29] While no one died, the slurry polluted one hundred miles of streams and obliterated any life forms in its path.[30] Eades noted that consultants who work for the coal companies are under a huge amount of pressure to provide data that is favorable to the consultant's client. He stated that "consultants must find the 'least-case scenario' of environmental risk, somewhere within their credible methods, to enable coal companies to do whatever they want to do." He did not make such assertions lightly, noting that he had been employed as a government and commercial consultant for sixteen years.[31] In fact, he worried that the mine's pillars did not have enough coal left in them to support the additional load from the Brushy Fork slurry impoundment, despite Eagle Fork Mine's claims that all was well. Constructing impoundments over underground mines could leave the impoundment vulnerable to breakthroughs, putting the communities near the impoundment in harm's way. Eades's concern stemmed from the fact that coal companies would never leave that much coal in a mine, and to suppose that the company had any prior knowledge that a slurry dam would be built above it is illogical.[32]

The fear of impoundments and dust problems grip the entire town of Sylvester. Even employees at the now-closed Sylvester grade school felt MTR's effects during the course of their workday. In the school cafeteria, coal dust blanketed the cooks' equipment so heavily that the cooks needed to wash it off before they could use it. Finally, the cooks decided to store their pots and pans in plastic bags to keep from having to wash them twice.[33] Just a year before the elementary school closed, it had conducted emergency evacuations of the students in case of a slurry-impoundment break. Officials at the school timed the children as they moved from the school to the tallest knoll in the area, which most of the town would be clamoring to reach if an actual slurry-impoundment break occurred.[34] The problems associated with the stoker plant above Sylvester and the dust it created were severe enough to cause the Boone County Board of Education to close the Sylvester school, rather than maintain its newer facilities, and consolidate it with Whitesville Elementary, an older school prone to repeated flooding.[35]

Unfortunately, Sylvester and Whitesville are not isolated cases in southern West Virginia. Companies have constructed huge coal-slurry impoundments above schools elsewhere, and in some instances, local districts have built schools in valleys below a dam. In Wyoming County, for example, the county built its new high school less than two miles from the high-hazard Itmann Preparation Plant impoundment (formerly the Joe Branch impoundment), which had existed for twelve years. Like the Martin County, Kentucky, impoundment that dumped more than 300 million gallons of coal slurry into tributaries that flow into the Tug Fork,[36] the Itmann Preparation Plant impoundment sits partially over underground mines. The March 14, 2003, emergency evacuation plan that Consolidation Coal Company submitted to the DEP indicated that if a "fair weather break" were to occur, the slurry would crest at 21 feet at New Richmond (home of the school), 11 feet at Pineville, 11.4 feet at Mullensville, and 11.4 feet at Marianna. These communities are 2.4, 7.7, 13.7, and 17.9 miles, respectively, from the impoundment. The high school consists of nearly six hundred students and employees; is downriver from any potential breakthroughs; and, as illustrated by the company's own evacuation plan, would be devastated should a breakthrough ever occur. (The company is quick to note in its evacuation plan the unlikelihood of this ever occurring, in spite of the dam's categorization as a high-hazard impoundment.) It is unclear if the Board of Education did not realize the danger existed or simply chose to ignore the fact when it decided to construct the school in its present location. Also downstream from the impoundment is a retirement home in New Richmond and several small communities. It is likely that Pineville, the seat of Wyoming County, would be hurt drastically by any breakthroughs, since it is a mere seven miles from the site.[37]

The danger associated with these impoundments being so close to communities is real, as is the potential for a disaster resulting in the deaths of schoolchildren. On October 21, 1966, a similar impoundment in Wales spilled over its boundaries and landed in the coal town of Aberfan below. The disaster resulted in a loss of 144 people, 116 of them schoolchildren who met their death after the rushing sludge completely covered three classrooms of their school.[38] Disasters such as Aberfan should be cautionary tales for those constructing coal dams above communities. The British government created and distributed warning documents to interested parties both inside and outside Britain. One of those interested parties was the coal company operating above Buffalo Creek in Logan County, West Virginia. The company even consulted with British experts. Yet in 1972, the collapse of a coal dam above Buffalo Creek destroyed the town and killed 125 people. Only after this loss of life would coal-dam failures receive attention in the U.S.[39]

Community Impact of Coal's Mono-Economy
All of the problems associated with MTR notwithstanding, there was an unquestionable need for jobs in an area where the unemployment rate ran as high as 11 percent, representing some of the highest numbers in the state. The loss of thousands of residents in search of work has resulted in a population made up mostly of the elderly and disabled. The few remaining working-age individuals lucky enough to have jobs work for the coal companies, the school system, or the supporting welfare system. Well-paying jobs are sparse in the coalfields, and while many southern West Virginians oppose MTR, others staunchly support it. Some of the most vocal protectors of the practice are the workers whose livelihood depends on the continuation of MTR.

In 1998, when Arch Coal's Dal-Tex mine tried to secure a controversial expansion permit, workers at the operation showed up in droves at a public DEP hearing. They complained about the high unemployment rate in their area and spoke of the desperate need for good jobs. Dal-Tex encouraged their outspokenness: before the hearing, the company enclosed notices with their employees' paychecks that read, "There will be people there who don't want this permit issued. They don't care about your job. Please attend this hearing and show that you support the future of our jobs here at Dal-Tex. Encourage your family and friends to join you. Arrive early to get your 'I'm proud to work at Dal-Tex' T-shirts while supplies last." At the hearing, one miner, a resident of Boone County who was employed at the Logan County mine, asked, "What are we going to give the next generation to live on? How are they going to make it? What are we going to do for jobs for our families?"[40]

Others whose livelihoods depended on MTR were also quite vocal in their support and were suspicious of the negative environmental-impact studies. Stephen Walker, the president of Walker Machinery, said, "Do not blame the modern coal industry for water-quality problems in Southern West Virginia today. Modern coal mining does not pollute." Coal-industry representatives were indignant. At another hearing about mountaintop mine permits in October 1998, Bill Raney, lead lobbyist for the West Virginia Coal Association, told EPA representatives at the Logan County hearing, "Today's hearing isn't about streams. It's about jobs, and families and kids, and a way of life."[41] At this hearing, opponents of MTR may have far outnumbered proponents, but proponents at the meeting were still especially vocal.

On the other side of the issue were citizens in the local community. Carlos Gore, a resident of Blair, asked the supporters of the permit how many of them lived in the area where the MTR was taking place. No one in the audience resided in Blair. Gore then asked all of the audience members who did not live in the area to raise their hand. His request was met with a flurry of hands in the air. Gore then emphasized, "We're not trying to shut you people down. We've got rules and regulations that these [DEP] people are supposed to enforce. That's all we want."[42] The president of the local UMWA spoke in support of the community. He tried to calm the audience and explain that the residents wanted the mining to be done according to law and that the company had an obligation to either mine around the residents or buy them out at a reasonable price.[43]

Still, those depending on the mine to support their families were not easily consoled. Noting that the mines provided one of the few avenues to secure a livable wage, a worker at Hobet described how his household consisted of ten people, including a son and daughter-in-law who traveled to Charleston every day for $7-an-hour jobs. He asked a very good question: "What are we going to do for jobs?"[44] It was a question that neither state nor federal politicians have adequately addressed. Community leaders in the affected towns were just as adamant as the MTR employees. An administrator for the Logan County Commission declared that there was no other way for the county to support itself except through the coal industry, stating, "If the mining process is stopped or impeded, Logan County would suffer devastating consequences. The county commission is not saying coal mining is perfect. But we cannot lessen the degree of dependence on coal that currently exists."[45]

Rather than addressing legitimate concerns posed by opponents of MTR, one union member working for Arch Coal lashed out at opponents and questioned their legitimacy as productive community members when he said, "Most of the people who are doing all the talking couldn't tell a dozer from a loader. Most of them are on a check or too old to have a family to raise." A company manager added, "All we have are Chicken Little environmentalists claiming the sky is falling, and they have a sympathetic press to help their cause."[46] Opponents implored the EPA to ignore the pleas about jobs and to stick to their duty of enforcing the law and protecting the environment.

As in the previous meeting, Carlos Gore was vehement about his right to protect his home, even though he understood the workers' desire to keep their jobs: "You put a pond and valley fill in my hollow. I had two streams running, and I had well water. Now I don't have anything. I've got a right to live there. I lived there before the mountaintop removal came in, and I'll be there long after it's gone."[47] Supporters of MTR expressed anger with what they perceived as flawed priorities. UMWA member Terry Vance, a vocal proponent and employee at an MTR mine, stated, "You need to take a good look around at what you're impacting. We're people, not crawdads or spotted salamanders . . . We're not going to go into the ranks of the unemployed quietly."[48]

In January 2007, the U.S. Army Corps of Engineers issued a Clean Water Act permit for Arch Coal's Spruce No. 1 Mine. While scaled back from 3,113 acres to 2,278 acres, this was essentially the same permit that was at the core of the 1998 Dal-Tex controversy, which resulted in a series of protests by UMWA members. Since that time, Arch Coal transferred the operation to its Mingo Logan Coal Co. subsidiary and will operate the mine with non-union workers.[49]

Workers and others who depended on MTR to make a living, sometimes used violent rhetoric when speaking of judicial attempts to monitor it. One operator of a local trucking company commented on Judge Haden's decision to halt the expansion of the Dal-Tex site: "It could get ugly. I'm surprised that some of these guys that have lost their jobs haven't taken it into their own hands with this judge."[50] Tensions boiled over in September 1999. When trying to commemorate the Battle of Blair Mountain, a group including long-time West Virginia political mainstay Ken Hechler was attacked by proponents of MTR who erroneously blamed the re-enactors for lost jobs. Hechler served as a United States congressman from West Virginia from 1959 to 1977 and as West Virginia's secretary of state from 1985 to 2001. Throughout his tenure as a congressman, Hechler was a true friend of the coal miner and, typically, enjoyed an excellent relationship with the UMWA. He was such a staunch ally for the miners that he had confronted not only the coal interests but also UMWA leadership when he felt its actions were not in the best interest of the miners it represented. Such was the case during the corrupt UMWA presidency of Tony Boyle when Hechler stood up for stronger safety measures and accountability from coal companies. Hechler's public service saw him repeatedly speaking up for the blue-collar worker even as others remained silent.[51]

History did not seem to matter to those lashing out at the re-enactors. Acting more like the company thugs, who historically strong-armed the union, than actual union members, laid-off UMWA workers and others kicked members of the re-enactment group, ripped the signs from their hands, and pelted them with eggs. Some re-enactors even had their eyeglasses broken. Hechler himself was slightly injured. The protestors would have been better served to lay their anger with the company who failed to obey the law and used quick extraction methods that would hasten the end of coal.[52]

Clearly, those who stand to lose their jobs if government regulation halts or curbs MTR have a vested interest and will do whatever is in their power to ensure its continuation. Still, while MTR does provide a few high-paying jobs, the jobs typically are not long-lasting. Companies obtain the coal at MTR sites so quickly that it dwindles at a faster rate than mining using traditional methods. Coal communities have always struggled to strike a balance between their need to maintain jobs and their need to preserve the environment, but with MTR, the balancing act has become even more delicate.

By Any Means Necessary

Julia (Judy) Bonds, a past director of Coal River Mountain Watch (CRMW), a Boone County grassroots organization fighting MTR, has become nationally and internationally recognizable among American environmental activists, winning the Goldman Prize—the Nobel Prize for the environment—in 2003. Her passionate fight against MTR climaxed in 2001 when her family history came full circle. It was then that she became the last of six generations to leave Marfork, a casualty of MTR. Holding out was sometimes a dangerous venture for her. She remembers being "run off" the narrow hollow roads into ditches by supporters of the mine. Bonds spent the first forty-eight years of her life in what she found to be an ideal place to raise children, to live, and to die. According to her, there had been continuous mining on Marfork Hollow for decades, but it had not been so intensive and its effects not so devastating environmentally. In 1993, Massey began actively moving people out of the hollow and started its rigorous MTR operations. Bonds could only watch as her small town died:

> When they first moved in there, we had a thriving community. It wasn't as thriving and prosperous as it was when I was a child . . . but it was still a thriving community with children and, of course, people that were retired lived there and it was still a thriving community. We had our little store that was always there—the Pantry Store—but I noticed people started moving out . . . the houses at the head of the hollow first from Marfork and Birch . . . Old Man Pop Aliff was the last house in Marfork, and he did not want to move. They moved him out because he was living on company land. He had a lifetime lease. He lived six months after he moved. . . . He was heartbroken. Certain

people there that didn't own their own land that was just leasing land, they were the first people to be moved out.[53]

Bonds witnessed the annihilation of her small community as one by one the families sold out to the coal company as the MTR operation came closer and closer to their homes. The majority of people in the hollow owned their own land, but they trickled out of town as the incessant blasting, noise, and coal dust worsened, driving them to quieter, more stable locations. Marfork Hollow no longer exists except in the memories of its former residents. Except for one family cemetery, MTR has consumed the rest of the hollow. The company quietly accumulated acreage, expanding its presence before residents realized what was happening. While some changes seemed subtle, they were actually drastic. The mountains provided protection from the sun, wind, and floods. As the mountains have disappeared, so has that protection. In late 1997 and early 1998, Bonds first became aware that the mining had begun its slow encroachment on her home. She related how the company put up cameras along the mine to monitor the activities near the company's property:

> The camera they had pointed at the house sitting in front of me and one right up above my house. Legitimately they could say that they had the camera on the one in front of [me] and on [their] property. They bought up around me . . . So it was an intimidation factor.[54]

Less than four years later, she moved from her ancestors' home-place. Bonds had not wanted to leave, but, as she noted, "the last blackwater spill . . . came right up to the bank of our creek that was right in front of our house."[55] The blackwater spill—combined with the noise, the safety issues, the dust problems, and her family members frantically encouraging her to leave—was enough to make Bonds pack up and move out of the hollow in 2001.

When Massey first arrived, company officials held a town meeting with community residents and assured them that Massey would be a good corporate neighbor. Many of the residents were not convinced since they had already heard otherwise from neighboring communities. Residents recalled that Armco Inc., the previous underground coal company in the area, was more sensitive to its workers and the community than Massey. Armco managers lived in the area, which helped to create a sense of community between the company and the town and made the company more mindful of the effects that mining had on the area. In contrast, Massey had no high-level officials living in either Whitesville or Sylvester. Prior to the onslaught of MTR, the underground coal companies like Armco had an unwritten policy of doing more neighborly things such as sending cards and food to employees and their spouses when a loved one died. With large companies such as Massey, civic gestures were no longer part of the corporate culture.[56] Instead, the companies were large, multinational corporations directed from distant headquarters and too removed from the community to entertain such ideas.

In a written response to the Draft Environmental Impact Statement (DEIS), the Ohio Valley Environmental Coalition collected responses from coalfield residents to present to the government. One man commented, "We live in fear. The whole hollow is in a state of anxiety now every time it floods."[57] The same OVEC document noted how people in these affected communities had lost insurance on their homes. In some cases, local officials condemned the homes; other residents have houses with cracked foundations, walls, and ceilings; destroyed water wells; and overall devaluations. The value of Mary Miller's home fell from $144,000 to $12,000. Hers is a beautiful, large brick house with hardwood floors that in a different setting would certainly be appraised for more than $144,000. Miller blamed the lower appraisal on the extensive mining occurring near her home, particularly the huge preparation plant, complete with nylon dome, that sits just behind her home, visible from her well-maintained lawn.[58] Additionally, her town has experienced a decrease in population and has recently suffered the closing of its elementary school.

Nevertheless, many of the people continue to live near the coalfields, even if it means putting their property, and themselves, at risk. They do so largely because of ties to the land, their communities, and their families. On the surface, it would seem far easier for the companies to simply buy out these homeowners, securing complete control of the entire area. In many places where significant MTR has occurred, this tactic has been used. The town of Blair, for instance, has seen a sharp decline in its population. By 1998, fewer than thirty families still remained in the area, down from 180 families just years earlier. In Blair, Arch Coal bought out both residents and businesses;

the businesses were purchased first, resulting in a loss of taxes. Residents soon found themselves traveling miles for milk and bread. Then, massive buyouts of the residents in the area took place, and population decline forced the closure of school systems, often the death knell for a small community.[59]

From the standpoint of the company, it made perfect sense. David Todd, vice president and spokesman for Arch, provided the company's philosophy in a court deposition: "Our philosophy is not to impact people and if there are no people to impact, that is consistent with our philosophy."[60] In its quest to limit the adverse effect MTR has on communities, the best solution equaled removal of the communities. In truth, only by eliminating the communities can corporations expand mining operations. Whether this elimination comes from paying someone to leave or by creating nuisances so severe that they force people to sell out is not the issue. The end result is the same: depopulation of the coalfields and easy access to the coal. Arch Coal and its Logan County operations usually did not offer to buy out residents. Rather, they relied on the increased activity at the mines to force residents to ask them for a buyout, thus ensuring that Arch could pay less money for the residents' property. Residents then signed an agreement stating that they would not come back to the communities, would not speak out against the mining activity, and would withdraw any previous complaints they had made about the mining. Such wording is illegal, but the people who signed the agreement often were not aware.[61] These buyout plans affected eleven hollows near Blair Mountain.

Arch Coal is not the only entity to embrace depopulation as a solution. Some industry lawyers openly applaud and encourage the removal of people. In 2000, an attorney in Charleston, West Virginia, made multiple presentations about that very subject. Employed by a well-respected Charleston law firm that handles labor, government, environment, and energy litigation, the firm boasts of its wide-ranging experience in identifying and minimizing environmental liabilities for its clients. The lawyer, who asked to remain anonymous, represents corporations in obtaining permits, in penalty negotiations, and in appeal hearings and rulemaking proceedings. One of the presentations detailed the problems facing MTR in southern West Virginia. The attorney presented the idea of ridding the area of its inhabitants for the sake of the company's growth as worthwhile—an example of the ends justifying the means. The effects on wildlife were inconsequential; he asserted that saving wildlife, particularly any endangered species that might be affected, was not worth the social or economic cost: "People will always be more important than insignificant species whose only value is spiritual."[62] The attorney did not address the environmental problems that would remain regardless of population numbers. The hundreds of miles of streams affected, for instance, run into other streams, negatively affecting those water outlets as well. In the course of arguing in favor of MTR, the attorney lauded reclamation, noting the "hardwood saplings up to three feet tall" that are the result of successful reforestation practices on reclaimed land.[63] Left unmentioned was the scientific evidence that such endeavors have rarely been successful.

The starkest argument in the presentation, however, was the claim that MTR-assisted depopulation was a good thing, and the only way to solve the long-term poverty found in the coalfield counties. Rather than encouraging economic development in the area, the presentation noted the huge financial drain on the miners who, it asserted, pay the majority of the taxes, with two-thirds of the taxpayers dependent on the one-third that worked as coal miners. The "core problem," as the presenter saw it, "is too many people. Way too many people."[64]

The attorney set out various proposals to handle this problem. One proposal encouraged the state to eradicate dilapidated coalfield homes. Another proposal suggested that the state provide grant money to help people settle outside of West Virginia and to revoke the grant if they moved back to the state. Yet another idea entailed offering free college education for coalfield kids whose families relocated, as well as to single adults and childless couples. One final suggestion concluded: if "stubborn people" refused to move, their land could be condemned, taken over by the state, and sold to companies who wanted it for MTR. The companies would then reimburse the state for any expenses incurred in securing the land.[65] In the end, the main goal focused on getting people to move.

In the presentation, the attorney admitted that a West Virginia government would likely never embrace such suggestions, declaring that the government lacked the political fortitude to depopulate the southern coalfields. The attorney nevertheless believed the coalfields would be depopulated, but in a more agonizing way than necessary, and without the assistance that the attorney outlined in the presentation. It is certain that such depopulation would allow the companies total control of the coal-rich southern counties. It is also certain that the tactics employed thus far by many coal operators seem to embrace the idea of depopulation, which increases the power of a coal corporation.

Conclusion

As companies currently practice MTR, three distinct stages of power relations can be discerned among communities affected by MTR. Stage one is the infancy/beginning stage. In this stage, community members often trust that the company has their best interests at heart. They welcome the company for the employment and tax revenues it will generate. Initially, the community sees the company as the savior who will deliver it from its economic plight.

Stage two is the intermediate or middle stage. Community members become shocked, dismayed, and angered as MTR begins to directly affect them in a negative way. Those with jobs at the site may still praise the company for providing them with work and see the company as protecting their homes. Many others leave in search of work or to escape the effects of MTR, and steady migration from the area begins. As the company begins to offer to buy out households, residents band together in an effort to save their communities. The role of savior begins to crumble, and the company instead finds itself in the dual role of protector of jobs and robber baron of the land.

In stage three, the final stage, massive buyouts of homeowners and businesses take place, and intense depopulation and out-migration occurs. The coal company essentially gobbles up the community as MTR expands, consumes the surrounding land, and displaces residents. Once established, stage three cements the company in the role of destroyer. The stage is complete when all members of the community have moved and the community itself is dissolved, with all associated local businesses and schools closed.

As the stages progress, the number of residents plummet, MTR acreage increases, and employment begins to slightly increase, followed by a tapering off of employment as MTR expands. The need for manpower is replaced by the need for huge draglines. It is expected that more and more southern West Virginia communities will experience these stages as MTR becomes more widespread there. Throughout all three stages, the power relationship remains the same; coal corporations enjoy the upper hand.

The old argument pitting jobs against the environment remains. Five generations have passed since railroads began to take West Virginia's bountiful natural-resource treasures of coal and timber to places outside the region, decimating the state's hardwood forests and diminishing its coal reserves. By 1920, West Virginia's bountiful hardwood forests had nearly disappeared. It took eighty years for the forest to replenish. Coal is a finite resource and will not replenish. The repercussions of constantly extracting resources with no thought of the future consumed those within the region. There were jobs, but at what cost? At the turn of the nineteenth century, West Virginia's inhabitants learned a difficult lesson about what such a "dance with the devil" could do. In this present struggle, once again balancing economics with community needs for a safe and functioning environment, this generation has now done the same.

NOTES

This essay is based on Chapter 3 of my earlier study, *Bringing Down The Mountains: The Impact of Mountaintop Removal on Southern West Virginia Communities* (Morgantown: West Virginia University Press, 2007).

1   Ronald Lewis, *Transforming the Appalachian Countryside: Railroads, Deforestation, and Social Change in West Virginia, 1880–1920* (Chapel Hill: University of North Carolina Press, 1998), 5, 9.

2   Christiadi Hammond and George Hammond. "How well did we retain and attract highly educated workers?" *West Virginia University College of Business and Economics, West Virginia Business and Economic Review* 9 (2003): 1–6.

3   West Virginia Coal Association, *West Virginia Coal Facts 2001* (Charleston: West Virginia Coal Association, 2001); West Virginia Bureau of Employment Programs, "West Virginia Nonfarm Employment by Industry, 1939–1999," http://www.wvbep.org/bep/lmi/e&e/nf_39-99.htm (accessed October 14, 2006).

4   West Virginia Bureau of Employment Programs, "West Virginia Coal Mining Employment, 1950–2004," http://www.wvbep.org/bep/lmi/e&e/nf_39-99.htm (accessed October 14, 2006).

5   U.S. Census Bureau, *Profile of General Demographic Characteristics: 2000, Sylvester, West Virginia,* generated by the author using American Fact Finder, available from http://factfinder.census.go (accessed October 14, 2006).

6   Pauline Canterberry, interview with the author, Sylvester, West Virginia, October 9, 2003, tape recording; Mary Miller, interview with the author, Sylvester, West Virginia, October 9, 2003, tape recording; Boone County, West Virginia, "Certificate of Incorporation of the town of Sylvester," 1952.

7   Mary Miller interview; Pauline Canterberry interview.

8   Massey Energy Company, *2003 Annual Report*, 9–10.

9   *All Shaken Up*, VHS, interviews with Russell Elkins, Dickie Judy, and Larry Brown (Charleston, WV: Omni Productions, 1998).

10   Ken Ward Jr., "State Environmental Officials Limit Blasting at Dal-Tex Mine: Rocks Sailing into Yards Upset Blair Residents," *Charleston (WV) Gazette*, August 23, 1997; Ken Ward Jr., "Strip-mining Battle Resurfaces in State," *Charleston Gazette*, March 22, 1998; personal conversations between the author and Carlos Gore, various dates.

11   Ward, "Strip-mining Battle Resurfaces in State"; personal conversations between the author and Carlos Gore, various dates.

12   United States Environmental Protection Agency, "Final Programmatic Environmental Impact Statement," Alternatives, Part 2 (Washington, D.C.: U.S. Environmental Protection Agency, 2005), II.A-6. (Hereafter, the "Final Programmatic Environmental Impact Statement" will be referred to as FPEIS.) According to the FPEIS signature page, the "FPEIS was prepared in accordance with the provision set forth in 40 CFR 1503.4(c) of the regulations implementing NEPA, which allow the agencies to attach an errata sheet to the statement instead of rewriting the draft statement and to circulate the errata, comments, responses, and the changes, rather than the entire document." In essence, much of the FPEIS remains the same document as its precursor, the "Draft Environmental Impact Statement" (DEIS).

13   Ken Ward Jr., "Mining Study: Blasts not 'Significant,'" *Charleston Gazette*, August 3, 2003; Ohio Valley Environmental Coalition, *Comments on Draft Programmatic Environmental Impact Statement on Mountaintop Removal/Valley Fill Activities in Appalachia: The Social and Cultural Effects of Mountaintop Removal/Valley Fill Coal Mining* (Huntington, WV: Ohio Valley Environmental Association, January 5, 2004), 6, available from http://www.ohvec.org/issues/mountaintop_removal/articles/EIS_social_cultural.pdf.

14   Ken Ward Jr., "Mining Study," *Charleston Gazette*, August 3, 2003.

15   Ward, "Mining Study," *Charleston Gazette*, August 3, 2003; Mark Squillace, *Strip Mining Handbook: A Coalfield Citizens' Guide to Using the Law to Fight Back Against the Ravages of Strip Mining and Underground Mining* (Washington, D.C.: Environmental Policy Institute and Friends of the Earth, 1990), 70.

16   For examples of fatal accidents see "2 Die in Collision with Coal Truck," *Charleston Gazette*, July 11, 2005; Deanna Wrenn, "Wreck May Revive Coal Truck Debate; Lawmaker Says Fatal Crash a Reminder of Ongoing Problem," *Charleston (WV) Daily Mail*, September 30, 2002; Toby Coleman, "Three Die in Accident: Car Skids into Path of Loaded Coal Truck, Officials Say," *Charleston Daily Mail*, September 28, 2002; Dan Radmacher, "House Finance Chairman Michael Should Take Heat in Coal Truck Death," *Charleston Gazette*, May 3, 2002; "Driver Cited Twice Over Truck Weight," *Charleston Daily Mail*, April 19, 2002; "Father, Son Die in Welch Wreck," *Charleston Gazette*, December 20, 1997; "Truck Driver Charged in Fatal Wreck," *Charleston Daily Mail*, August 6, 1994.

17   Sam Truman, "Coal Haulers See Little Difference," *Charleston Daily Mail*, July 29, 2002; Dan Radmacher, "Coal Truck Bill is an Economic Disaster for West Virginia," *Charleston Gazette*, February 28, 2003. I have had numerous close calls with coal trucks in the coalfields of southern West Virginia. I've been both a driver and a passenger in cars that were run off the road by huge trucks heaped with coal. Longer reaction times could have ended disastrously. It should be noted that the truck drivers did not appear to be doing anything malicious. The roads are simply not wide enough to accommodate the large coal trucks and any other vehicle simultaneously. The coal trucks, quite simply, must use more than just one lane in order to maneuver the steep, mountainous terrain.

18   Martha Bryson Hodel, "DOH Shows its Plans for Overweight Trucks on Southern W. Va. Roads," *Associated Press*, August 5, 2003; "Monster Truck Companies Can Run Legally," *Charleston Gazette*, November 8, 2001; Tara Tuckwiller, "Lighten Up, Haulers Say: Coal Truck Owners Hoping Legislators Ease Weight Limits; Latest State Crackdown Could Close Their Doors, Company Officials Fear," *Charleston Gazette*, November 4, 2001; Dan Radmacher, "Monster Mash," *Charleston Gazette*, February 9, 1994.

19   Tuckwiller, "Lighten Up," *Charleston Gazette*, November 4, 2001; Gavin McCormick, "Wise Names Overweight Truck Panel Members," *Charleston Gazette*, April 4, 2002.

20   McCormick, "Wise Names Overweight Truck Panel Members," *Charleston Gazette*, April 4, 2002.

21   Paul Nyden, "Both Sides Make a Stand on Hauling Bill," *Charleston Gazette*, March 2, 2002; Brian Bowling, "Coal Trucks Roll on State Capitol; Both Sides of Issue Take Message to Lawmakers," *Charleston Daily Mail*, March 1, 2002.

22   Mary Miller interview.

23   "Replacement Dome Rips at Massey Plant," *Charleston Gazette*, April 24, 2003.

24   Mary Miller interview; Pauline Canterberry interview.

25   Robert Meyers, *Coal Handbook* (New York: M. Dekker, 1981), chapter 6; Jann Vendetti, "Mining: Storing Coal Slurry," (Alexandria, VA: American Geological Society), http://www.agiweb.org/geotimes/dec01/NNcoal.html (accessed September 24, 2006).

26   Robert C. Byrd National Technology Transfer Center (NTTC), Wheeling Jesuit University, West Virginia University, and the National Energy Technology Laboratory, "Coal Impoundment Location and Warning System," http://www.coalimpoundment.com/locate/impoundment.asp?impoundment_id=1211-WV04-40234-02. (accessed September 24, 2006).

27   Rick Eades, *Brushy Fork Slurry Impoundment: A Preliminary Report* (Huntington, WV: Ohio Valley Environmental Coalition, 2000), 17, 13; "UMWA Launches Community Education Campaign about Massey's Brushy Fork Impoundment," *United Mine Workers of America Journal*, January–February 2004. (Hereafter cited as *UMWJ*.)

28   Mary Miller interview.

29   Ken Ward Jr., "Criminal Probe Is Ongoing in Coal Mining Dam Failure in Kentucky," *Charleston Gazette*, May 12, 2004.

30   Jack Spadaro, "Mountaintop Removal Mining Practices Must Change or Ecosystem will be Destroyed," *Charleston Gazette*, February 21, 2005.

31   Eades, *Brushy Fork Slurry Impoundment*, 11; "UMWA launches community education campaign," *UMWJ*, January–February 2004.

32   Eades, *Brushy Fork Slurry Impoundment*, 11.

33   Mary Miller interview; Pauline Canterberry interview.

34   Mary Miller interview.

35   Boone County Board of Education, *Reasons Supporting Data—Executive Summary of Consolidation of Sylvester Elementary and Whitesville Schools for Boone County*

Schools (Madison, WV: Boone County Board of Education, 2002), 37; Pauline Canterberry interview; Mary Miller interview.

36   Ken Ward Jr., "Mine Spill on Tug Underscores Dam Concerns," *Charleston Gazette,* October 13, 2000; Ward, "Criminal Probe Is Ongoing," *Charleston Gazette,* May 12, 2004.

37   Consolidation Coal Company, *Monitoring and Emergency Warning Plan and Procedures for the Joe Branch Coal Refuse Dam* (Welch, WV: Department of Environmental Protection, 2003), 12, http://www.coalimpoundment.org/EmergencyPlans/1211-WV4-0709-01.pdf. GIS mapping also illustrates this.

38   Iain McLean and Martin Johnes, *Aberfan: Government and Disasters* (Cardiff, Wales: Welsh Academic Press, 2000), 22.

39   McLean and Johnes, *Aberfan*, 90, 234–235.

40   Ken Ward Jr., "Miners Pack Hearing to Support Strip Permit: Area Needs the Jobs, UMW Member Says," *Charleston Gazette,* May 6, 1998.

41   Ken Ward, Jr., "EPA Gets Earful on Mountaintop Mining," *Charleston Gazette*, October 25, 1998.

42   "Miners Outnumber Others at Mountaintop Removal Hearing," *Charleston Daily Mail,* May 6, 1998.

43   Ward, "Miners Pack Hearing to Support Strip Permit," *Charleston Gazette*, May 6, 1998.

44   "Miners Outnumber Others," *Charleston Daily Mail,* May 6, 1998.

45   Ward, "EPA Gets Earful on Mountaintop Mining," *Charleston Gazette,* October 25, 1998.

46   Ward, "EPA Gets Earful on Mountaintop Mining," *Charleston Gazette,* October 25, 1998.

47   Ward, "EPA Gets Earful on Mountaintop Mining," *Charleston Gazette,* October 25, 1998.

48   Ken Ward Jr., "UMW Members Protest Mine Ruling," *Charleston Gazette,* March 6, 1999. While it is true that the lawsuits brought against the coal companies rely largely on arguments that focus on damage afforded the environment, plaintiffs have realized that the most effective arguments are *not* those that deal with the destruction of their personal property but those that deal with the destruction of the environment in general. Any one of a number of federal laws against such damage may be invoked.

49   Ken Ward, Jr., "Corp Gives Final OK to Record Strip Mine in Logan," *Charleston Gazette*, January 30, 2007.

50   Steve Meyers, "Tensions Over Mining Could Grow; Those At Rally Say They Need Jobs to Support Families," *Charleston Daily Mail,* March 13, 1999.

51   Charles H. Moffat, *Ken Hechler: Maverick Public Servant* (Charleston, WV: Mountain State Press, 1987), 209–231.

52   "Violence Logan, Ravenswood," *Charleston Gazette,* September 2, 1999; Dan Radmacher, "Union Members Become Blair Thugs," *Charleston Gazette,* September 3, 1999.

53   Judy Bonds, interview with author, Sylvester, West Virginia, October 8, 2003, tape recording.

54   Judy Bonds interview.

55   Judy Bonds interview.

56   Judy Bonds interview.

57   Ohio Valley Environmental Coalition, *Comments on Draft Programmatic Environmental Impact Statement,* 4.

58   Ohio Valley Environmental Coalition, *Comments on Draft Programmatic Environmental Impact Statement,* 53; Mary Miller interview.

59   Martha Bryson Hodel, "Residents Praise, Vilify the Coal Industry," *Charleston Gazette,* September 23, 1998; "Mine Guts Town, Some Say; Blair Mountain Residents Upset over Shrinking Population," *Charleston Daily Mail,* August 18, 1997.

60   "Buying Blair: Arch Removes Community," *Charleston Gazette,* November 27, 1998.

61   This type of wording takes away a person's first amendment rights to free speech. For coverage of this, see "Buying Blair: Arch Removes Community," *Charleston Gazette*, November 27, 1998.

62   Presentation information in possession of author. During the course of research for this document, I was able to obtain a copy of one of the presentations given in January 2000. I was asked to maintain confidentiality from where the document was originally obtained. In order to do so, no pronouns have been used so the gender of the presenter is not revealed, nor the presenter's employer.

63   Quote from January 2000 presentation.

64   Quote from January 2000 presentation.

65   Quote from January 2000 presentation.

Publications

By Ronald L. Lewis

Authored Books

Coal, Iron, and Slaves: Industrial Slavery in Maryland and Virginia, 1715–1865. Westport, Conn.: Greenwood Press, 1979. 283 pages.

Black Coal Miners in America: Race, Class, and Community Conflict, 1780–1980. Lexington: University Press of Kentucky, 1987. 239 pages.

Transforming the Appalachian Countryside: Railroads, Deforestation, and Social Change in West Virginia, 1880–1920. Chapel Hill: University of North Carolina Press, 1998. 348 pages.

Welsh Miners, American Coal, and the Loss of Memory, 1840–1920. Chapel Hill: University of North Carolina Press, 2008. 395 pages.

Edited Books

The Other Slaves: Mechanics, Artisans and Craftsmen. Boston: G. K. Hall, 1978. Co-edited with James E. Newton. 245 pages.

The Black Worker: A Documentary History From Colonial Times to the Present. Philadelphia: Temple University Press, 1978–84. Co-edited with Philip S. Foner. 8 vols., 4,102 pages.

    Vol. 1 *The Black Worker to 1869* (1978)

    Vol. 2 *Era of the National Labor Union* (1978).

    Vol. 3 *Era of the Knights of Labor* (1978).

    Vol. 4 *Era of the AFL and the Railway Brotherhoods* (1979).

    Vol. 5 *The Black Worker from 1900 to 1919* (1980).

Vol. 6 *Era of Prosperity and the Great Depression* (1981).

Vol. 7 *From the Founding of the CIO to the AFL-CIO Merger, 1936–1954* (1983).

Vol. 8 *Since the AFL-CIO Merger, 1955–1980* (1984).

Black Workers: A Documentary History from Colonial Times to the Present. Philadelphia: Temple University Press, 1989. One volume, abridged hardcover and papercover editions. Co-edited with Philip S. Foner. 733 pages.

The Transformation of Life and Labor in Appalachia. Guest Editor. *Journal of the Appalachian Studies Association,* 2 (1989). 198 pages.

West Virginia History: Critical Essays on the Literature. Dubuque, Iowa: Kendall-Hunt Pub. Co., for WV Humanities Council, 1991. Co-edited with John C. Hennen, Jr. 247 pages.

West Virginia: Documents in the History of a Rural-Industrial State, 2nd. rev. ed. Dubuque, Iowa: Kendall-Hunt Publishing Co., l996. Co-edited with John C. Hennen, Jr. 371 pages.

Transnational West Virginia: Ethnic Communities and Economic Change, 1840–1940. Morgantown: West Virginia University Press, 2002. Co-edited with Ken Fones-Wolf. 350 pages.

Articles, Chapters, Essays

"Slavery on Chesapeake Iron Plantations before the American Revolution," *The Journal of Negro History,* 59 (July 1974): 242–254.

"Slave Labor in the Chesapeake Iron Industry: The Colonial Era," *Labor History,* 17 (Summer 1976): 388–405.

"The American Dream and the Rationalization of Slavery," *The Crisis,* 83 (August-September 1976): 253–254.

"Slave Labor in the Virginia Iron Industry: The Ante-Bellum Era," *West Virginia History,* 38 (January 1977): 141–156.

"Race and the United Mine Workers in Tennessee: The Letters of William Riley, 1892–1895," *Tennessee Historical Quarterly,* 36 (Winter 1977): 524–536.

"Cultural Pluralism and Black Reconstruction: The Career of Richard H. Cain," *The Crisis,* 85 (February 1978): 57–65.

"Black Coal Miners in the Eastern Virginia Coal Field, 1765–1860," *The Other Slaves* (August-September 1976): 87–108.

"Slave Families at Early Chesapeake Ironworks," *The Virginia Magazine of History and Biography*, 86 (April 1978): 169–179.

"The Darkest Abode of Man: Black Miners in the First Southern Coal Field," *The Virginia Magazine of History and Biography*, 87 (April 1979): 190–202.

"The Rev. T. G. Steward and 'Mixed' Schools in Delaware, 1882," *Delaware History*, 19 (Spring-Summer 1980): 53–58.

"Rev. T. G. Steward and Black Education in Reconstruction Delaware," *Delaware History*, 20 (Spring-Summer 1981): 156–178.

"Job Control and Race Relations in the Coal Fields, 1870–1920," *The Journal of Ethnic Studies*, 12 (Winter 1985): 36–64.

"Blacks in the Paint-Cabin Creek Strike, 1912–1913," *West Virginia History*, 46 (1985–86): 59–71.

"From Peasant to Proletarian: The Migration of Southern Blacks to the Central Appalachian Coalfields," *The Journal of Southern History*, 55 (February 1989): 77–102.

"'Why Don't You Bake Bread?' Franklin Trubee and the Scotts Run Reciprocal Economy," *Goldenseal: Traditional Life in West Virginia*, 15 (Spring 1989): 34–41.

Chapter. "Black Coal Miners in Southwestern Pennsylvania Coal Fields During the Labor Disputes of 1924–28," in *The Early Coal Miner* (1992), ed. by Dennis F. Brestensky, 63–67.

"Appalachian Restructuring in Historical Perspective: Coal, Culture, and Social Change in West Virginia," *Urban Studies: International Journal of Urban and Regional Studies* (Glasgow, Scotland), special issue on Economic Restructuring in Older Industrial Regions, 30 (March 1993): 299–308.

"Conflict at Coal River Collieries: The United Mine Workers of America vs. The Brotherhood of Locomotive Engineers," *West Virginia History*, 52 (1993): 73–90. Co-authored with Tom Robertson.

"Scotts Run: An Introduction," *West Virginia History*, 53 (1994):1–5.

Chapter. "Railroads, Deforestation, and the Transformation of Agriculture in the West Virginia Back Counties, 1880–1920," in *Appalachia in the Making: The Mountain South in the Nineteenth Century*, ed. by Mary Beth

Pudup, Dwight B. Billings, and Altina L. Waller (Chapel Hill: University of North Carolina Press, 1995), 297–320.

Chapter. "Scotts Run, Symbol of the Great Depression in the Coal Fields," in *A New Deal for America: Arthurdale and the New Deal Homesteads*, ed. by Bryan Ward (Morgantown, WV: Arthurdale Heritage, Inc., 1995), 1–23.

Chapter. "Coal Miners and the Social Equality Wedge in Alabama, 1880–1908," in *A Model of Industrial Solidarity?: The United Mine Workers of America, 1890–1990*, ed. by John H. M. Lasslett (University, Pa.: Penn State University Press, 1996), 297–319.

"Cutting West Virginia's Virgin Forest," *WVU Alumni Magazine,* 19 (Summer 1996): 10–12.

Co-authored with Dwight Billings, "Appalachian Culture and Economic Development," *Journal of Appalachian Studies,* 3 (Spring 1997): 3–36.

Chapter. "Beyond Isolation and Homogeneity: Diversity and the History of Appalachia," in *Recycling Appalachia: Back-Talk from An American Region*, ed. by Dwight B. Billings (Lexington: University Press of Kentucky, 1999), 21–43.

Chapter. "African American Convicts in the Coal Mines of Southern Appalachia," in *Appalachia in Black and White: Race Relations in the Nineteenth-Century South*, ed. by John C. Inscoe (Lexington: University Press of Kentucky, 2000), 259–283.

Author's Response. Forum on *Transforming the Appalachian Countryside: Railroads, Deforestation, and Social Change in Appalachia, 1880–1920* (Chapel Hill: University of North Carolina Press, 1998), in *West Virginia History,* 58 (1999–2000): 45–67.

Essay. "The Coal Industry," *Encyclopedia of the United States in the Nineteenth Century* (New York: Charles Scribner's Sons, 2001). 2000 words.

Chapter. "Industrialization in Appalachia," in *High Mountains Rising: Appalachia in Time and Place,* ed. by Tyler Blethan and Richard Straw (Urbana: Univ. of Illinois Press, 2004), 59–73.

Chapter. "Industrial Slavery: Linking the Periphery and the Core," in *African American Urban Studies: Historical, Contemporary, and Comparative Perspectives,* ed. by Joe William Trotter, Tera Hunter, and Earl Lewis (New York: Palgrave Macmillan, 2004), 35–57.

Chapter. "Appalachian Myths and the Legacy of Coal," in *The Appalachians: America's First and Last Frontier,* ed. by Mari-Lynn Evans, Robert Santelli, and Holly George-Warren (New York: Random House, 2004), 75–83. Companion book to the National PBS documentary of the same title.

Chapter. "Americanizing Immigrant Coal Miners in Northern West Virginia: Monongalia County Between the Wars, 1920–1940," in *Transnational West Virginia: Ethnic Communities and Economic Change, 1840–1940,* ed. by Ken Fones-Wolf and Ronald L. Lewis (Morgantown: West Virginia University Press, 2002), 261–296.

Essay. "Networks Large and Small." Introduction to *Transnational West Virginia: Ethnic Communities and Economic Change,* ed. by Ken Fones-Wolf and Ronald L. Lewis (Morgantown: West Virginia University Press, 2002), ix–xvii. Co-authored with Ken Fones-Wolf.

Chapter. "Networking Among Welsh Coal Miners in Nineteenth Century America," in *Toward a Comparative History of Coalfield Societies,* ed. by Stefan Berger, Andy Croll, and Norman LaPorte (London: Ashgate Publishers, 2005), 191–203.

Foreword. *Afflicting the Comfortable* (Morgantown: West Virginia University Press, 2005), ix–xiv.

Essay. "Labor in Appalachia." 4000 word introduction to the Labor section in the *Encyclopedia of Appalachia* (Knoxville: University of Tennessee Press, 2006), 551–557. Co-authored with John C. Hennen.

Article. "Gender and Transnationality Among Welsh Tinplate Workers in Pittsburgh: The Hattie Williams Affair, 1895." Co-authored with William D. Jones. *Labor History,* 48 (May 2007): 175–194.

Report

Report. "Appalachian Culture and Economic Development," Report 5, in *Thirty Years Later: A Socio-Economic Profile of the Appalachian Region* (Washington, DC: Appalachian Regional Commission, 1996). Co-authored with Dwight Billings.

Articles Reprinted

"Race and the United Mine Workers in Tennessee: The Letters of William Riley, 1892–1895," *Tennessee Historical Quarterly,* 36 (Winter 1977): 524–

536, reprinted in William Turner and Edward Cabbell, eds., *Blacks in Appalachia* (Lexington: University Press of Kentucky, 1985), pp. 173–182.

"Job Control and Race Relations in the Coal Fields, 1870–1920," *The Journal of Ethnic Studies,* 12 (Winter 1985): 36–64, reprinted in Donald G. Nieman, ed., *African Americans and Non-Agricultural Labor in the South, 1865–1900* (New York: Garland Publishing Inc., 1994), 245–274.

"From Peasant to Proletarian: The Migration of Southern Blacks to the Central Appalachian Coalfields," *Journal of Southern History,* 55 (February 1989): 77–102, reprinted in Donald G. Nieman, ed., *African Americans and Non-Agricultural Labor in the South, 1865-1900* (New York: Garland Publishing Inc., 1994), 245–275. Also reprinted by the Schomburg Center for Research, New York Public Library, *African American Migration Experience* website.

"Slave Families at Early Chesapeake Ironworks," *Virginia Magazine of History and Biography,* 86 (1978): 169–179, reprinted in Paul Finkelman, ed., *Women and the Family in a Slave Society,* Vol. 9 in the series *Articles on American Slavery* (New York: Garland Publishing, Inc., 1989), 299–310.

"'The Darkest Abode of Man': Black Miners in the First Southern Coal Field, 1780–1865," *Virginia Magazine of History and Biography,* 87 (1979): 190–202, reprinted in Paul Finkelman, ed., *Articles on American Slavery* Vol. 10 (New York: Garland Publishing, Inc., 1989), 288–202.

"Beyond Isolation and Homogeneity: Diversity and the History of Appalachia," *Confronting Appalachian Stereotypes* (Lexington: University Press of Kentucky, 1999), 21–43, Dwight B. Billings, ed., reprinted in Philip Obermiller, ed., *Appalachia in Social and Political Context* (Dubuque: Kendall Hunt Publishing, 2003).

Article. "The Forest Primeval." West Virginia Highlands Conservancy, *The Highlands Voice,* 36 (December 2003): 8–11. This article was excerpted and edited from Ronald L. Lewis's book, *Transforming the Appalachian Countryside,* and reprinted, with permission, with his name as author.

Reprints. Eleven excerpts from *The Black Worker,* Vol. 2, reprinted in *Making Freedom,* Vol. 4. (New York: Heinemann, forthcoming).

Contributors

Rebecca Bailey, a native of rural northern Virginia, but with family roots in McDowell and Mercer Counties of West Virginia, is a graduate of the College of William & Mary. Specializing in public and Appalachian history, she earned both graduate degrees at West Virginia University. *Matewan Before the Massacre* grew out of her doctoral dissertation, which was inspired by her participation in the West Virginia Humanities Council-funded Matewan Oral History projects of 1989–1990. Now an assistant professor at Northern Kentucky University, Bailey is also the public history coordinator for the NKU History & Geography Department. She lives in Crestview, Kentucky, with her dogs, Mickey and Ellie Grace.

Dwight B. Billings, a professor of sociology at the University of Kentucky, studies Appalachia and the American South. He is a past president of the Appalachian Studies Association and a past editor of the *Journal of Appalachian Studies*.

Shirley Stewart Burns holds a B.S. in news editorial journalism, a master's degree in social work, and a PhD in history, with an Appalachian focus, from West Virginia University. She is the author of *Bringing Down the Mountains: The Impact of Mountaintop Removal on Southern West Virginia Communities*, the first academic treatment of the topic. She is also the editor of *Coal Country*, the companion book for the documentary of the same name. A native of Wyoming County in the southern West Virginia coalfields and the daughter of an underground coal miner, she has a passionate interest in the communities, environment, and histories of the southern West Virginia coalfields. Shirley Stewart Burns lives in Charleston, West Virginia, with her husband, Matthew.

JEFFERY B. COOK is an associate professor of history and the chairman of the Department of History and Political Science at North Greenville University. He received a B.A. from Fairmont State University, and earned a master's and PhD from West Virginia University. Professor Cook is a contributor to several encyclopedias and he has written several scholarly articles and reviews for academic publications. His first book, *The Missouri Compromise: The Presidency of Harry S. Truman*, will be published in 2010. His wife, Laura, is a Kentucky native, and they have three daughters, Margaret Anne (Maggie), Sara Elizabeth, and Samantha Joy.

JENNIFER EGOLF is currently a visiting assistant professor at Indiana University of Pennsylvania. Her teaching and research focus on nineteenth and twentieth century history. She was a co-organizer of the Rush Holt Conference that honored Dr. Ronald L. Lewis, to whom this book is dedicated.

KEN FONES-WOLF is the Stuart and Joyce Robbins Professor of History at West Virginia University, where he teaches American social and working-class history. After earning his PhD from Temple University in 1986, he taught at the University of Massachusetts and for the Institute for Labor Studies and Research at WVU before joining the History Department in 2000. Fones-Wolf is the author or editor of five previous books and numerous articles on American labor and social history. His most recent works are: *Glass Towns: Industry, Labor and Political Economy in Appalachia, 1890–1930s* (University of Illinois Press, 2007) and *Transnational West Virginia: Ethnic Groups and Economic Change, 1840–1940* (West Virginia University Press, 2002). He is currently working on a coauthored book with Elizabeth Fones-Wolf, tentatively titled, *Struggle for the Soul of the Postwar South: Protestantism and the CIO's Operation Dixie*.

CONTRIBUTORS

JOHN HENNEN is a professor of history and associate director of Appalachian Studies at Morehead State University. A 1993 PhD from WVU, he is writing a book on Local 1199 in West Virginia, Kentucky, and Ohio from 1970 to 1989. He is the author of *The Americanization of West Virginia: Creating a Modern Industrial State, 1916–1925* (University Press of Kentucky, 1996) and several articles on Appalachian working-class history. Hennen co-edited the Labor section of *The Encyclopedia of Appalachia* with Ron Lewis.

LOUIS C. MARTIN is an assistant professor of history at Chatham University. He offers several courses in American, Latin American, African American, and labor and working-class history. His research interests include labor, working-class politics, Appalachian history and culture, and twentieth century political economy. He has researched workers in the steel and pottery industries in West Virginia and published multiple works on the steel industry of West Virginia. He has presented at the North American Labor History Conference, the Working-Class Studies Association Conference, and the Appalachian Studies Association Conference. Originally from West Virginia, he earned his B.A. from West Virginia University, his M.A. from Carnegie Mellon University, and his PhD from West Virginia University.

RICHARD P. MULCAHY is professor of history and political science with the University of Pittsburgh's Titusville Regional Campus in Titusville, Pennsylvania. A graduate of West Virginia University, he worked under Dr. Ronald L. Lewis from 1985 to 1988. The author of *A Social Contract for the Coal Fields*, Dr. Mulcahy is an active scholar in the areas of labor, Appalachia, and health care reform.

MARK MYERS is currently an instructor of history at the Indiana Academy, a gifted high school at Ball State University. He received his PhD from West Virginia University in December 2008, where he wrote a dissertation on deindustrialization in McDowell County, West Virginia.

After twenty years as a coal miner, PAUL H. RAKES acquired his PhD in Appalachian history at West Virginia University and is now associate professor at WVU Institute of Technology. His scholarly journal contributions include such articles as "Technology in Transition: The Dilemmas of Early-Twentieth Century Coal Mining" and "West Virginia Coal Mine Fatalities: The Subculture of Danger and a Statistical Overview of the Pre-enforcement Era."

CONNIE PARK RICE has a PhD in American history with an emphasis on Appalachian regional history. Rice has published several essays in journals and books and is in the process of completing two book projects, *For Men and Measure: The Life and Legacy of Civil Rights Pioneer J. R. Clifford* documenting the life of West Virginia's first black editor and practicing attorney John Robert Clifford, and *Daughters of Appalachia: A History of Women in the Mountain South*, an edited collection of essays on women in Appalachia. She is currently a lecturer in the History Department at West Virginia University and the assistant editor of *West Virginia History: A Journal of Regional Studies*.

DEBORAH R. WEINER'S book, *Coalfield Jews: An Appalachian History*, was published in 2006 by the University of Illinois Press. In 2007 it won the Southern Jewish Historical Society Prize for the book that has made "the most significant contribution to Southern Jewish history over the past four years." Weiner received a PhD in history from West Virginia University in 2002 and currently serves as research historian and family history coordinator at the Jewish Museum of Maryland, in Baltimore.

MICHAEL E. WORKMAN was born and raised at Falls View in Fayette County, West Virginia. After working in the coal industry, he attended West Virginia University, where he earned degrees in political science, public history, and a doctorate in history in 1995. He worked at the Institute for History of Technology from 1990 to 2006, and now teaches at Fairmont State University.

Index

Absentee owners: xiv, 9, 14
African Americans: x, xii, 16–18, 70, 118, 120–22, 124, 127, 131–134, 143, 334
Agriculture: 11, 13, 92, 280n96, 331. *See also* farming
Alternative fuels: 283, 285, 299
American Federation of Labor (AFL): 158–159, 238–239, 241, 294, 329–330
American Friends Service Committee: 109, 126
American Medical Association (AMA): 206–208, 210–211
American Mining Congress (AMC): 262–265
Americanization: xv, 20, 91, 98–100, 102, 105, 150, 156
Anti-union tactics: 38–39, 97–98, 102–103, 145, 175, 185, 187, 189–190, 192, 229–232, 240–243
Animals (in mines): 65–66

Appalachian Agreement: 291–292
Appalachian Regional Commission: 216, 333
Appalachian Studies: viii, 1–4, 6–10, 16, 18–19, 21, 23
Appalachian Studies Association: viii, 1, 6
Arch Coal's Dal Tex mine: 314–316, 319–320
Area Medical Officers (AMO): 210, 213
Arthurdale (WV): 130, 332

Baldwin-Felts Detective Agency: 39, 166–168, 174–176, 181–182, 184, 188, 191, 193n2, 293, 303n38
Battle of Blair Mountain: 181, 187, 316
Berwind White Coal Mining Company: 94–95
Bituminous Coal Commission: 170–172

Black Hand: 151, 153–154
Blankenhorn, Heber: 88, 94, 101, 104, 192
Blankenship, G. T.: 166–167, 169, 174–175, 180–181, 183–185
Bolshevism: 20, 91, 110. *See also* communism
Bonds, Julia: 317–318
Boone County (WV): 307–308, 312, 314, 317
Boone Report, The: 212, 226
Boyle, W. A. "Tony": 215, 245n3, 316
Brophy, John: 88, 90, 95, 96, 103, 104, 108, 109, 110, 111
Bureau of Mines (1910): xv, 76–77, 254, 260–269

Cabell-Huntington Hospital: 229, 239
Captive mines: 292, 300n2
Central Competitive Field: 89, 91, 95, 111, 284, 291, 302n30
Central West Virginia Coal Operators Association: 156, 158
Chambers, E. B.: 180–181, 184
Child labor: 56, 59, 66–67, 74, 79n2, 102, 254, 261
Children: 23, 98–100, 102, 104, 108–109, 118–119, 121, 124, 126–128, 130, 132, 167, 224, 235, 288, 312–313, 317
Civil liberties: 111n1, 189
Cleveland Agreement: 89, 99
Clinch Valley Strike: 233–235, 238
Coal Age: 34, 36, 40–41, 157

Coal code: xvi, 283–284, 289–292, 294, 297–299
Coal industry: xi, xiv, 8, 11, 14, 17–18, 31, 33, 38–39, 42, 50, 77–78, 103, 108, 111, 122, 143, 170, 172, 185, 186–187, 205, 207, 213–214, 222, 258, 260, 263, 266, 283–300, 306, 314–315, 332
Coal miners: viii, x–xvi, 5, 17–19, 22, 31, 33–39, 41–43, 45–47, 50, 56, 58–65, 67–74, 76–78, 79n2, 83n32, 88–91, 93–110, 122, 124–127, 131, 142, 144–146, 148–150, 152, 155–160, 6n161, 166–168, 171–176, 178–179, 181–192, 207–208, 212, 215, 234, 238, 259, 261, 263–264, 267–268, 271n16, 276n62, 278n82, 283–288, 290–300, 302n27, 305, 316, 321, 329, 331–334
Coal mining: xi, 5, 8, 9, 13–15, 17–18, 23, 32, 38, 40, 41, 50, 57, 58–59, 62, 65–71, 73, 75–79, 82n25, 91, 92, 102, 112n3, 130, 229, 262–264, 284, 288, 306–307, 314–315, 317. *See also* surface mining and mountaintop removal
Coal operators: xi–xiii, xv–xvi, 17–18, 31–34, 36–40, 44–45, 55n53, 59–61, 66–69, 73–76, 82n18, 88–89, 91, 94–98, 100–107, 109–111, 124, 130–131, 144, 146, 149, 156–158, 169–170, 172, 174–179, 181–182, 184–190, 192, 207, 215–216, 254–260, 263–269, 276n62, 278n82, 279n90, 284–285, 287, 289–291,

293–296, 298–299, 302n25, 307, 322
Coal River Mountain Watch (CRMW): 317
Coal towns: xvi, 19, 39, 50, 288, 299. *See also* company towns
Collective bargaining: 227, 229, 232, 238–239, 241–242, 244n3, 284, 289, 291
Collins, Justus: 37–40, 44, 48, 60, 255, 264
Communism: 15, 19, 91–92, 97, 101, 103, 105–106, 236–237, 240. *See also* Bolshevism
Communist Party: 236
Company physicians (doctor): 54n36, 68, 73, 205, 207–208, 210–212, 214
Company store: xi, 31–36, 39–51, 53n30, 54n42, 103–104, 124, 130, 173, 176, 273n31, 291, 302n27
Company towns: 46, 89, 104, 123–124, 145, 284. *See also* coal towns
Congress of Industrial Organizations (CIO): 238, 239, 241, 330
Conservatism: xv, 15, 22, 90–92, 97, 101, 106, 109–111, 144, 149, 155, 177, 191, 236, 254, 257–258, 263, 266, 269, 287
Consolidation Coal Company: 14, 62, 78, 94, 96, 104, 108, 144, 146, 149, 155–157, 164n62, 297, 312
Convict labor: 17, 70
Cornwell, John J.: 157, 172, 177, 184, 188

Cultural modernization theory: 2, 5, 7–9, 15

Davis, Henry Gassaway: 12, 74, 257
Davis, Leon: 230, 235–237, 239, 241, 243
Dawson, William M. O.: 75–76, 256–257, 259
Deadwork: 95, 103–104
Debs, Eugene V.: 148
Deindustrialization: 112n3, 214, 218, 222n45, 283–284, 299–300
Democracy: 17, 92, 99, 106–110, 238
Democratic unionism: 5, 225, 232, 238
Democrats: 14, 16, 131, 170, 177, 179–181, 184, 192, 288
Department of Environmental Protection (DEP): 08, 311–312, 314
Depopulation: 23, 307, 320–322. *See also* out-migration
Discrimination: 129–133
District 2 (UMWA): 88, 90, 95, 108, 114n30
District 17 (UMWA): 148, 155, 157–159, 172–173, 175–176, 179, 181, 184–187, 189, 292
Dixon, Samuel: 74–75, 80, 81–82, 274n42
Drake, Wendell: 232–233, 235
Draper, Warren F.: 209–210, 220

Education: xii, 3, 14, 16, 74, 91, 111, 118–119, 121–122, 124, 126–129,

131, 133–134, 170, 206, 225, 237, 265, 274n41, 312–313, 321, 331
Edwin, James: 144, 158
Elections: 110, 148, 158, 170, 177–178, 180–181, 184, 187, 283, 288 Union 215, 224, 232–234, 239–240, 241–242
Elkins, Stephen B.: 12, 74, 257, 267, 274n42
Ethnicity: vii, ix, 8, 15, 21–22, 47, 76, 93, 143–144, 149, 152–156, 159, 237, 330, 333
Evictions: 89, 103, 109, 124, 130–131, 166–167, 174–176, 196n21

Fair Labor Standards Act (1967): 227
Fairmont Coal Company: 74–75, 144, 146–147, 253, 257, 259–260, 265–268, 274n41, 279n90
Farm Security Administration (FSA): 207, 209
Farmers: vii, x–xi, 10–12, 22, 37, 89, 92–94, 101–102, 120, 123, 305
Farming: 10–11, 13, 32, 130–131. *See also* agriculture
Farmington (WV): 78, 149–155
Fayette County (WV): 56, 65, 74, 276n62
Federal Bureau of Mines: 76–77, 254, 260–262, 264–266, 268–269, 280n94
Federal Coal Mine Health and Safety Act (1969): 78
Federal Emergency Relief Administration (FERA): 127, 131

Federal government: xiv, xvi, 89, 102, 127, 170, 187, 189–192, 207–208, 211, 214, 259, 262, 267, 269, 279n90, 289–290, 292, 294, 298
Felts, Albert C.: 166–168, 174, 176
Felts, Lee: 168, 176
Fleming, A. Brooks: xv, 74–75, 144, 253–260, 262–269
Foreclosures: 131–133
Fox, Dewey William: xii, 118–135

Glass industry: 13–14, 22, 130, 148, 158, 226
Grange: 101–102
Great Depression: xv, 126, 129, 185, 192, 207, 212, 283–284, 286–288, 297, 330, 332
Green, William: 294

Harding, Warren G.: 90–91, 107, 109, 177
Harlan County (KY): 19, 35, 37, 40
Harless, Larry: 225–226, 231–235
Hatfield, Anderson "Ance": 168, 183, 194n6
Hatfield, Greenway: 179, 184
Hatfield, Sid: 18, 166–168, 181–183, 188, 190
Health care: xiii–xiv, 18, 21, 68–69, 73, 204–205, 207–209, 211–216, 218, 224, 226, 228, 238, 240, 242
Health care workers: 225–231, 234, 237–240, 243
Health clinics: 204, 212–214, 218, 244n3

Health insurance: 207–209, 215, 217–218

Holmes, Joseph: 260, 262, 264, 266–268, 280n94, 280n96, 281n115, 282n118

Homogeneity: x–xii, 4, 9, 21, 92, 141–142, 332, 334

Immigrants: x–xii, 17, 31, 38–39, 42, 47, 74, 92, 97, 99–100, 103, 105, 141, 143–145, 149–153, 333

Immigration: ix; restriction on: 92, 102, 142

Industrialization: vii–viii, xi, xv, 4, 13, 15, 21–22, 141–144, 332

Injunctions: 89, 96, 146–147, 190, 234–235

Isolation: x, 2–4, 8, 206, 332, 334

Jacksonville Agreement: 111

Jewish merchants: xi, 22–23, 31–32, 35, 43, 47, 49–50, 54n42, 55n53

Jim Crow: 16, 118, 235

Keeney, Frank: 148, 157–160, 187

Ku Klux Klan (KKK): 132, 134, 154

Labor movement: viii, xiii, 19, 21–22, 37, 39

Labor relations: xii–xiii, 9–10, 15–17, 22, 37

Laing, James: 34, 41, 48, 54n36

Laing, John: 70, 76, 268, 281n110

Landrum-Griffin amendment (1959): 225

Lane, Winthrop: 34, 36–37, 44

Lewis, John L.: 89–90, 95, 106, 110–111, 171–172, 179, 185–187, 208, 291, 294, 296–297

Lewis, Ronald L.: viii–xi, xvi, 1, 3, 7, 10, 15, 17–18, 21,

Liberalism: 90, 177, 191–192, 257, 266, 273n32

Local 1199: xiv, 224–225, 227–244, 245n5, 247n21, 248n31, 250n51

Logan (WV): 37, 39, 47, 145, 172

Logan County (WV): 147, 186, 188, 247n21, 313–315, 320

Lyons, F. R.: 96–98

Machinery: 13, 59, 62, 77–78, 253–254, 256, 259, 266, 269, 284–285, 295–299, 306, 309. *See also* mechanization

Management consultants: 239–242

Manufacturing: 13–14, 22, 130, 214, 229

Marshall University: 236–237

Masculinity: 69–72

Massey Energy Company: 307–308, 310–311, 317–319

Matewan (WV): xiii, 18, 166–170, 173–176, 180–185, 191–193

McDowell County (WV): xv, 37, 46, 178–179, 186, 188, 283–284, 286–288, 291–293, 295, 297–300

Mechanization: 23, 254, 256, 259, 283–284, 295–299. *See also* machinery

Medicaid: 214, 216–217, 226

Medicare: 214–215, 226
Merchants: xi, 23, 32–38, 40, 43–48, 50, 175, 183. *See also* retailers
Methane gas: 56, 59–60, 66–67, 77, 79n2, 84n44, 255, 264, 266
Miller, Arnold: 215–216, 234, 248
Mine accidents: xi, 18, 64, 77–78, 149, 224, 261–262, 265
Mine casualties: xi, 58–63, 67–69, 72, 77, 109, 192, 317. *See also* mine fatalities
Mine guard system: 19, 39, 109, 124–125, 171, 184, 293, 303n38
Mine explosions: xiii, 57–58, 60, 63–64, 67, 69, 74–78, 79n5, 80n6, 80n7, 81n16, 84n44, 253-255, 258–261, 263–264, 266, 268–269, 271n16, 272n31
Mine inspections: 261, 265, 268
Mine fatalities: 58–59, 61, 65, 75–76, 78, 254, 256. *See also* mine casualties
Mine rescues: 66, 69–70, 76, 78
Mine safety: 58, 60–61, 63–64, 67, 75, 77–79, 142, 253–257, 259–262, 264–266, 268–269, 298
Mine safety laws (or legislation): 60, 75, 78–79, 256–257, 259–262, 264–266
Mingo County (WV): 54n42, 166, 170, 172, 174–177, 179–185, 189, 191– 192, 196n21, 307
Mitchell, John: 77, 260, 267
Monongah Mine Disaster (WV): xv, 58, 60, 74–75, 77, 82n19, 253–255, 258–261, 268–269, 273n31, 274n41
Monongahela Valley: 4, 13–14, 18, 141, 143
Monongalia County: xii, 118, 121–123, 127–134, 143, 157, 333
Montgomery, Samuel B.: 155–157, 177–181
Mooney, Fred: 148, 159, 173, 175–176, 187
Morgantown (WV): viii, 123–129, 132, 141, 148, 150, 212, 221n24, 274n41
Mother Jones: 159–160
Mountaintop Removal (MTR): ix, xvi, 7, 11, 19, 306–309, 312–322. *See also* surface mining and strip mining
Muehlenkamp, Robert: 232–233, 237–238, 243

National Association for the Advancement of Colored People (NAACP): 123, 128–129, 131–134
National coal strike of 1919: 91, 109, 170–172, 285
National Industrial Recovery Act (1933): xv, 192, 283–284, 289–291, 294, 297, 299
National Labor Board: 170
National Labor Relations Act (1935): 225, 227, 238–239, 242
National Labor Relations Board: 232, 241
National Origins Act (1924): 92
National Recovery Administration

(NRA): xv–xvi, 283–284, 289–290, 292–295, 297–299
National Union of Hospital and Health Care Employees: 224
Nativism: 91, 99, 101–102, 112n10
New Deal: xv, 111, 184–185, 209, 216–217, 226, 283, 289, 296, 332
New River Coal Company: 34–36, 46, 74
Nixon, Richard: 238, 242

O'Toole, Edward: 178, 296
Open shop: 110, 157–158, 190–191, 293
Out-migration: viii, 322. *See also* depopulation
Overproduction: 11, 42, 284–285, 295

Paint Creek-Cabin Creek Strike: 146, 148, 177
Paternalism: 290, 293, 299
Patriotism: xii, 20, 92, 105–108, 156, 160, 256, 291
Paul, James W.: 79n5, 257, 273n31
Paul, John W.: 74–76
Peerless Coal and Coke: 284, 286, 288, 295, 297–298, 300n2
Pennsylvania State Police: 89, 95–96
Pocahontas (VA): 46, 48
Pocahontas Field: 188–189, 284
Press: 89, 94, 97–107, 149, 238–239, 261–262, 268, 315
Progressive Era: 61, 64, 76, 92, 148, 177, 255, 258

Prohibition: 151, 153, 180
Property damage: 100, 190, 309

Radicalism: xii, 19, 75, 91, 96–102, 105, 111, 144, 148–149, 154, 160, 236, 238, 257–258, 264
Railroads: vii, 8, 11–13, 16, 19, 22, 39, 111n3, 118, 122, 125, 141–142, 145, 138, 153, 158, 166, 168, 173, 177, 183, 261, 263, 285, 305, 323, 329, 331, 332
Regulation: 75, 110, 169–171, 189, 255, 257, 261, 265, 279n88, 314n12, 317, 324
Relief: strike relief : 90, 109, 166, 189; work relief: 127, 131, 287–289, 294–295
Religion: 19, 102, 105–106, 109, 126, 132, 134, 142–143, 148, 159, 173, 180, 182–183, 190, 212, 244
Republicans: 14, 90, 109, 131, 148, 170, 176–181, 184, 257, 264, 288
Resettlement Administration: 128, 130
Retail, Wholesale, Department Store Union (RWDSU): 225, 236
Retailers: xi, 23, 31–40, 42–50
Roads: 13, 16, 36–38, 124, 129, 133–134, 152, 171, 309, 317, 325n17
Rockefeller, John D.: 14, 94, 128
Roosevelt, Eleanor: 129–130, 133
Roosevelt, Franklin: 184, 192, 209, 283, 288–290, 297

Scotts Run (WV): 122–125, 130–131,

331, 332

Scrip: 33, 42–46, 49, 54n36, 55n53, 104, 146, 291, 293, 302

Segregation: racial: 18, 118–119, 128, 134–135; workplace: 228

Slavery: 17, 118

Social equality: 17, 118, 122, 134, 237, 332

Socialist Party: 148–149

Socialists: 96, 147–149

Somerset County (PA): xi, 88–93, 95–102, 104, 106–107, 109–110

Speculators: 12, 305

Steel industry: 20, 55n53, 68, 185–187, 226, 239, 244, 292, 300n2

Stewart, Danie Joe: 235–238, 240

Strike for Union (1922–1923): xi, 88–89, 100, 103, 106–107, 110

Strikebreakers: 100, 105, 122, 124–125, 146

Strikes: xi–xii, 32, 38, 43, 88–111, 122, 124–126, 145–148, 151–157, 159–160, 170–173, 177, 181–190, 196, 208, 224–225, 228–232, 234–235, 237–238, 241, 243–244, 245, 248, 285, 292.

Strip mining: 295, 306, 309. *See also* mountaintop removal and surface mining

Stuart Mine: 56, 66, 74–75, 80, 81

Surface mining: 5–6, 306–307. *See also* mountaintop removal and strip mining

Taft, William Howard: 267–368

Taft-Hartley (1947): 225, 234, 237

Tent colonies: 89, 99, 109, 176, 182–183, 188–189, 192

Testerman, Cabell C.: 166–169, 180–181, 183

Timbering: vii, 8, 10–13, 305, 323

Underground mining: xi, xvi, 58–62, 66, 68–69, 71, 74–76, 82n18, 83n32, 295, 298, 305, 307, 311–312, 318–319

Unemployment: 103, 131, 214, 232, 294, 300, 313–315

Unionization: 34, 38–40, 46, 88–92, 94–111, 130, 144–147, 150, 155–160, 171–176, 179–181, 184–189, 192, 224–243, 264, 285, 289–296

United Mine Workers of America (UMWA): xiii–xiv, 17–19, 88–89, 91, 94–101, 103–111, 145–149, 152, 155–160, 164n62, 169–172, 175–181, 184–185, 187, 189–190, 205, 207–209, 213, 215, 224, 226, 234, 238, 244n3, 260–262, 264, 267–268, 284–285, 290–294, 296–299, 314–316, 331, 332

United Mine Workers of America Health (Welfare) and Retirement Fund: xiii, 18, 205, 209, 244n3

United States Coal and Coke Company: 284, 286, 288, 292, 295–298, 300n2

United States Coal Commission: 35–36, 43, 46–49, 111n1

United States Immigration

Commission: 37, 46, 48
United States Public Health Service (USPHS): 208–209, 214, 218
United States Steel Corporation (U.S. Steel): 68, 185–187, 292, 300n2
United States Supreme Court: 190, 298–299
United Steel Workers of America: 226, 239, 244n3

Violence: xiii, 17, 96–97, 99–101, 106, 151–155, 170, 174, 176, 184, 186, 191, 236
Volunteers in Service to America (VISTA): 3, 214

Watson, C. W.: 14, 144–145, 147, 158
Watson, George T.: 14, 156–158, 265
Watson, James Otis: 14, 144
Welfare: 18, 127, 313
Welfare capitalism: 20, 146, 191
West Virginia Board of Trade: 257–258
West Virginia State Federation of Labor: 160, 175–177
West Virginia Supreme Court of Appeals: 178, 180, 188, 190
Weyland, Clarence: 144, 158
White, M. Z.: 168, 178–180, 190
Wilson, Woodrow: 91, 107, 156, 177
Wolfe, George: 37–40, 44, 48–49, 60
Women: 35, 124–125, 133, 159, 168–169, 193, 209, 229–231, 317–318, 333
Wood, Kenneth: 229–230, 240–241

Woodruff, Tom: 235–239
Works Progress Administration (WPA): 127, 129, 131–133
World War I: xi, xiii, 20, 59, 73, 109, 143, 186, 189, 192, 284
World War II: 43, 226, 299

Yellow dog contracts: 125, 175, 289, 301n22

www.ingramcontent.com/pod-product-compliance
Lightning Source LLC
Chambersburg PA
CBHW050200240426
43671CB00013B/2189